Composites in Biomedical Applications

Composites in Biomedical Applications

S. M. Sapuan
Y. Nukman
N. A. Abu Osman
R. A. Ilyas

CRC Press
Taylor & Francis Group
Boca Raton London New York

CRC Press is an imprint of the
Taylor & Francis Group, an **informa** business

First edition published 2020
by CRC Press
6000 Broken Sound Parkway NW, Suite 300, Boca Raton, FL 33487-2742

and by CRC Press
2 Park Square, Milton Park, Abingdon, Oxon, OX14 4RN

© 2021 Taylor & Francis Group, LLC
CRC Press is an imprint of Taylor & Francis Group, LLC

ISBN: 978-0-367-27168-8 (hbk)
ISBN: 978-0-367-54513-0 (pbk)
ISBN: 978-0-429-32776-6 (ebk)

Typeset in Times LT Std
by Cenveo® Publisher Services

Contents

Preface

Biomedical engineering can be defined as the application of engineering principles and techniques into the medical field. In the past 20 years, the field of biomedical engineering has expanded successfully due to the supportive advances in biological science and other related disciplines. This phenomenon has created new opportunities for development of tools for therapy, ergonomics, and diagnosis for human disease. The book *Composites in Biomedical Engineering Application* presents a comprehensive overview on recent developments in composites and their use in biomedical applications. The information contained in this book is critical for the more efficient use of composites, as detailed up-to-date information is a pre-requirement. The information provided brings cutting-edge developments to the attention of young investigators to encourage further advances in the field of composite material research. Further, chapters discuss advanced techniques for the development of synthetic-based composites as well as biopolymer-based composites. By introducing these topics, the book highlights a totally new research theme in advanced composites in biomedical engineering applications. This book covers advanced materials, conceptual design, classification, ergonomic design applications, selection of biocomposites in advanced biomedical and electronic systems applications, anthropometrics, dentistry resin-based composites, bone substitute material, rubber glove waste fiber reinforced polymer composites, piezoelectric/electroactive polymeric composites, marine skeletal nanocomposites, nanocomposite tissue engineering, and regenerative as well as nanocellulose composites in biomedical application. This book also discusses the latest developments of the synthesis, preparation, characterization, and applications of nano/microstructure-based materials in the biomedical field. The latest results in material evaluation for targeted applications are also presented. As such, it offers a valuable reference guide for scholars interested in synthesis polymers and biopolymers and evaluation of nano/microstructure-based materials, as well as their biomedical applications. It also provides essential insights for graduate students and scientists pursuing research in the broad fields of biomedical engineering, composite materials, polymers, organic/inorganic hybrid materials, nano-assembly, etc.

Editors

Prof. Ir. Dr. Mohd Sapuan Salit (S. M. Sapuan) is a professor of composite materials at Universiti Putra Malaysia. He earned his B.Eng degree in Mechanical Engineering from University of Newcastle, Australia, in 1990, MSc from Loughborough University, UK, in 1994, and PhD from De Montfort University, UK, in 1998. His research interests include natural fiber composites, materials selection, and concurrent engineering. To date he has authored or co-authored more than 1,521 publications (730 papers published/accepted in national and international journals, 16 authored books, 25 edited books, 153 chapters in books, and 597 conference proceedings/seminar papers/presentation [26 of which are plenary and keynote lectures and 66 of which are invited lectures]). S. M. Sapuan was the recipient of the Rotary Research Gold Medal Award 2012; the Alumni Medal for Professional Excellence Finalist, 2012 Alumni Awards, University of Newcastle, NSW, Australia; and the Khwarizmi International Award (KIA). In 2013, he was awarded the 5 Star Role Model Supervisor award by UPM. He has been awarded "Outstanding Reviewer" by Elsevier for his contribution in reviewing journal papers. He received the Best Technical Paper Award in the UMIMAS STEM International Engineering Conference in Kuching, Sarawak, Malaysia. S. M. Sapuan was recognized as the first Malaysian to be conferred Fellowship by the US-based Society of Automotive Engineers International (SAE International) in 2015. He was the 2015/2016 recipient of the SEARCA Regional Professorial Chair. In the 2016 ranking of UPM researchers based on the number of citations and h-index by SCOPUS, he is ranked at 6th of 100 researchers. In 2017, he was awarded with IOP Outstanding Reviewer Award by Institute of Physics, UK, National Book Award, The Best Journal Paper Award, UPM, Outstanding Technical Paper Award, Society of Automotive Engineers International, Malaysia, and Outstanding Researcher Award, UPM. He also received in 2017 a Citation of Excellence Award from Emerald, UK, SAE Malaysia the Best Journal Paper Award, IEEE/TMU Endeavour Research Promotion Award, Best Paper Award by Chinese Defence Ordnance, and Malaysia's Research Star Award (MRSA) from Elsevier. In addition, in 2019, he was awarded with Top Research Scientists Malaysia (TRSM 2019) and Professor of Eminence Award from AMU, India.

Professor Ir. Dr. Nukman Yusoff (Y. Nukman) is currently working in the Department of Mechanical Engineering, University of Malaya (UM), Malaysia. He worked for 1 year as visiting Professor in Qassim University, Saudi Arabia, for developing research and teaching in the field of Advanced Manufacturing Process. Being a professional engineer registered under the Board of Engineers Malaysia, he has pioneered the Computer Aided Design/Computer Aided Manufacturing (CAD/CAM) field in Malaysia. He acquired his PhD from Loughborough University, UK (field of research: Laser Materials Processing) in 2009 and Master in Engineering Science, MSc (Mechatronics) from De Montfort University, Leicester, UK, in 1998. He acquired his first degree in CAD/CAM from the University of Central England in Birmingham, UK. Since then he has been teaching undergraduate courses in the field of CAD/CAM and manufacturing and graduate level of teaching in Manufacturing Automation and Control and other fields of subjects in manufacturing. His administration experience in UM was recognized when he was appointed as Head of Department of Engineering Design & Manufacture from 2007–2009. Since working in UM, Nukman was actively doing research in the field of Laser Materials Processing, CAD/CAM, CNC, and Manufacturing Process and Management. Being a Principal Researcher in UM, he has secured several millions of grants from internal and international sources, and he manages more than handful of PhD graduates and quite a number of postgraduates who are currently working under his supervision. He has published more than 50 articles in many ISI Journals in the field of manufacturing and in different international, regional, and national journals, conference proceedings and bulletins, and he is co-author of several technical books in CAD/CAM and Manufacturing Process.

Prof. Ir. Dr. Noor Azuan Abu Osman (N. A. Abu Osman) graduated from University of Bradford, UK, with a B.Eng Hons. in Mechanical Engineering, followed by MSc and PhD in Bioengineering from University of Strathclyde, UK. Azuan's research interests are quite wide-ranging under the general umbrella of biomechanics. However, his main interests are the measurements of human movement, prosthetics design, development of instrumentation for forces and joint motion, and the design of prosthetics, orthotics, and orthopedic implants. In 2004, he received the BLESMA award from ISPO UK NMS in recognition of his significant contribution to the development of the prosthetic socket. He is currently the President of Malaysian Society of Biomechanics, which is affiliated with the International Society of Biomechanics, Council Member of World Congress of Biomechanics and the Deputy Vice-Chancellor (Academic and International), Universiti Malaysia Terengganu (UMT). Prior to seconded to UMT, he was a Dean of Engineering and also the Director of Centre for Applied Biomechanics and the Coordinator of Motion Analysis Laboratory, Faculty of Engineering, University of Malaya, Malaysia. He has published more than 200 articles in ISI high impact journals and authored 3 engineering academic books. He is also a Professional Engineer (P.Eng) with Board of Engineer, Malaysia, Chartered Engineer (C.Eng.) and Fellow (FIMechE) with The Institute of Mechanical Engineers, United Kingdom, Chartered Professional Engineer (C.PEng) and Fellow (FIEAust) with Engineers Australia, and also Fellow for Academy of Sciences, Malaysia (FASc).

Dr. Ahmad Ilyas Rushdan (R. A. Ilyas) received his Diploma in Forestry at Universiti Putra Malaysia, Bintulu Campus (UPMKB), Sarawak, Malaysia, from May 2009 to April 2012. In 2012, he was awarded the Public Service Department (JPA) scholarship to pursue his Bachelor's Degree (BSc) in Chemical Engineering at Universiti Putra Malaysia (UPM). Upon completing his BSc program in 2016, he was again awarded the Graduate Research Fellowship (GRF) by the Universiti Putra Malaysia to undertake a PhD degree in the field of Biocomposite Technology & Design at Institute of Tropical Forestry and Forest Products (INTROP) UPM. R. A. Ilyas was the recipient of Gold Medal Malaysian Vaccines & Pharmaceuticals

(MVP) Doctor of Philosophy 2019, and Top Student Award (Institute of Tropical Forest and Forest Products (INTROP), Universiti Putra Malaysia, Malaysia). In 2018, he was awarded with Outstanding Reviewer by Carbohydrate Polymers, Elsevier, United Kingdom; Best Paper Award (11th AUN/SEED-Net Regional Conference on Energy Engineering); Best Paper Award (Seminar Enau Kebangsaan 2019, Persatuan Pembangunan dan Industri Enau Malaysia); and National Book Award 2018. His main research interests are: (1) polymer engineering (biodegradable polymers, biopolymers, polymer composites, polymer-gels) and (2) material engineering (natural fiber reinforced polymer composites, biocomposites, cellulose materials, nanocomposites). He has authored and published more than 39 citation-indexed journals on green materials related subjects. To date he has authored and co-authored more than 148 publications (45 papers published/accepted/submitted in national and international journals, 1 authored book, 7 edited books, 58 chapters in books, 10 conference papers, 2 research bulletins, 3 conference/seminar proceedings, and 22 conference proceedings/seminar papers/presentations.

Contributors

S. Abdullah
Faculty of Applied Sciences and
  Technology
Universiti Tun Hussein Onn Malaysia
Pagoh Educational Hub
Pagoh, Johor, Malaysia

A. S. Abu
Department of Mechanical and
  Manufacturing Engineering
Universiti Malaysia Sarawak
Kota Samarahan, Sarawak, Malaysia

N. A. Abu Osman
Centre for Applied Biomechanics
Department of Biomedical Engineering
Faculty of Engineering
University of Malaya
Kuala Lumpur, Malaysia

The Chancellery
Universiti Malaysia Terengganu
Kuala Lumpur, Malaysia

S. Adzila
Faculty of Mechanical and
  Manufacturing Engineering
Universiti Tun Hussein Onn Malaysia
Parit Raja, Johor, Malaysia

H. A. Aisyah
Institute of Tropical Forestry and Forest
  Products (INTROP)
Universiti Putra Malaysia
Serdang, Selangor, Malaysia

Faris M. AL-Oqla
Department of Mechanical
  Engineering
Hashemite University
Zarqa, Jordan

S. Asman
Faculty of Applied Sciences and
  Technology
Universiti Tun Hussein Onn Malaysia
Pagoh Educational Hub
Pagoh, Johor, Malaysia

A. Ataollahi
Centre for Applied Biomechanics
Department of Biomedical Engineering
Faculty of Engineering
University of Malaya
Kuala Lumpur, Malaysia

N. A. Badarulzaman
Faculty of Mechanical and
  Manufacturing Engineering
Universiti Tun Hussein Onn
  Malaysia
Parit Raja, Johor, Malaysia

N. Bano
Department of Chemistry
Government Postgraduate College
  for Women
Raiwind, Lahore, Pakistan

H. Basri
Faculty of Applied Sciences and
  Technology
Universiti Tun Hussein Onn
  Malaysia
Pagoh Educational Hub
Pagoh, Johor, Malaysia

V. K. Bommala
Department of Mechanical
  Engineering
Acharya Nagarjuna University
Guntur, Andhra Pradesh, India

R. A. Diab
University of Malaya
Kuala Lumpur, Malaysia

M. A. G. Gonzalez
University of Malaya
Kuala Lumpur, Malaysia

M. Gopi Krishna
Department of Mechanical
   Engineering
Acharya Nagarjuna University
Guntur, Andhra Pradesh, India

L. He
School of Mechanical Engineering
Dongguan University
   of Technology
Songshan Lake, Dongguan, China

R. A. Ilyas
Advanced Engineering Materials and
   Composites Research Centre
Department of Mechanical and
   Manufacturing Engineering
Faculty of Engineering
Universiti Putra Malaysia
Serdang, Selangor, Malaysia

Laboratory of Biocomposite Technology
Institute of Tropical Forestry and Forest
   Products (INTROP)
Universiti Putra Malaysia
Serdang, Selangor, Malaysia

M. Jawaid
Institute of Tropical Forest and Forest
   Products (INTROP)
Universiti Putra Malaysia
Serdang, Selangor, Malaysia

S. S. Jikan
Faculty of Applied Sciences and
   Technology
Universiti Tun Hussein Onn Malaysia
Pagoh Educational Hub
Pagoh, Johor, Malaysia

K. Karmegam
Department of Environmental and
   Occupational Health
Faculty of Medicine and Health Sciences
Universiti Putra Malaysia
UPM Serdang, Selangor, Malaysia

A. Khalina
Institute of Tropical Forestry and Forest
   Products (INTROP)
Universiti Putra Malaysia
Serdang, Selangor, Malaysia

Faculty of Engineering
Universiti Putra Malaysia
Serdang, Selangor, Malaysia

I. J. Macha
Department of Mechanical and
   Industrial Engineering
University of Dar es Salaam
Dar es Salaam, Tanzania

A. Mahmood
Laboratory of Biocomposite
   Technology
Institute of Tropical Forestry
   and Forest Products
Universiti Putra Malaysia
UPM Serdang, Selangor, Malaysia

Department of Mechanical Engineering
Politeknik Port Dickson
Port Dickson, Negeri Sembilan, Malaysia

M. T. Mastura
Fakulti Teknologi Kejuruteraan
   Mekanikal dan Pembuatan (FTKMP)
Universiti Teknikal Malaysia
Melaka, Malaysia

R. Nadlene
Fakulti Teknologi Kejuruteraan
   Mekanikal dan Pembuatan
Universiti Teknikal Malaysia Melaka
   (UTeM)
Melaka, Malaysia

M.N.F. Norrrahim
Research Centre for Chemical
  Defence
Universiti Pertahanan Nasional
  Malaysia
Kuala Lumpur, Malaysia

Y. Nukman
Centre for Applied Biomechanics
Department of Biomedical
  Engineering
Faculty of Engineering
University of Malaya
Kuala Lumpur, Malaysia

N. Mohd. Nurazzi
Center for Defence Foundation
  Studies
National Defence University of
  Malaysia
Kuala Lumpur, Malaysia

M. Nuzaimah
Advanced Engineering Materials
  and Composites Research Centre
  (AEMC)
Department of Mechanical and
  Manufacturing Engineering
Universiti Putra Malaysia
Serdang, Selangor, Malaysia

Fakulti Teknologi Kejuruteraan
  Mekanikal dan Pembuatan
Universiti Teknikal Malaysia Melaka
  (UTeM)
Melaka, Malaysia

M. T. Paridah
Institute of Tropical Forestry and Forest
  Products (INTROP)
Universiti Putra Malaysia
Serdang, Selangor, Malaysia

Z. Radzi
University of Malaya
Kuala Lumpur, Malaysia

N. N. Ruslan
Faculty of Mechanical and
  Manufacturing Engineering
Universiti Tun Hussein Onn Malaysia
Parit Raja, Johor, Malaysia

S. M. Sapuan
Advanced Engineering Materials and
  Composites Research Centre
Department of Mechanical and
  Manufacturing Engineering
Faculty of Engineering
Universiti Putra Malaysia
Serdang, Selangor, Malaysia

Laboratory of Biocomposite Technology
Institute of Tropical Forestry and Forest
  Products (INTROP)
Universiti Putra Malaysia
Serdang, Selangor, Malaysia

F. N. Shafiqa
Advanced Engineering Materials and
  Composites Research Centre (AEMC)
Department of Mechanical and
  Manufacturing Engineering
Faculty of Engineering
Universiti Putra Malaysia
Serdang, Selangor, Malaysia

N.S. Sharip
Laboratory of Biopolymer and
  Derivatives
Institute of Tropical Forestry and Forest
  Products (INTROP)
Universiti Putra Malaysia
Serdang, Selangor, Malaysia

S. S. Shazleen
Laboratory of Biopolymer and
  Derivatives
Institute of Tropical Forestry and Forest
  Products (INTROP)
Universiti Putra Malaysia
Serdang, Selangor, Malaysia

S. Sivasankar
Department of Environmental and
    Occupational Health
Faculty of Medicine and Health
    Sciences
Universiti Putra Malaysia
UPM Serdang, Selangor, Malaysia

S. H. M. Suhaimy
Faculty of Applied Sciences and
    Technology
Universiti Tun Hussein Onn
    Malaysia
Pagoh Educational Hub
Pagoh, Johor, Malaysia

C. Wang
Department of Mechanical Engineering
The University of Hong Kong
Pokfulam Road, Hong Kong

School of Mechanical Engineering
Dongguan University of Technology
Songshan Lake, Dongguan, China

M. Wang
Department of Mechanical Engineering
The University of Hong Kong
Pokfulam Road, Hong Kong

N. A. Yahya
University of Malaya
Kuala Lumpur, Malaysia

U. J. Yap
National University of Singapore
Singapore

T.A.T. Yasim-Anuar
Department of Bioprocess Technology
Faculty of Biotechnology and
    Biomolecular Sciences
Universiti Putra Malaysia
Serdang, Selangor, Malaysia

Z. M. Yunus
Faculty of Applied Sciences and
    Technology
Universiti Tun Hussein Onn Malaysia
Pagoh Educational Hub
Pagoh, Johor, Malaysia

Weiwei Zhao
School of Mechanical and Electronic
    Engineering
Wuhan University of Technology
Wuhan, Hubei, P. R. China

1 The Hip Joint and Total Hip Replacement

N. A. Abu Osman,[1,4] *A. Ataollahi,*[1]
S. M. Sapuan,[2,3] *Y. Nukman,*[1] *and R. A. Ilyas*[2,3]

[1]Centre for Applied Biomechanics, Department
of Biomedical Engineering, Faculty of Engineering,
University of Malaya, Malaysia

[2]Advanced Engineering Materials and Composites Research
Centre, Department of Mechanical and Manufacturing
Engineering, Universiti Putra Malaysia, Malaysia

[3]Laboratory of Biocomposite Technology,
Institute of Tropical Forestry and Forest Products
(INTROP), Universiti Putra Malaysia, Malaysia

[4]The Chancellery, Universiti Malaysia Terengganu, Malaysia

CONTENTS

1.1   INTRODUCTION

This chapter presents a brief on hip joint and total hip replacement (THR). The hip joint is composed of soft and hard tissues. A joint comprises the femoral head, acetabulum, cartilage, and ligaments (Figure 1.1). The hip joint is classified as a ball-and-socket joint (Polkowski & Clohisy, 2010). The ball-and-socket joint provides three rotational movements, namely, flexion–extension, abduction–adduction, and internal–external rotation. The femoral head is connected to the femur via the femoral neck. The cartilage supplies a frictionless joint. The stability of the hip joint is supplied by the ligaments and muscles. This structure provides optimal stability for the stance and bipedal locomotion, but the hip joint endures complex dynamic and static loads (Bowman Jr et al., 2010).

Mechanical injury, chemical process, and/or their combination can cause degeneration and dysfunction in the articular hip joint (Bougherara et al., 2011). The most common causes of hip joint degeneration are osteoarthritis, fracture of the hip, inflammatory arthritis, femoral head necrosis, and rheumatoid arthritis (Figure 1.2).

The final recourse but the most successful procedure to remedy a severely degenerated hip joint is THR (Caeiro et al., 2011). This procedure alleviates the pain and

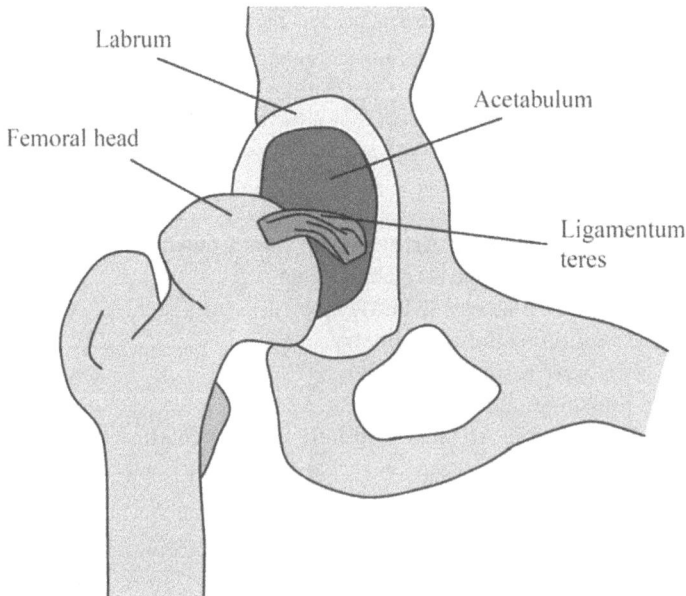

FIGURE 1.1   The hip joint (Stops et al., 2011).

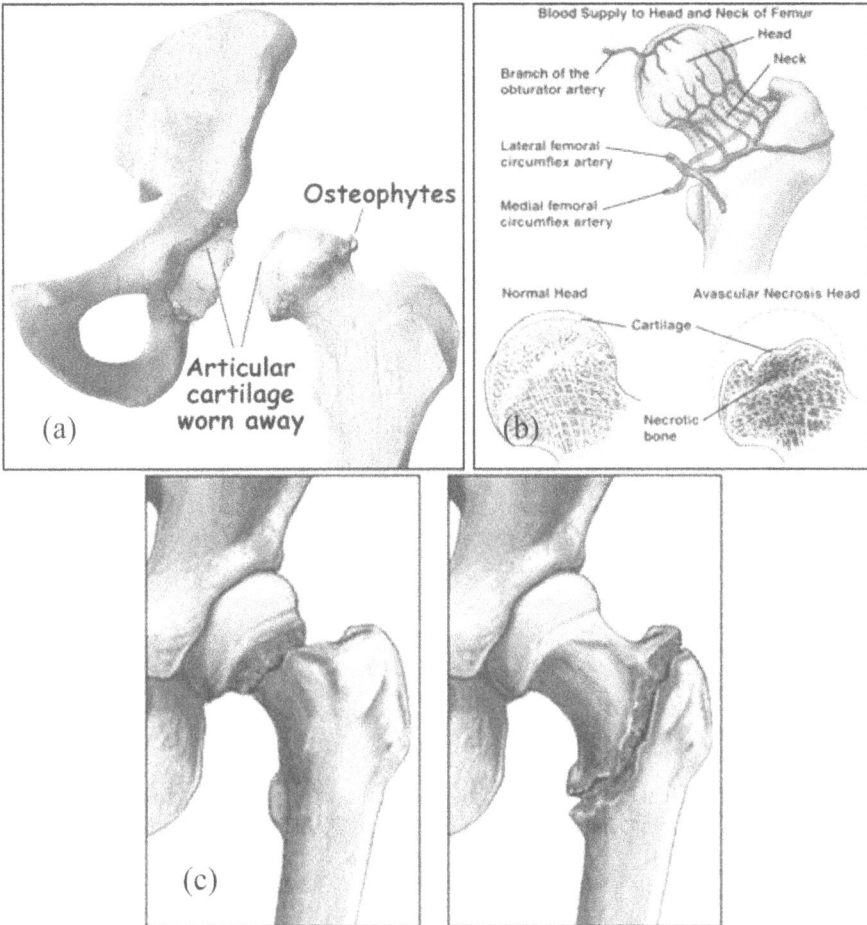

FIGURE 1.2 Three typical hip joint diseases: (a) osteoarthritis, (b) necrosis, and (c) neck fracture. (Reproduced with permission from Dunne and Ormsby (2011) and Ilesanmi (2010), Creative Commons Attribution 3.0 License 2012, IntechOpen.)

restores hip joint function. In THR, the natural hip joint is replaced with an artificial hip joint, which consists of the femoral head, acetabular cup (acetabular shell and liner), and femoral prosthesis (stem) (Figure 1.3). The artificial hip joint components are formed in a modular or monoblock structure. A femoral head may also be included in a femoral prosthesis in a monoblock structure.

1.2 IMPLANT FIXATION METHODS

The implants are fixed inside the bone with or without cement (Figure 1.4). Cemented prosthesis fixation secures an orthopedic cement prosthesis within the bone. An orthopedic cement is made of polymethylmethacrylate, which is a self-curing and nonadhesive polymeric material (Pal et al., 2013). Therefore,

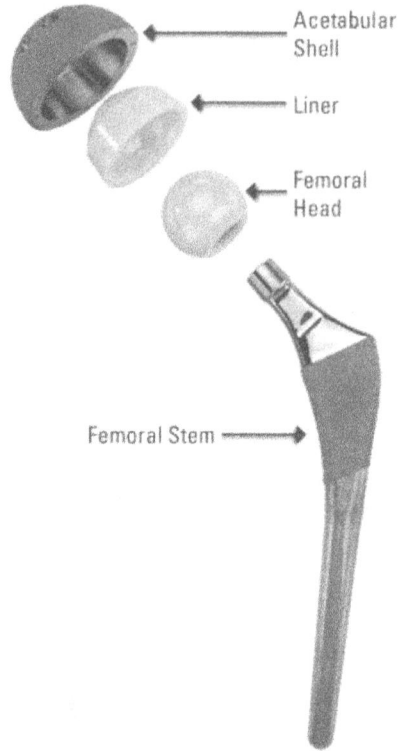

FIGURE 1.3 A typical artificial hip prosthesis. (Reproduced with permission from Li et al. (2014), Creative Commons Attribution 3.0 License 2012, IntechOpen.)

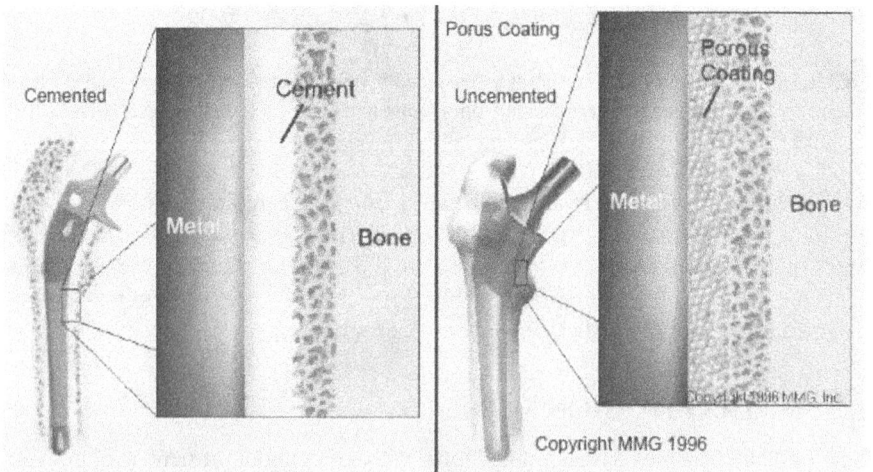

FIGURE 1.4 Typical cemented and uncemented fixation. (Reproduced with permission from Izzo (2012), Creative Commons Attribution 4.0 License 2012, IntechOpen.)

interlocking the spongy bone–cement and cement–implant features provides fixation (Pal et al., 2013). However, in a cementless prosthesis, fixation is performed by press fitting or screwing the components in the bone. This procedure guarantees primary stability for the in-growth and on-growth of the bone to the implant surfaces, thus providing secondary fixation and long-term durability. Porous and hydroxyapatite (HA) coatings are applied on the surface of a cementless prosthesis to strengthen primary and secondary fixation.

Moreover, a hybrid THR is a process in which cementless and cemented methods are used to fix the artificial hip joint components in THR. Bone quality is the most influential criterion in selecting a fixation procedure. Young and more active patients have better bone quality than old and less active patients. Accordingly, a cementless prosthesis is more appropriate for young patients, whereas a cemented prosthesis is more suitable for older patients. Each implant fixation method has advantages and disadvantages. For example, cement provides instant fixation, but a cementless prosthesis bone must grow to secure the prosthesis in the bone. In addition, a cemented prosthesis requires a bigger hole or more reaming inside the bone than a cementless prosthesis. The revision rate of patients who underwent THR with cemented prosthesis is lower than that of patients with cementless prosthesis.

1.3 TOTAL HIP REPLACEMENT FAILURE

Developments in the design, technology, and technical operation increased the success rate of THR. However, THR failure remains a problem, so revision surgery is essential and unpreventable. For example, 10% of all THR surgeries in the USA per year undergo THR revision surgery (Brown & Huo, 2002). Accordingly, the components of the old artificial joint are partially or totally replaced with new components. Mechanical factors are more common causes of THR failure than infection. Aseptic loosening is the most important cause of THR failure (Gross & Abel, 2001). The mechanisms leading to aseptic loosening remain ambiguous. Osteolysis, lack of sufficient primary stability, stress shielding, cement failure, and debonding are some of the main factors that contribute to the development of aseptic loosening and ultimately destruction of THR (Boyle & Kim, 2011; Sivarasu et al., 2011).

1.3.1 Osteolysis

The fretting of the THR joint components against each other releases debris in the joint environment, and the released debris activates the immune system, which causes bone resorption in a biological process known as osteolysis (Figure 1.5) (Bourghli et al., 2010; Fabbri et al., 2011; Suárez-Vázquez et al., 2011). Osteolysis is the main biological factor that causes aseptic loosening. Young patients have higher risk of osteolysis than old patients because of their higher range of activities that release more frictional debris (Beldame et al., 2009; Canales et al., 2010; Yoo et al., 2013). Thus, engineers have developed new designs, coatings, alloys, and bearing surfaces to cope with osteolysis (Canales et al., 2010). The risk of osteolysis in THR with polyethylene component is higher than those with ceramic-on-ceramic (CoC) or metal-on-metal (MoM) joints because of the size and amount of debris (Yoo et al., 2013).

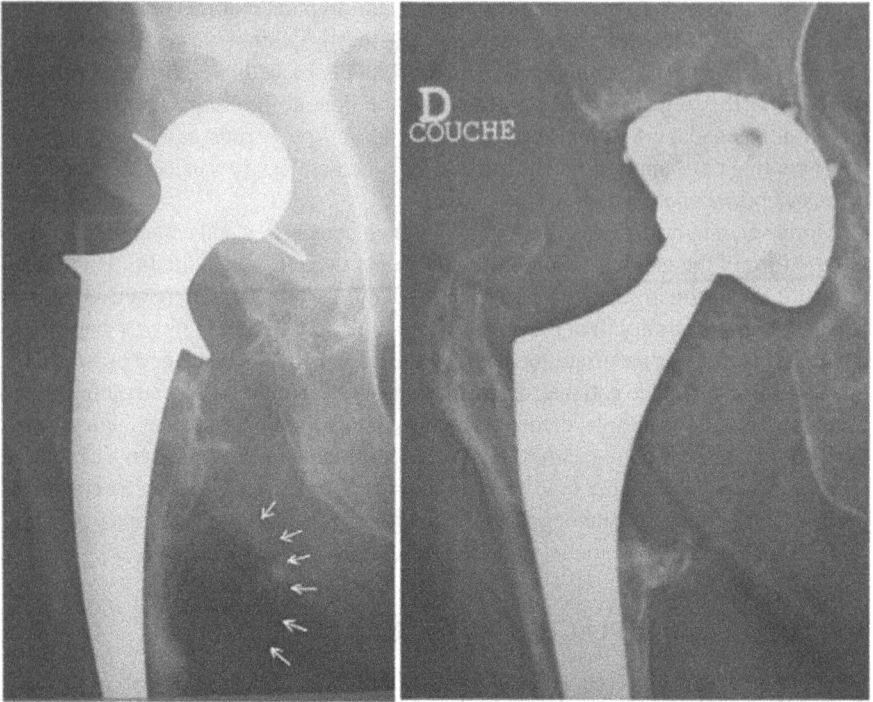

FIGURE 1.5   Osteolysis after total hip replacement (Bourghli et al., 2010).

A cross-linked polyethylene has been developed to improve its property against wear and decrease the amount of released particles.

1.3.2   PRIMARY STABILITY

Primary stability refers to the stability of prosthesis after surgery (Viceconti et al., 2006). This stability is necessary to ensure the short- and long-term THR survival and is more crucial for the cementless prosthesis than the cemented one (Abdul-Kadir et al., 2008). Primary stability is a prior condition resulting in osseointegration and reduced movement at the interfaces of THR (bone–prosthesis and cement–bone) (Cristofolini et al., 2006). Insufficient primary stability will ultimately lead to THR failure because of the excessive motion at its interfaces. Excessive motion prevents the good biological fixation between the bone and the prosthesis by decreasing the bone in-growth into the prosthesis (Hao et al., 2010). Press fitting and proper rasping procedures provide primary stability for cementless prostheses (Varini et al., 2008). Additionally, the expertise of surgeons in selecting femoral prostheses with proper size is crucial to achieve good primary stability (Varini et al., 2008). In vitro tests and numerical methods have been employed to measure the primary stability of different femoral prostheses (Viceconti et al., 2006). Moreover, intraoperative devices help surgeons to immediately examine the prosthesis stability after surgery (Varini et al., 2008).

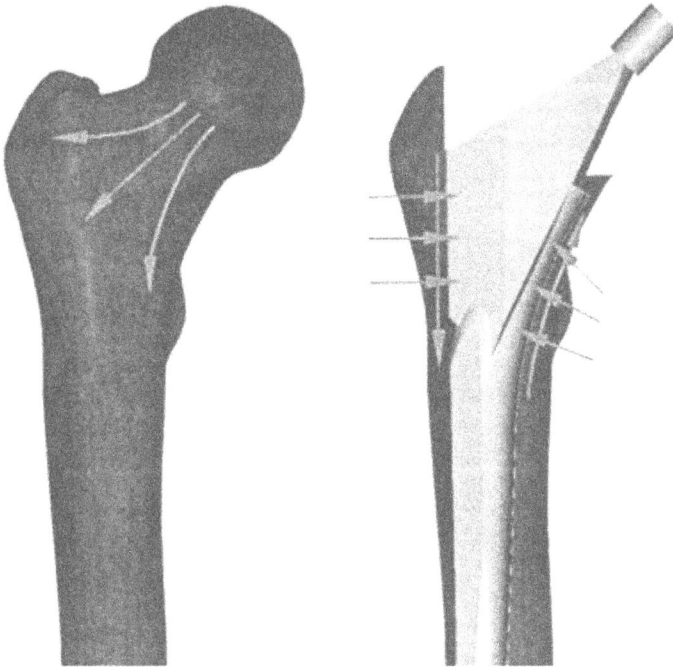

FIGURE 1.6 Load transfer before and after total hip replacement. (Reproduced with permission from Joshi et al. (2000), Elsevier.)

1.3.3 STRESS SHIELDING

The stress distribution in the femur at the hip joint is altered after implanting a femoral prosthesis because load transfer changes from the joint to the bone, as shown in Figure 1.6 (Joshi et al., 2000). The change in stress distribution is due to the mismatch between the prosthesis and bone stiffness (rigidity) (Behrens et al., 2008; Katoozian et al., 2001). Thus, some portions of the bone in THR tolerate less stress compared with the natural bone. This phenomenon is called stress shielding, and rigid prostheses can shield more load transfer from the hip joint to the femur at the proximal metaphysis (Gross & Abel, 2001). Contrary to engineering materials, the bone is a living tissue that can adapt to its mechanical and chemical environment, and it loses its strength because of load absence and stress shielding (Doblaré et al., 2004; Katoozian et al., 2001). Accordingly, excessive stress can develop at the interface of the bone–prosthesis and bone–cement (Gross & Abel, 2001) and cause aseptic loosening, which ultimately results in THR collapse. Therefore, stress shielding should be minimized after THR, and stress distribution should be similar to the physiological condition to increase THR durability (Behrens et al., 2008).

1.3.4 CEMENT FAILURE

Bone cement is a brittle material that provides stability and fixation to the prostheses cemented to the host bone (Janssen et al., 2008; Lewis, 1997). Therefore, cement layer failure results in aseptic loosening (Lai et al., 2009). The strengths of the mechanical bone cement in compressive, tensile, and bending are 75–105 MPa, 50–60 MPa, and 65–75 MPa, respectively. Moreover, the recommended thickness of the cement layer in THR ranges from 2 mm to 5 mm, whereas a cement layer with 5–10 mm thickness is deleterious for the THR life span; more cracks are also detected in thinner cement layers (Scheerlinck & Casteleyn, 2006). Cement endures dynamic mechanical repetitive loadings during daily activities (De Santis et al., 2000), and the amplitude of such loadings depends on the type of activity, such as walking, running, or stair climbing. Cyclic loads cause fatigue, which accumulates in the cement layer (Verdonschot & Huiskes, 1997). These loads cause crack initiation and propagation (Waanders et al., 2011; Zivic et al., 2012). In addition, cracks could be initiated during polymerization because of porosities or internal tension and then propagated in the cement mantle caused by the fatigue loads of daily activities, which ultimately result in cement failure (Achour et al., 2010; Zivic et al., 2012).

1.3.5 DEBONDING

Debonding at the interfaces is another factor that causes aseptic loosening and THR failure. This condition can occur at the cement–prosthesis or cement–bone interface, although most studies have shown the former case (Pérez et al., 2005). Debonding at cement–prosthesis causes higher hoop cement stress and increases the crack densities at the cement (Verdonschot & Huiskes, 1997).

1.3.6 IMPLANT FRACTURE

Fracture rarely occurs in the femoral component of THR. The number of fractures in a cemented femoral prosthesis is higher than that in a cementless prosthesis. However, fractures in the ceramic head and acetabular cup are frequently detected because of the brittleness of ceramics (Figure 1.7).

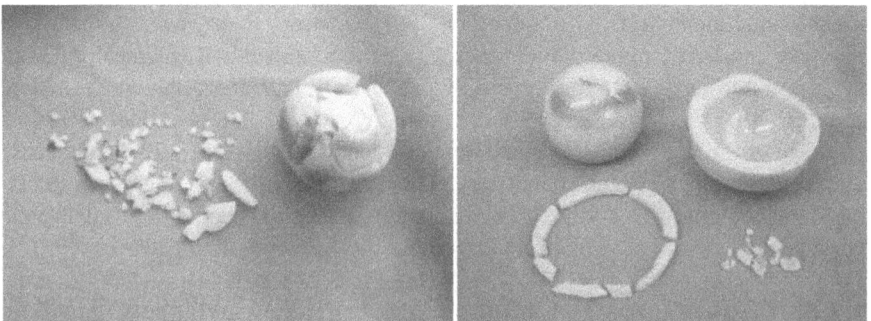

FIGURE 1.7   Fractures in a ceramic ball and acetabular cup. (Reproduced with permission from Jenabzadeh et al. (2012), Elsevier.)

1.4 MATERIAL AND GEOMETRY OF ARTIFICIAL HIP JOINT CONSTITUENTS

The main factors in THR failure are briefly presented in previous sections. Reducing the effect of these factors is the initial step to create a successful design of artificial joint components. According to the literature, the geometry, materials, and surface finishing of the prosthesis are the possible characteristics that should be adjusted to achieve optimal designs. Therefore, the following sections briefly review the materials used and the geometries of artificial hip joint constituents.

1.4.1 FEMORAL HEAD AND ACETABULAR CUP

After THR, the femoral head and acetabulum are replaced with MoM, metal-on-polymer (MoP), ceramic-on-polyethylene (CoP), or CoC bearing couples (Figure 1.8). The most commonly used couple joints are MoP and MoM (Catelas et al., 2011). The main criteria for selecting the design and materials for the hip joint bearing are fracture toughness, wear resistance, and frictional properties. Different bearings exhibit varying strengths and weaknesses. In MoP and CoP couple joints, the polymer against ceramic and metal is soft. Therefore, wear occurs in the polymer part of the joint couple. The wearing of polymer and the release of debris into the joint environment primarily cause joint luxation and osteolysis (Tudor et al., 2013). However, the developments in new cross-linked polyethylene can decrease the wear rate and particle sizes (Catelas et al., 2011). Moreover, in designing a process for the MoP couple joint to decrease friction, the artificial femoral head size should be approximately 28–36 mm, which is considerably smaller than the intact natural femoral head. MoM and CoC have been developed to prevent the release of debris in the joint environment. The second generation of the MoM joint couple with a large head and low wear rate is more preferred for THR than

FIGURE 1.8 Typical femoral heads and acetabular cups (Heimann, 2010).

the first generation, which shows very weak performance because of poor design features and surgical techniques (Molli et al., 2011; Naudie et al., 2004). Improving and optimizing metallurgical approaches (carbon content, method of fabrication, and heat treatment) and geometries (clearance, sphericity, surface finish, functional arc, fixation surface, and head size) enable the second generation of MoM to be superior to the first generation (Molli et al., 2011). Regardless of these advances in producing the MoM couple joint, its exposure to released metal ions because of the articulation wear in the joint remains unsolved (Vendittoli et al., 2011). Thus, CoC couple joints, which have outstanding wear resistance, have been developed as an alternative couple joint for MoP, MoM, and CoP (Al-Hajjar et al., 2013). The CoC couple joint provides a joint with negligible wear because of wettability and wear resistance, thus reducing periprosthetic osteolysis and the release of metal ions in the joint environment (Traina et al., 2013). Contrary to the MoM, MoP, and CoP couple joints, the wear rate in the CoC couple joint does not increase along with the femoral head size (Al-Hajjar et al., 2013). However, the intrinsic brittleness of ceramic materials is the main disadvantage of CoC couple joints.

1.4.2 FEMORAL PROSTHESIS (STEM)

A femoral prosthesis is secured within the femur and connects the upper and lower limbs (Figure 1.9). Accordingly, loads are transferred from the upper limb and hip joint to the lower limb through the femoral prosthesis, so the geometry and material for femoral prosthesis are crucial in the life span of THR.

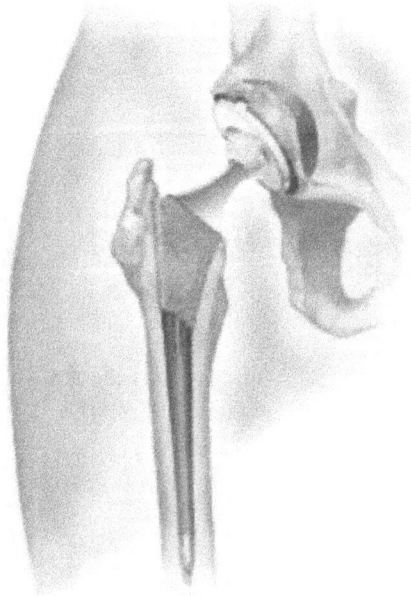

FIGURE 1.9   A typical total hip replacement. (Reproduced with permission from Jenabzadeh et al. (2012) and Jun and Choi (2010), Elsevier.)

1.4.3 FEMORAL PROSTHESIS GEOMETRY

Four different commercial femoral prostheses are presented in Figure 1.10. The optimal femoral prosthesis geometry can transfer axial and torsional loads without causing destructive stress and excessive micromotion (Scheerlinck & Casteleyn, 2006). In addition to the angle and length of the neck, the geometry of a femoral prosthesis consists of its cross section, profile, and length. Moreover, prosthesis stiffness (rigidity) is a function of the prosthesis geometry and could be optimized by altering the geometry to decrease stress shielding and bone resorption, thus prolonging the THR life span. In addition, the developed stress in the cement layer depends on the prosthesis geometry (Simpson et al., 2009). The stress in the cement and prosthesis can also be reduced by increasing the prosthesis cross section (Gross & Abel, 2001). However, anatomic factors limit the development of a new geometry for femoral prostheses (Ruben et al., 2007).

The initial stability and type of fixation within the cement and bone of the prosthesis are affected by the prosthesis geometry (Kleemann et al., 2003). The two designs used to fix a cemented prosthesis inside the cement are shape-closed (composite-beam) and force-closed (loaded-taper) (Scheerlinck & Casteleyn, 2006). In the shape-closed design, stability is provided by interlocking the cement and prosthesis through the rough surface, collars, flanges, and grooves. By contrast, in

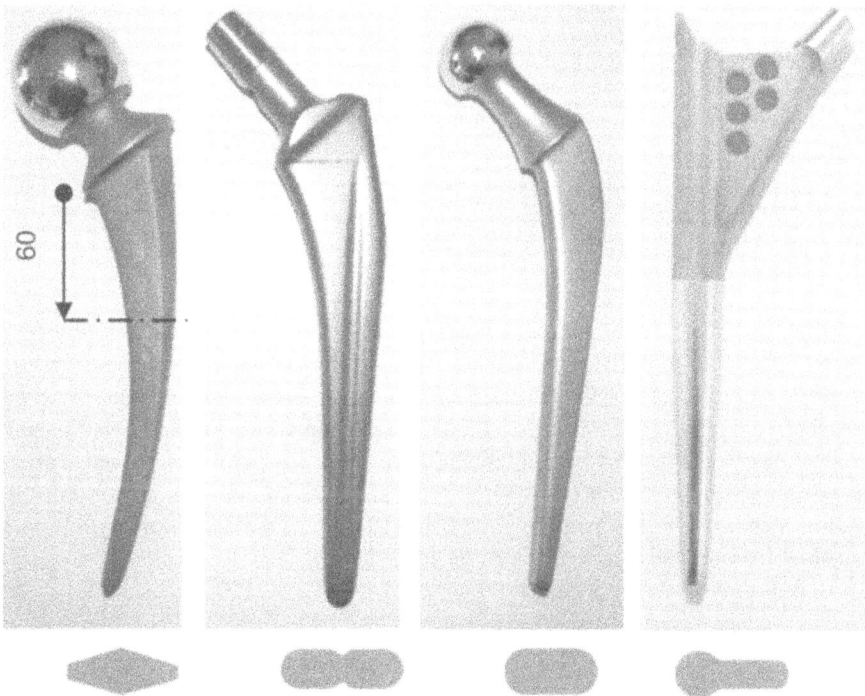

FIGURE 1.10 Typical prosthesis geometries with different cross sections and profiles. (Reproduced with permission from Ramos et al. (2012), Elsevier.)

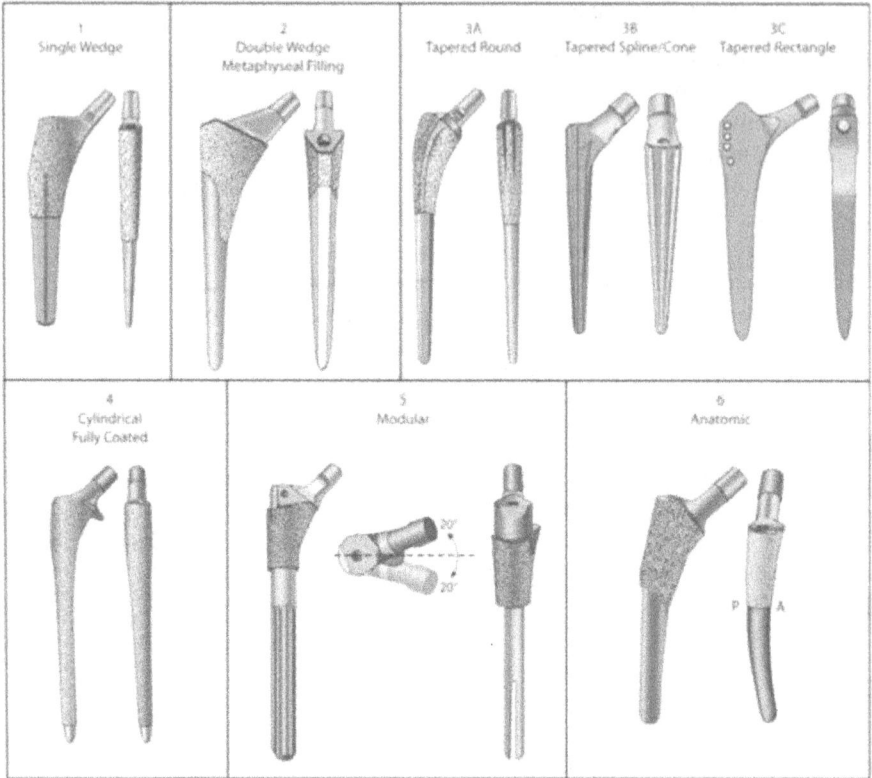

FIGURE 1.11 Schematic illustration of the different classifications of cementless femoral stem designs. Type 1 is a single wedge, Type 2 is a double wedge, Type 3A is tapered and round, Type 3B is tapered and splined, Type 3C is tapered and rectangular, Type 4 is cylindrical and fully coated, Type 5 is modular, and Type 6 is anatomic. P = posterior and A = anterior. (Reproduced with permission from Khanuja et al. (2011), Elsevier.)

force-closed design, the friction and the transfer of forces across the interface maintain the tapered prosthesis into the cement. Moreover, cementless prostheses can be categorized into six groups based on their distinct geometries (Figure 1.11) (Khanuja et al., 2011). Types 1–4 are straight femoral prostheses. Types 1 (single-wedge prosthesis), 2, and 3 are tapered with more proximal fixation, whereas Type 4 is fully coated with more distal fixation. Type 5 is a modular prosthesis, and Type 6 is a curved femoral prosthesis with anatomic designs.

1.4.4 Femoral Prosthesis Materials

Selecting materials for a femoral prosthesis is a complex task, because the implant that would be introduced into the aggressive physiological environment of the human body would be exposed to various biological and mechanical stresses (Enab & Bondok, 2013). The implant material should be biocompatible and resistant to corrosion and wear (Enab & Bondok, 2013). Moreover, the Young's modulus of the prosthesis material directly affects its stiffness and stress shielding. The Young's modulus values of the conventional materials (Ti alloy, Cr–Co, and St alloy) applied in femoral prostheses are ten times higher than the Young's modulus value of cortical bone. Thus, the risk of THR failure caused by stress shielding is high.

1.5 SURFACE FINISHING

Surface finishing is one of the main factors in the design of implants that significantly affects the longevity of THR (Jamali et al., 2006). The surface finishing (surface roughness) of the head and cup is required to provide good function, whereas the surface finishing of the stem remains debatable (Zhang et al., 2008). The finish of a metallic stem can be smooth-polished surfaces, roughened-blasted surfaces, or geometrically textured surfaces (Crowninshield, 2001).

1.6 MATERIALS UTILIZED IN ARTIFICIAL HIP JOINT COMPONENTS

The following section presents a brief review on the materials used in artificial hip joint components. The materials can be classified into four main groups, namely, metals, polymers, ceramics, and composites. Each group has strengths and weaknesses.

1.6.1 METALS

St, Co–Cr–Mo alloys, and Ti alloys are the most commonly used metals for implant designs (Khanuja et al., 2011). St is advantageous in terms of cost and processing availability (Long & Rack, 1998). However, given that St-based prostheses are prone to corrosion and fracture, Co–Cr–Mo alloys and Ti alloys are more frequently used in prostheses (Musolino et al., 1996). Co–Cr alloys are stronger than St and Ti alloys and have better corrosion resistance than St (Manicone et al., 2007). Ti alloys have lower Young's modulus, better biocompatibility, and more corrosion resistance than St- and Co-based alloys (Long & Rack, 1998). However, Ti alloys have poor shear strength and wear resistance (Long & Rack, 1998).

1.6.2 POLYMERS

Polymers are long-chain high molecular weight materials that consist of repeating monomer units (Löser & Stropp, 1999). Orthopedic implants made of polymeric material can be classified into two groups: temporary (bioresorbable or biodegradable) and permanent (long-term implant). Permanent polymeric implants are commonly produced using polyethylene, urethane, and polyketone, whereas temporary polymeric implants consist of polycaprolactone, polylactide, and polyglycolide. Sir John Charnley developed a low-friction joint with a polymeric acetabular cup made of polytetrafluoroethylene (PTFE) and a small metallic femoral head; however, PTFE has been replaced by ultrahigh molecular weight polyethylene, which has excellent energy absorption and low coefficient of friction (Long & Rack, 1998; Slouf et al., 2007).

1.6.3 CERAMICS

Ceramics are inorganic materials composed of metallic and nonmetallic elements (Asthana et al., 2006; Mackenzie, 1969). Ceramics are widely used in engineering, particularly in the aviation and automotive industries. In addition, ceramic materials have good biocompatibility, and thus they are suitable for medical devices and hard tissue replacement. Ceramics, including HA, alumina, and zirconia, have orthopedic applications.

HA, with the chemical formula of $Ca_{10}(PO_4)_6(OH)_2$, is a crystalline molecule that consists of phosphorus and calcium (Saithna, 2010). HA is the main mineral component (65%) of human bone (Havlik, 2002). This compound exhibits significant properties such as excellent biocompatibility, bioactivity, nontoxicity, and unique osteoinductivity for orthopedic applications (Ohgaki & Yamashita, 2003; Pramanik & Kar, 2013). The brittleness of HA and its lack of mechanical strength limit its application in implants (Aminzare et al., 2012). Therefore, HA can be used as a composite material by reinforcing with other materials or can be applied as a coating on the surface of implants (Aminzare et al., 2012). The HA coat creates a firm fixation by forming a biological bond between the host bone and implant (Singh et al., 2004). Thus, cementless implants coated with HA have higher survival rate than the uncoated implants (Singh et al., 2004).

Calcium silicate (CS) ($CaSiO_3$) is a highly bioactive material that induces the formation of an HA layer on its surface after soaking in simulated body fluid or human saliva. Hence, CS is an appropriate material for bone filling, implant, and bone tissue regeneration because of its osseointegration properties. However, similar to HA, CS has low fracture toughness and load-bearing capacity, thus limiting its application in the human body. Therefore, numerous studies have endeavored to enhance the load-bearing capacity and toughness of CS by reinforcing it with other materials such as alumina (Shirazi et al., 2014), carbon nanotube (Borrmann et al., 2004), graphene oxide (Xie et al., 2014), and reduced graphene oxide (Mehrali et al., 2014). In addition, CS is applied as a coating layer on metallic implants to increase their surface bioactivity and to provide a good bond with the bone and a firm fixation.

Alumina is the most stable and inert ceramic material that has been utilized in orthopedic implants (Shikha et al., 2009). Alumina is a polycrystalline ceramic that contains aluminum oxide, which is extremely hard and ranks third after diamonds and silicon carbide, and is also a scratch resistant material (Jenabzadeh et al., 2012). Alumina has a Young's modulus of 380 GPa, which is approximately twice as much as that of St (Hannouche et al., 2005). Zirconia, a crystalline dioxide (ZrO_2) of zirconium, has good chemical and dimensional stability, wear resistance, mechanical strength, and toughness, in addition to the following characteristics: Young's modulus similar to that of St; tensile strength, between 900 MPa and 1,200 MPa; and compressive strength, 2,000 MPa (Piconi & Maccauro, 1999). A molecularly stable zirconia can be achieved by mixing it with other metallic oxides, such as MgO, CaO, or Y_2O_3 (Manicone et al., 2007). Despite the difficulty in stabilizing zirconia with Y_2O_3 sintering, this combination presents better mechanical properties than other combinations (Manicone et al., 2007). A biomedical grade of zirconia that was

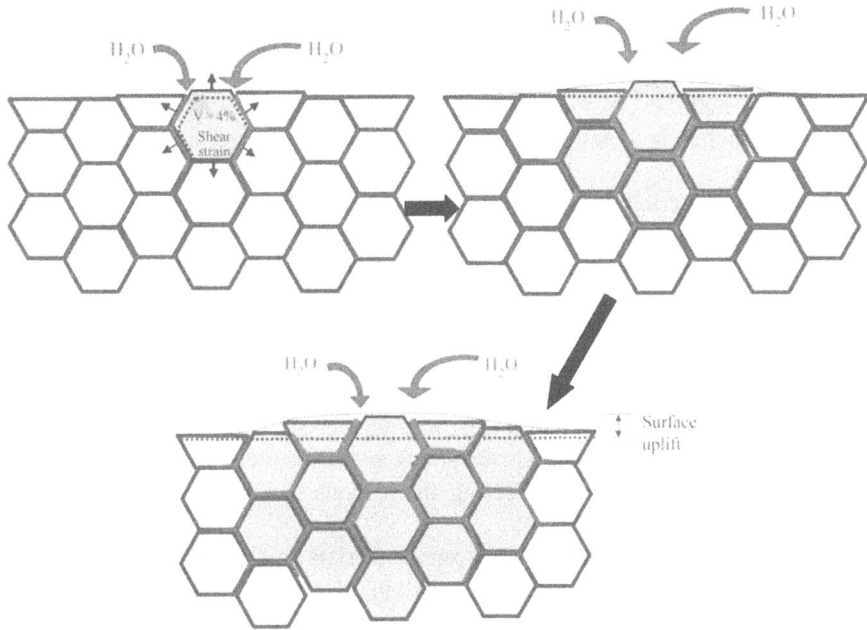

FIGURE 1.12 Scheme of the aging process. (Reproduced with permission from Chevalier (2006), Elsevier.)

proposed in 1969 for orthopedic implants and for replacement of Ti and alumina implants has comparable brittleness with that of alumina, thus preventing implant failure (Chevalier, 2006). However, zirconia aging and surface grinding have detrimental effects on its properties and toughness (Figure 1.12) (Kosmač et al., 1999; Luthardt et al., 2002).

1.6.4 COMPOSITES

Composites are engineered materials composed of two or more constituents. Currently, composite materials have been used in different fields of engineering, such as biomedical engineering, to produce new devices and implants (De Oliveira Simoes & Marques, 2001). The properties of composites can be modified according to different requirements; moreover, composites overcome the limitation of using single-phase material with the use of combined materials (Evans & Gregson, 1998). Therefore, these materials have better biological and mechanical compatibilities with body tissues and optimal strength and durability (Evans & Gregson, 1998). Orthopedic composites can be classified into polymer composites, ceramic composites, metal composites, and functionally graded materials (FGMs). In polymer composites, biocompatible polymers are applied as matrix with the reinforced materials (particulates, short or continuous, woven fibers (fabric), and nanofillers), regardless of the curing process (thermoset and thermoplastic). Thermoset polymer composites with low Young's modulus and high strength have been implemented in femoral prostheses and fixation devices (Scholz et al., 2011). However, their performance in fixation devices is better than that in femoral prostheses (Evans & Gregson, 1998). Moreover, thermoplastic polymer composites have been used in acetabular cups and artificial knee joint bearing.

Composite materials made of ceramics and metals are categorized based on the matrix and reinforcing materials into ceramic–metal composites (CMCs) and metal–ceramic composites (MCCs). The significant change in mechanical properties is caused by the inclusion of ceramic or metal particles into the metal or ceramic matrixes (Rodriguez-Suarez et al., 2012). Therefore, CMCs and MCCs possess superior stiffness, fracture, fatigue, tribological, and thermal properties to their monolithic ceramic and metal counterparts because of the overlapped strengths and weaknesses of the ceramics and metals (Mattern et al., 2004). Accordingly, conventional and monolithic materials (ceramics and metals) can rapidly change with these composites in various engineering applications such as in the aerospace and automobile industries (Sahin, 2005).

FGMs are special groups of composite materials that incorporate continuous change (gradient) or stepwise change (graded) in their microstructure and properties as shown in Figure 1.13 (Miao & Sun, 2009). This concept was obtained from their natural biological structures (Pompe et al., 2003). Adapting materials with specific structural, compositional, morphological, and mechanical properties has emphasized that FGMs can be utilized in the design of new prostheses. The mechanical properties of FGMs can be optimized and controlled by adjusting the volume fraction of each material phase (Nie & Batra, 2010). In addition, the FGM-based implants provide better load-bearing capacity, fracture toughness, and wear resistance than

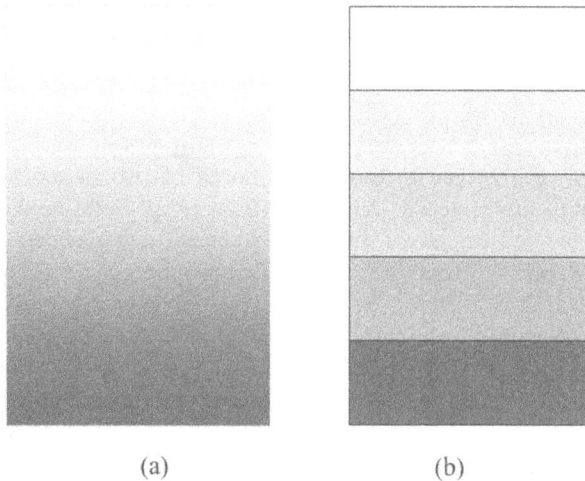

(a)   (b)

FIGURE 1.13   A typical FGM structure: (a) gradient and (b) graded.

their monolithic ceramic or metallic counterparts (Miao & Sun, 2009; Mishina et al., 2008; Zhang et al., 2012).

1.7 NUMERICAL METHODS IN HIP JOINT BIOMECHANICS AND IMPLANT STUDY

Numerical (computational) methods, such as the finite element method (FEM), extended FEM, and boundary element method, are powerful mathematical analysis tools that are widely used in different fields of engineering. Numerical methods are well accepted in biomedical engineering and biomechanics. In the numerical study of implant design, various implant design configurations are considered in a computer rather than performing expensive and destructive experimental tests (Asgari et al., 2004). Accordingly, many researchers have used FEM in analyzing hip joint, hip joint biomechanics, and hip implants. FEM has been used by prosthetists and engineers in hip implant design to address problems such as implant failure, stress shielding, and bone resorption, which are related to the prosthesis material and design. The materials for prosthesis were discussed by El-Sheikh et al. (2002), Akay and Aslan (1996), Kaddick et al. (1997), Katoozian et al. (2001), Simões and Marques (2005), Kuiper and Huiskes (1997), and Hedia et al. (2004, 2006). Moreover, El-Sheikh et al. (2002) examined stress distribution in the implanted hip components to select the optimal material for femoral prosthesis; the study was conducted by inserting a femoral prosthesis with four different Young's moduli: 25, 100, 196, and 400 GPa. The prostheses with lower Young's modulus tolerate less stress compared with those with higher Young's modulus, as shown in Figure 1.14. Moreover, the developed stresses increase in the bone and cement.

Akay and Aslan (1996), Kaddick et al. (1997), Katoozian et al. (2001), and Simões and Marques (2005) used FEM to determine whether composite materials can replace the conventional materials used in femoral prosthesis. Katoozian et al. (2001) investigated the effect of the fiber orientations in the composites on the stress distribution in the implanted femur components. Moreover, Simões and Marques (2005) used composite materials to construct a prosthesis with a metal core and variable stiffness, as shown in Figure 1.15. High strain energy was observed in the proximal metaphysis of the bone, indicating the less stress-shielding effect of the prosthesis with tailorable stiffness. However, more principal stress was detected in the bone because of the implantation of the developed femoral prosthesis compared with those of the prostheses composed of conventional materials.

Kuiper and Huiskes (1997) and Hedia et al. (2004, 2006) evaluated the FGM performance in femoral prosthesis using two-dimensional (2D) FEM. The stress shielding, interface stress, and developed stress in the implant decline when FGMs are utilized in the femoral prosthesis. However, the developed stress in the bone and cement increases. Moreover, designing and optimizing the geometry of a femoral prosthesis using FEM were performed by Gross and Abel (2001), Sabatini and Goswami (2008), Bennett and Goswami (2008), and Ramos et al. (2012). Gross and

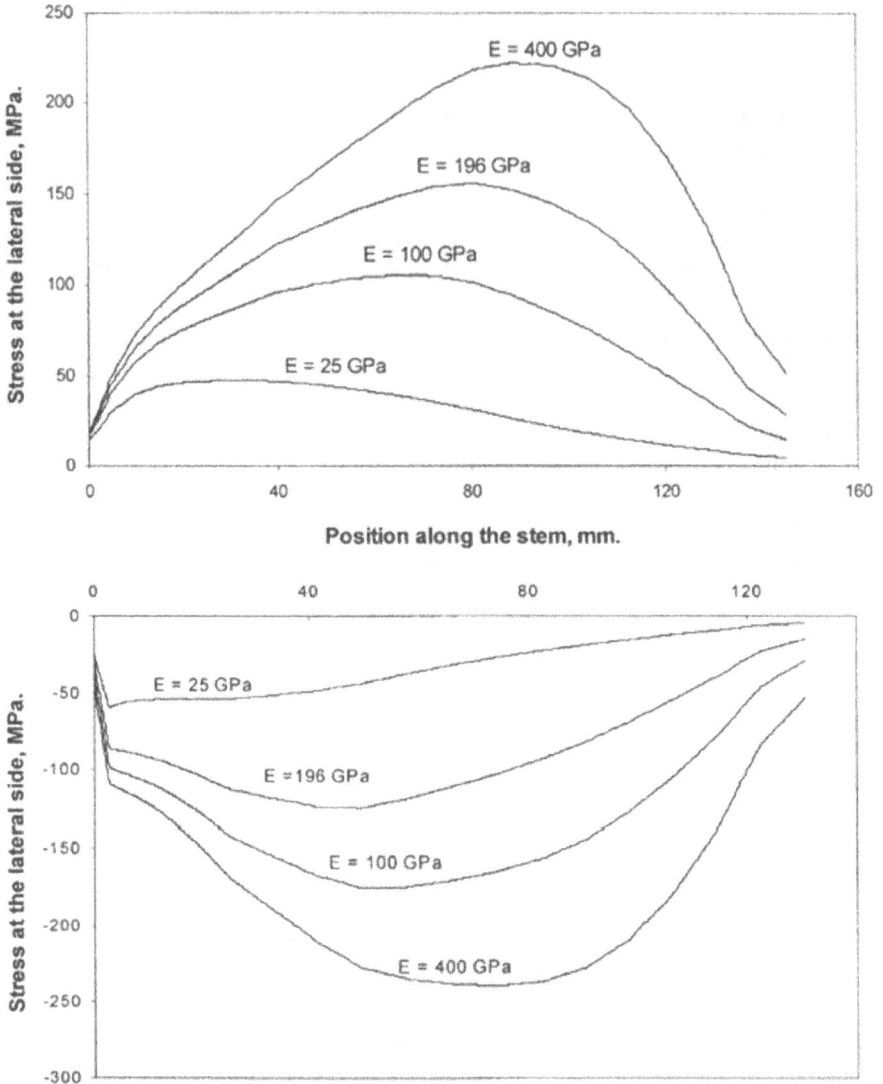

FIGURE 1.14 Minimum and maximum principal stress distributions on the lateral and medial sides of the stem as a function of the prosthesis Young's modulus. (Reproduced with permission from El-Sheikh et al. (2002), Elsevier.)

Abel used a hollow stem to decrease prosthesis stiffness (rigidity) and stress shielding as shown in Figure 1.16.

Sabatini and Goswami (2008) and Bennett and Goswami (2008) investigated the effect of the different geometries of prostheses on stress distribution in the implanted femur components. The different cross sections and profiles used in their studies are presented in Figure 1.17.

Ramos et al. (2012) numerically examined various cemented femoral prostheses with different cross sections and developed a new cemented femoral prosthesis

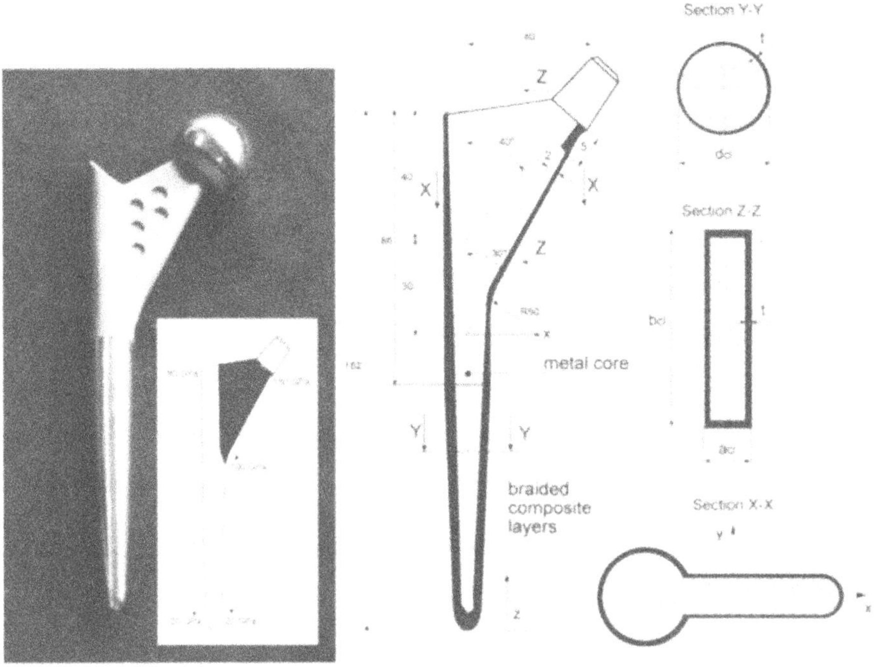

FIGURE 1.15 A femoral prosthesis with metal core and variable stiffness. Unit in mm. (Reproduced with permission from Simões and Marques (2005), Elsevier.)

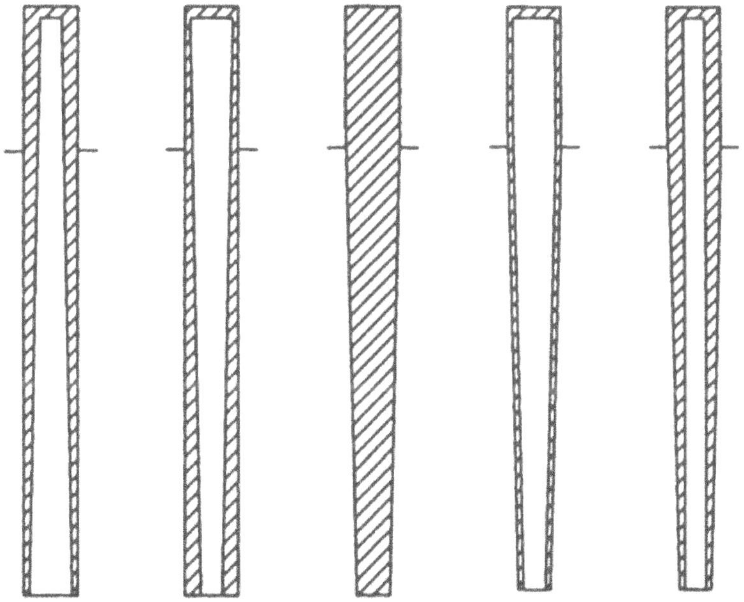

FIGURE 1.16 Hollow stems introduced by Gross and Abel (2001). (Reproduced with permission from Gross and Abel (2001), Elsevier.)

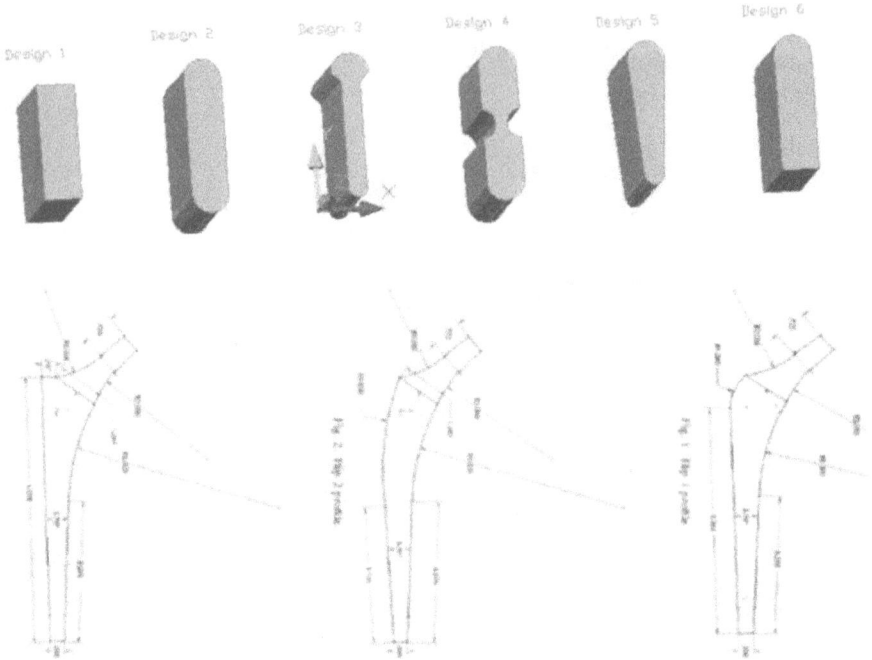

FIGURE 1.17 Different cross sections and profiles. (Reproduced with permission from Bennett and Goswami (2008); Sabatini and Goswami (2008), Elsevier.)

geometry as shown in Figure 1.18. The new design provides 25% less stress in the cement compared with that of conventional prostheses.

1.8 LOAD TRANSFER IN THE PROXIMAL FEMUR

All biological tissues have composite structures. This assumption is true for bone–implant combinations. The bone and implant have different material properties. Thus, an interface is required at which the two materials are integrated. One of the most important issues in the bone–implant interfaces is the mechanism of transferring loads from the implant to the surrounding bone. If the two materials are bonded and equal forces are applied with equal strains, Hook's law and some simple algebra should be used to determine the load shared on each part of the composite structure as shown in Equations (1.1) and (1.2).

$$F_i = \frac{A_i E_i F}{A_i E_i + A_b E_b} \tag{1.1}$$

$$F_b = \frac{A_b E_b F}{A_i E_i + A_b E_b} \tag{1.2}$$

where the subscript i denotes implant, and b represents the bone.

FIGURE 1.18   Stem geometry developed by Ramos et al. (2012). (Reproduced with permission from Ramos et al. (2012), Elsevier.)

As indicated by these analogies for the composite structures and bone–implant configurations, the load transfer mechanism in the femoral hip component exhibits some basic characteristics, regardless of the stem shape and the precise joint load. These basic characteristics can be illustrated by a simplified model of a straight implant mounted on a straight bone tube (Figure 1.19) (Huiskes, 1988).

Figure 1.19 shows that a large load is initially supported by the stem axially and in bending. Afterward, the bone and implant share the load. A high shear stress exists at the interface when the load is transferred. Similar to the previous results of composite bar analysis, a load is shared between the bone and implant with the ratio of implant stiffness to bone stiffness. Higher implant stiffness results in more loads supported by the implant as shown in Equations (1.3) and (1.4).

$$\frac{F_i}{F_{\text{normal}}} = \frac{A_i E_i}{A_i E_i + A_b E_b} \tag{1.3}$$

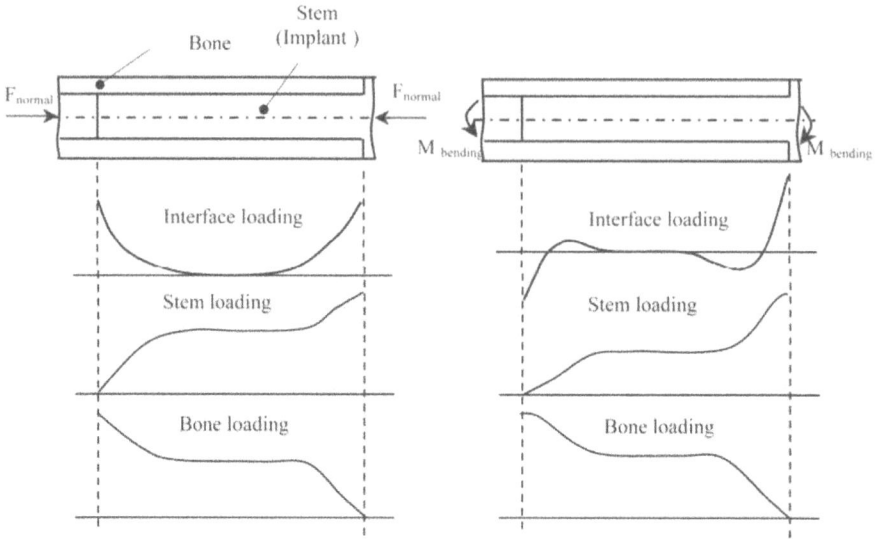

FIGURE 1.19 Principles of the load transfer mechanism explained with a simplified intramedullary fixation model (Huiskes, 1988).

$$\frac{M_i}{M_{\text{bending}}} = \frac{I_i E_i}{I_i E_i + I_b E_b} \tag{1.4}$$

where F_i/F_{normal} is the normal load shared under axial load, and M_i/M_{bending} is the transverse load shared under bending loads. This finding demonstrates that stress shielding is attributed to high implant stiffness relative to bone stiffness.

The mismatch between the bone and implant stiffness is important for determining stress shielding and interface stress. Higher mismatches between the implant and bone stiffness result in a higher degree of stress shielding because more load is supported by the implant. However, as the implant carries more stress, lower loads should be transferred to the bone, resulting in lower interface stress.

1.9 BONE

The skeleton is mostly composed of bony components. Unlike engineering materials, bones are living tissues that can adapt to their mechanical and hormonal environment. Bones are composite materials composed of minerals and collagen with complex and unique mechanical properties. The bone functions are related to age, disease, and use. Moreover, bones are considered FGMs with composition and property dependent on direction and location. Long bones, such as the femur, consist of two different bony structures: spongy (cancellous or trabecular) and cortical (compact bone) (Figure 1.20).

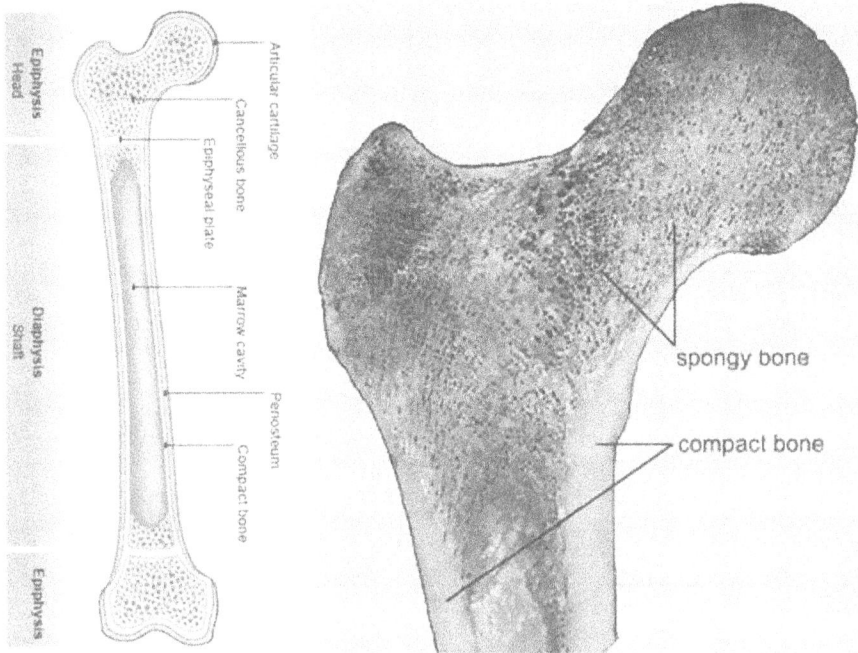

FIGURE 1.20   Bone structure (Juillard, 2011).

1.10   CONCLUSION

This chapter presented a brief review regarding the hip joint and THR. Several studies have been performed to determine the optimal material and geometrical design for femoral prosthesis. However, an optimal hip design remains ambiguous and should be further investigated. Previous studies have focused only on the materials or geometrical design for prosthesis. The increasing incidence of THR involving young patients has motivated the use of FGMs in femoral prosthesis to prolong the life span of THR.

REFERENCES

Abdul-Kadir, M. R., Hansen, U., Klabunde, R., Lucas, D., & Amis, A. (2008). Finite element modelling of primary hip stem stability: The effect of interference fit. *Journal of Biomechanics*, 41(3), 587–594.

Achour, T., Tabeti, M. S. H., Bouziane, M. M., Benbarek, S., Bachir Bouiadjra, B., & Mankour, A. (2010). Finite element analysis of interfacial crack behaviour in cemented total hip arthroplasty. *Computational Materials Science*, 47(3), 672–677.

Akay, M., & Aslan, N. (1996). Numerical and experimental stress analysis of a polymeric composite hip joint prosthesis. *Journal of Biomedical Materials Research*, 31(2), 167–182.

Al-Hajjar, M., Jennings, L. M., Begand, S., Oberbach, T., Delfosse, D., & Fisher, J. (2013). Wear of novel ceramic-on-ceramic bearings under adverse and clinically relevant hip simulator conditions. *Journal of Biomedical Materials Research Part B: Applied Biomaterials*, 101(8), 1456–1462.

Aminzare, M., Eskandari, A., Baroonian, M., Berenov, A., Razavi Hesabi, Z., Taheri, M., & Sadrnezhaad, S. (2012). Hydroxyapatite nanocomposites: Synthesis, sintering and mechanical properties. *Ceramics International*, 39(3), 2197–2206.

Asgari, S. A., Hamouda, A., Mansor, S., Singh, H., Mahdi, E., Wirza, R., & Prakash, B. (2004). Finite element modeling of a generic stemless hip implant design in comparison with conventional hip implants. *Finite Elements in Analysis and Design*, 40(15), 2027–2047.

Asthana, R., Kumar, A., & Dahotre, N. B. (2006). 3 - Powder metallurgy and ceramic forming. *Materials processing and manufacturing science* (pp. 167–245). Burlington: Academic Press.

Behrens, B., Wirth, C., Windhagen, H., Nolte, I., Meyer-Lindenberg, A., & Bouguecha, A. (2008). Numerical investigations of stress shielding in total hip prostheses. *Proceedings of the Institution of Mechanical Engineers, Part H: Journal of Engineering in Medicine*, 222(5), 593–600.

Beldame, J., Carreras, F., Oger, P., & Beaufils, P. (2009). Cementless cups do not increase osteolysis risk in metal-on-metal total hip arthroplasty. *Orthopaedics & Traumatology: Surgery & Research*, 95(7), 478–490.

Bennett, D., & Goswami, T. (2008). Finite element analysis of hip stem designs. *Materials & Design*, 29(1), 45–60.

Borrmann, T., Edgar, K., McFarlane, A., Spencer, J., & Johnston, J. (2004). Calcium silicate–carbon nanotube composites. *Current Applied Physics*, 4(2), 359–361.

Bougherara, H., Zdero, R., Dubov, A., Shah, S., Khurshid, S., & Schemitsch, E. H. (2011). A preliminary biomechanical study of a novel carbon–fibre hip implant versus standard metallic hip implants. *Medical Engineering & Physics*, 33(1), 121–128.

Bourghli, A., Fabre, T., Tramond, P., & Durandeau, A. (2010). Total hip replacement pseudotumoral osteolysis. *Orthopaedics & Traumatology: Surgery & Research*, 96(3), 319–322.

Bowman Jr, K. F., Fox, J., & Sekiya, J. K. (2010). A clinically relevant review of hip biomechanics. *Arthroscopy*, 26(8), 1118–1129.

Boyle, C., & Kim, I. Y. (2011). Comparison of different hip prosthesis shapes considering micro-level bone remodeling and stress-shielding criteria using three-dimensional design space topology optimization. *Journal of Biomechanics*, 44(9), 1722–1728.

Brown, B. S., & Huo, M. H. (2002). Revision total hip replacement for osteolysis. *Current Opinion in Orthopaedics*, 13(1), 48–52.

Caeiro, J., Riba, J., & Gomar, F. (2011). Incidence and risk factors of dislocation after total hip replacement with a ceramic acetabular system. *Revista Española de Cirugía Ortopédica y Traumatología (English Edition)*, 55(6), 437–445.

Canales, V., Panisello, J. J., Herrera, A., Sola, A., Mateo, J. J., & Caballero, M. J. (2010). Extensive osteolysis caused by polyethylene particle migration in an anatomical hydroxyapatite-coated hip prosthesis: 10 years' follow-up. *The Journal of Arthroplasty*, 25(7), 1115–1124, 1124.e1.

Catelas, I., Wimmer, M. A., & Utzschneider, S. (2011). Polyethylene and metal wear particles: Characteristics and biological effects. *Seminars in Immunopathology*, 33(3), 257–271.

Chevalier, J. (2006). What future for zirconia as a biomaterial? *Biomaterials*, 27(4), 535–543.

Cristofolini, L., Varini, E., Pelgreffi, I., Cappello, A., & Toni, A. (2006). Device to measure intra-operatively the primary stability of cementless hip stems. *Medical Engineering & Physics*, 28(5), 475–482.

Crowninshield, R. (2001). Femoral hip implant fixation within bone cement. *Operative Techniques in Orthopaedics*, 11(4), 296–299.

De Oliveira Simoes, J., & Marques, A. (2001). Determination of stiffness properties of braided composites for the design of a hip prosthesis. *Composites Part A: Applied Science and Manufacturing*, 32(5), 655–662.

De Santis, R., Ambrosio, L., & Nicolais, L. (2000). Polymer-based composite hip prostheses. *Journal of Inorganic Biochemistry*, 79(1–4), 97–102.

Doblaré, M., García, J. M., & Gómez, M. J. (2004). Modelling bone tissue fracture and healing: A review. *Engineering Fracture Mechanics*, 71(13–14), 1809–1840.

Dunne, N., & Ormsby, R. W. (2011). MWCNT Used in Orthopaedic Bone Cements.

El-Sheikh, H., MacDonald, B., & Hashmi, M. (2002). Material selection in the design of the femoral component of cemented total hip replacement. *Journal of Materials Processing Technology*, 122(2), 309–317.

Enab, T. A., & Bondok, N. E. (2013). Material selection in the design of the tibia tray component of cemented artificial knee using finite element method. *Materials & Design*, 44, 454–460.

Evans, S., & Gregson, P. (1998). Composite technology in load-bearing orthopaedic implants. *Biomaterials*, 19(15), 1329–1342.

Fabbri, N., Rustemi, E., Masetti, C., Kreshak, J., Gambarotti, M., Vanel, D., Toni, A., & Mercuri, M. (2011). Severe osteolysis and soft tissue mass around total hip arthroplasty: Description of four cases and review of the literature with respect to clinico-radiographic and pathologic differential diagnosis. *European Journal of Radiology*, 77(1), 43–50.

Gross, S., & Abel, E. W. (2001). A finite element analysis of hollow stemmed hip prostheses as a means of reducing stress shielding of the femur. *Journal of Biomechanics*, 34(8), 995–1003.

Hannouche, D., Hamadouche, M., Nizard, R., Bizot, P., Meunier, A., & Sedel, L. (2005). Ceramics in total hip replacement. *Clinical Orthopaedics and Related Research*, 430, 62–71.

Hao, S., Taylor, J. T., Bowen, C. R., Gheduzzi, S., & Miles, A. W. (2010). Sensing methodology for in vivo stability evaluation of total hip and knee arthroplasty. *Sensors and Actuators A: Physical*, 157(1), 150–160.

Havlik, R. J. (2002). Hydroxyapatite. *Plastic and Reconstructive Surgery*, 110(4), 1176–1179.

Hedia, H., Shabara, M., El-Midany, T., & Fouda, N. (2004). A method of material optimization of cementless stem through functionally graded material. *International Journal of Mechanics and Materials in Design*, 1(4), 329–346.

Hedia, H., Shabara, M., El-Midany, T., & Fouda, N. (2006). Improved design of cementless hip stems using two-dimensional functionally graded materials. *Journal of Biomedical Materials Research Part B: Applied Biomaterials*, 79(1), 42–49.

Heimann, R. B. (2010). *Classic and advanced ceramics: From fundamentals to applications*, John Wiley & Sons, Wiley-VCH Verlag GmbH & Co., KGaA.

Huiskes, R. (1988). Stress patterns, failure modes, and bone remodeling. *Non-cemented hip arthroplasty* (pp. 283–302). New York: Raven Press.

Ilesanmi, O. O. (2010). Pathological basis of symptoms and crises in sickle cell disorder: Implications for counseling and psychotherapy. *Hematology Reports*, 2(1), e2.

Izzo, G. M. (2012). Support for total hip replacement surgery: Structures modeling, gait data analysis and report system. *European Journal of Translational Myology*, 22(1–2), 69–121.

Jamali, A. A., Lozynsky, A. J., & Harris, W. H. (2006). The effect of surface finish and of vertical ribs on the stability of a cemented femoral stem: An in vitro stair climbing test. *The Journal of Arthroplasty*, 21(1), 122–128.

Janssen, D., Mann, K. A., & Verdonschot, N. (2008). Micro-mechanical modeling of the cement–bone interface: The effect of friction, morphology and material properties on the micromechanical response. *Journal of Biomechanics*, 41(15), 3158–3163.

Jenabzadeh, A.-R., Pearce, S. J., & Walter, W. L. (2012). Total hip replacement: Ceramic-on-ceramic. *Seminars in Arthroplasty*, 23(4), 232–240.

Joshi, M. G., Advani, S. G., Miller, F., & Santare, M. H. (2000). Analysis of a femoral hip prosthesis designed to reduce stress shielding. *Journal of Biomechanics*, 33(12), 1655–1662.

Juillard, F. (2011). Effects of Tissue-Level Ductility on Trabecular Bone Strength. Paper No. 0135 • ORS 2012 Annual Meeting.

Jun, Y., & Choi, K. (2010). Design of patient-specific hip implants based on the 3D geometry of the human femur. *Advances in Engineering Software*, 41(4), 537–547.

Kaddick, C., Stur, S., & Hipp, E. (1997). Mechanical simulation of composite hip stems. *Medical Engineering and Physics*, 19(5), 431–439.

Katoozian, H., Davy, D. T., Arshi, A., & Saadati, U. (2001). Material optimization of femoral component of total hip prosthesis using fiber reinforced polymeric composites. *Medical Engineering & Physics*, 23(7), 505–511.

Khanuja, H. S., Vakil, J. J., Goddard, M. S., & Mont, M. A. (2011). Cementless femoral fixation in total hip arthroplasty. *The Journal of Bone & Joint Surgery*, 93(5), 500–509.

Kleemann, R. U., Heller, M. O., Stoeckle, U., Taylor, W. R., & Duda, G. N. (2003). THA loading arising from increased femoral anteversion and offset may lead to critical cement stresses. *Journal of Orthopaedic Research*, 21(5), 767–774.

Kosmač, T., Oblak, C., Jevnikar, P., Funduk, N., & Marion, L. (1999). The effect of surface grinding and sandblasting on flexural strength and reliability of Y-TZP zirconia ceramic. *Dental Materials*, 15(6), 426–433.

Kuiper, J., & Huiskes, R. (1997). Mathematical optimization of elastic properties: Application to cementless hip stem design. *Transactions-American Society of Mechanical Engineers Journal of Biomechanical Engineering*, 119, 166–174.

Lai, Y. S., Wei, H. W., Chang, T. K., & Cheng, C. K. (2009). The effects of femoral canal geometries, stem shapes, cement thickness, and stem materials on the choice of femoral implant in cemented total hip replacement. *Journal of the Chinese Institute of Engineers*, 32(3), 333–341.

Lewis, G. (1997). Properties of acrylic bone cement: State of the art review. *Journal of Biomedical Materials Research*, 38(2), 155–182.

Li, Y., Yang, C., Zhao, H., Qu, S., Li, X., & Li, Y. (2014). New developments of Ti-based alloys for biomedical applications. *Materials*, 7(3), 1709–1800.

Long, M., & Rack, H. J. (1998). Titanium alloys in total joint replacement—a materials science perspective. *Biomaterials*, 19(18), 1621–1639.

Löser, E., & Stropp, G. (1999). Chapter 38 - Polymers. In: H. Marquardt, S. G. Schäfer, R. O. McClellan, and F. Welsch (Eds.), *Toxicology* (pp. 919–936). San Diego: Academic Press.

Luthardt, R., Holzhüter, M., Sandkuhl, O., Herold, V., Schnapp, J., Kuhlisch, E., & Walter, M. (2002). Reliability and properties of ground Y-TZP-zirconia ceramics. *Journal of Dental Research*, 81(7), 487–491.

Mackenzie, J. D. (1969). Ceramics in ocean engineering. *Ocean Engineering*, 1(5), 555–571.

Manicone, P. F., Rossi Iommetti, P., & Raffaelli, L. (2007). An overview of zirconia ceramics: Basic properties and clinical applications. *Journal of Dentistry*, 35(11), 819–826.

Mattern, A., Huchler, B., Staudenecker, D., Oberacker, R., Nagel, A., & Hoffmann, M. (2004). Preparation of interpenetrating ceramic–metal composites. *Journal of the European Ceramic Society*, 24(12), 3399–3408.

Mehrali, M., Moghaddam, E., Shirazi, S. F. S., Baradaran, S., Mehrali, M., Latibari, S. T., Metselaar, H. S. C., Kadri, N. A., Zandi, K., & Osman, N. A. A. (2014). Synthesis, mechanical properties, and in vitro biocompatibility with osteoblasts of calcium silicate–reduced graphene oxide composites. *ACS Applied Materials & Interfaces*, 6(6), 3947–3962.

Miao, X., & Sun, D. (2009). Graded/gradient porous biomaterials. *Materials*, 3(1), 26–47.

Mishina, H., Inumaru, Y., & Kaitoku, K. (2008). Fabrication of ZrO_2/AISI316L functionally graded materials for joint prostheses. *Materials Science and Engineering: A*, 475(1), 141–147.

Molli, R. G., Lombardi, A. V., Berend, K. R., Adams, J. B., & Sneller, M. A. (2011). Metal-on-metal vs metal-on-improved polyethylene bearings in total hip arthroplasty. *The Journal of Arthroplasty*, 26(6), 8–13.

Musolino, M., Pettit, F., Burleigh, T., Rubash, H., & Shanbhag, A. (1996). Analysis of corrosion in stainless steel total hip prostheses. *Proceedings of the 1996 Fifteenth Southern Biomedical Engineering Conference* (pp. 5–6).

Naudie, D., Roeder, C. P., Parvizi, J., Berry, D. J., Eggli, S., & Busato, A. (2004). Metal-on-metal versus metal-on-polyethylene bearings in total hip arthroplasty: A matched case-control study. *The Journal of Arthroplasty*, 19(7), 35–41.

Nie, G., & Batra, R. (2010). Material tailoring and analysis of functionally graded isotropic and incompressible linear elastic hollow cylinders. *Composite Structures*, 92(2), 265–274.

Ohgaki, M., & Yamashita, K. (2003). Preparation of polymethylmethacrylate-reinforced functionally graded hydroxyapatite composites. *Journal of the American Ceramic Society*, 86(8), 1440–1442.

Pal, B., Puthumanapully, P. K., & Amis, A. A. (2013). Biomechanics of implant fixation. *Orthopaedics and Trauma*, 27(2), 76–84.

Pérez, M. A., García, J. M., & Doblaré, M. (2005). Analysis of the debonding of the stem–cement interface in intramedullary fixation using a non-linear fracture mechanics approach. *Engineering Fracture Mechanics*, 72(8), 1125–1147.

Piconi, C., & Maccauro, G. (1999). Zirconia as a ceramic biomaterial. *Biomaterials*, 20(1), 1–25.

Polkowski, G. G., & Clohisy, J. C. (2010). Hip biomechanics. *Sports Medicine and Arthroscopy Review*, 18(2), 56–62.

Pompe, W., Worch, H., Epple, M., Friess, W., Gelinsky, M., Greil, P., Hempel, U., Scharnweber, D., & Schulte, K. (2003). Functionally graded materials for biomedical applications. *Materials Science and Engineering: A*, 362(1), 40–60.

Pramanik, S., & Kar, K. (2013). Nanohydroxyapatite synthesized from calcium oxide and its characterization. *The International Journal of Advanced Manufacturing Technology*, 66(5–8), 1181–1189.

Ramos, A., Completo, A., Relvas, C., & Simões, J. A. (2012). Design process of a novel cemented hip femoral stem concept. *Materials & Design*, 33, 313–321.

Rodriguez-Suarez, T., Bartolomé, J., & Moya, J. (2012). Mechanical and tribological properties of ceramic/metal composites: A review of phenomena spanning from the nanometer to the micrometer length scale. *Journal of the European Ceramic Society*, 32(15) Special issue, 3887–3898.

Ruben, R., Folgado, J., & Fernandes, P. (2007). Three-dimensional shape optimization of hip prostheses using a multicriteria formulation. *Structural and Multidisciplinary Optimization*, 34(3), 261–275.

Sabatini, A. L., & Goswami, T. (2008). Hip implants VII: Finite element analysis and optimization of cross-sections. *Materials & Design*, 29(7), 1438–1446.

Sahin, Y. (2005). The effects of various multilayer ceramic coatings on the wear of carbide cutting tools when machining metal matrix composites. *Surface and Coatings Technology*, 199(1), 112–117.

Saithna, A. (2010). The influence of hydroxyapatite coating of external fixator pins on pin loosening and pin track infection: A systematic review. *Injury*, 41(2), 128–132.

Scheerlinck, T., & Casteleyn, P.-P. (2006). The design features of cemented femoral hip implants. *Journal of Bone & Joint Surgery, British Volume*, 88(11), 1409–1418.

Scholz, M.-S., Blanchfield, J., Bloom, L., Coburn, B., Elkington, M., Fuller, J., Gilbert, M., Muflahi, S., Pernice, M., & Rae, S. (2011). The use of composite materials in modern orthopaedic medicine and prosthetic devices: A review. *Composites Science and Technology*, 71(16), 1791–1803.

Shikha, D., Jha, U., Sinha, S., Barhai, P., Nair, K., Dash, S., Tyagi, A., Kalavathy, S., & Kothari, D. (2009). Microstructure and biocompatibility investigation of biomaterial alumina after 30 keV and 60 keV nitrogen ion implantation. *Surface and Coatings Technology*, 203(17), 2541–2545.

Shirazi, F., Mehrali, M., Oshkour, A., Metselaar, H., Kadri, N., & Abu Osman, N. (2014). Mechanical and physical properties of calcium silicate/alumina composite for biomedical engineering applications. *Journal of the Mechanical Behavior of Biomedical Materials*, 30, 168–175.

Simões, J., & Marques, A. (2005). Design of a composite hip femoral prosthesis. *Materials & Design*, 26(5), 391–401.

Simpson, D. J., Little, J. P., Gray, H., Murray, D. W., & Gill, H. S. (2009). Effect of modular neck variation on bone and cement mantle mechanics around a total hip arthroplasty stem. *Clinical Biomechanics*, 24(3), 274–285.

Singh, S., Trikha, S., & Edge, A. (2004). Hydroxyapatite ceramic-coated femoral stems in young patients: A prospective ten-year study. *Journal of Bone & Joint Surgery, British Volume*, 86(8), 1118–1123.

Sivarasu, S., Beulah, P., & Mathew, L. (2011). Novel approach for designing a low weight hip implant used in total hip arthroplasty adopting skeletal design techniques. *Artificial Organs*, 35(6), 663–666.

Slouf, M., Eklova, S., Kumstatova, J., Berger, S., Synkova, H., Sosna, A., Pokorny, D., Spundova, M., & Entlicher, G. (2007). Isolation, characterization and quantification of polyethylene wear debris from periprosthetic tissues around total joint replacements. *Wear*, 262(9), 1171–1181.

Stops, A., Wilcox, R., & Jin, Z. (2011). Computational modelling of the natural hip: A review of finite element and multibody simulations. *Computer Methods in Biomechanics and Biomedical Engineering*, 15(9), 963–979.

Suárez-Vázquez, A., Hernández-Vaquero, D., Del Valle López-Díaz, M., & Pérez-Coto, I. (2011). Distribution of periprosthetic osteolysis in the hip: A study using magnetic resonance. *Revista Española de Cirugía Ortopédica y Traumatología (English Edition)*, 55(3), 193–203.

Traina, F., De Fine, M., Di Martino, A., & Faldini, C. (2013). Fracture of ceramic bearing surfaces following total hip replacement: A systematic review. *BioMed Research International*, 2013.

Tudor, A., Laurian, T., & Popescu, V. M. (2013). The effect of clearance and wear on the contact pressure of metal on polyethylene hip prostheses. *Tribology International*, 63, 158–168.

Varini, E., Cristofolini, L., Traina, F., Viceconti, M., & Toni, A. (2008). Can the rasp be used to predict intra-operatively the primary stability that can be achieved by press-fitting the stem in cementless hip arthroplasty? *Clinical Biomechanics*, 23(4), 408–414.

Vendittoli, P.-A., Amzica, T., Roy, A. G., Lusignan, D., Girard, J., & Lavigne, M. (2011). Metal ion release with large-diameter metal-on-metal hip arthroplasty. *The Journal of Arthroplasty*, 26(2), 282–288.

Verdonschot, N., & Huiskes, R. (1997). The effects of cement-stem debonding in THA on the long-term failure probability of cement. *Journal of Biomechanics*, 30(8), 795–802.

Viceconti, M., Brusi, G., Pancanti, A., & Cristofolini, L. (2006). Primary stability of an anatomical cementless hip stem: A statistical analysis. *Journal of Biomechanics*, 39(7), 1169–1179.

Waanders, D., Janssen, D., Mann, K. A., & Verdonschot, N. (2011). The behavior of the micromechanical cement–bone interface affects the cement failure in total hip replacement. *Journal of Biomechanics*, 44(2), 228–234.

Xie, Y., Li, H., Zhang, C., Gu, X., Zheng, X., & Huang, L. (2014). Graphene-reinforced calcium silicate coatings for load-bearing implants. *Biomedical Materials*, 9(2), 025009.

Yoo, J. J., Yoon, P. W., Lee, Y.-K., Koo, K.-H., Yoon, K. S., & Kim, H. J. (2013). Revision total hip arthroplasty using an alumina-on-alumina bearing surface in patients with osteolysis. *The Journal of Arthroplasty*, 28(1), 132–138.

Zhang, H., Brown, L., Blunt, L., & Barrans, S. (2008). Influence of femoral stem surface finish on the apparent static shear strength at the stem–cement interface. *Journal of the Mechanical Behavior of Biomedical Materials*, 1(1), 96–104.

Zhang, Y., Sun, M.-J., & Zhang, D. (2012). Designing functionally graded materials with superior load-bearing properties. *Acta Biomaterialia*, 8(3), 1101–1108.

Zivic, F., Babic, M., Grujovic, N., Mitrovic, S., Favaro, G., & Caunii, M. (2012). Effect of vacuum-treatment on deformation properties of PMMA bone cement. *Journal of the Mechanical Behavior of Biomedical Materials*, 5(1), 129–138.

2 A Review of Biocomposites in Biomedical Application

H. A. Aisyah,[1] M. T. Paridah,[1] S. M. Sapuan,[1,2]
A. Khalina,[1,2] R. A. Ilyas,[1,2] and N. Mohd. Nurazzi[3]

[1]Institute of Tropical Forestry and Forest Products (INTROP), Universiti Putra Malaysia, Malaysia

[2]Faculty of Engineering, Universiti Putra Malaysia, Malaysia

[3]Center for Defence Foundation Studies, National Defence University of Malaysia, Malaysia

CONTENTS

2.1 INTRODUCTION

Recently, the dependence on synthetic or petroleum-based products has decreased due to higher concern on producing green materials. This awareness has led researchers to search for a solution to replace existing materials, and they have implemented biodegradable materials by utilizing natural fibers as a reinforcement agent (Aisyah et al., 2019; Asyraf et al., 2020; Atiqah et al., 2019; Norizan et al., 2020; Nurazzi et al., 2019). Natural fibers incorporated with polymer-based materials produce biocomposites, which generally have high strength and stiffness but low weight. Natural fibers are also abundantly available, renewable, noncorrosive, low in density per unit volume, and easily processed into many forms (Jumaidin et al., 2019a, 2019b, 2019c; Ticoalu et al., 2010).

Natural fibers are derived from three main sources: plants, animals, and minerals (Ilyas, Sapuan, Sanyang, et al., 2018). Among these sources, plant fibers have gained interest among scientists, researchers, and engineers due to their high availability, degradability, renewability, and environmental friendly nature. The utilization of natural fibers in composite material for industrial component, automotive industry, structural building, and furniture sectors improves environmental sustainability and leads to the development of alternative materials instead of using synthetic or humanmade fibers (Ilyas, Sapuan, Ibrahim, Atiqah, et al., 2019; Ilyas, Sapuan, Ishak, Zainudin, et al., 2018; Ilyas & Sapuan, 2020a, 2020b; Syafri et al., 2019).

Natural fibers from plants include biomass from agriculture crops such as oil palm fronds, empty fruit bunch (EFB), coir, straws, husks, and bagasse. There are also other fiber crops such as cotton, ramie, flax, bamboo, kenaf, jute, abaca, sisal, hemp, and sugar palm (Azammi et al., 2020; Ilyas et al., 2017; Ilyas, Sapuan, & Ishak, 2018; Ilyas, Sapuan, Ibrahim, Abral, et al., 2019a; Ilyas, Sapuan, Ishak, et al., 2019; Ilyas et al., 2019; Mazani et al., 2019). Plant fibers are lignocellulosic materials that mainly consist of cellulose microfibrils, hemicelluloses, and lignin. Natural fibers from plants possess high strength, thus they are suitable for many load-bearing applications and widely used as reinforcements.

2.2   VALUE OF FUELS AND LIGNOCELLULOSE AS RAW MATERIAL

2.2.1   PLANT-BASED NATURAL FIBERS

Generally, fibers can be classified into two main categories: natural and synthetic, based on their chemical origin. Three main groups fall under natural fibers, namely plant, mineral, and animal fibers. Natural fibers from plants are also denoted as cellulosic fibers, as they are composed mainly from cellulose, as well hemicellulose and lignin components. Normally, this type of fiber is derived from plant bast (outer layer of stem), stalk, leaf, and seed hair (e.g., cotton and kapok). Meanwhile, animal-based fibers are referred to as protein fibers as they are naturally derived from animal fur. On the other hand, synthetic fibers are chemically manufactured and do not occur naturally. There are several types of synthetic fibers, including organic and inorganic polymers; where organic can be classified into synthetic polymer (such as nylon, polyester, polyethylene, and polyurethane [PU]) and natural polymer (e.g., polylactide [PLA] and viscose [rayon]). In addition, inorganic synthetic fiber originates from minerals, such as carbon fiber and glass fiber.

Natural fibers are renewable, recyclable, and degradable, which are advantages that synthetic fiber cannot offer. Moreover, according to Jawaid and Abdul Khalil (2011), natural fibers are abundantly available, low in cost (because they are from plantation wastes), strong, and less hazardous to the environment. However, several researchers have pointed out the disadvantages of using natural fibers, such as having low durability, poor fiber–matrix bonding, poor water and heat/fire resistance, and nonhomogeneous properties (Jawaid & Abdul Khalil, 2011; Romanzini et al., 2012; Sreekumar, 2008). However, these drawbacks can be overcome by treating the fiber and modifying its structures (Ali et al., 2018; Ibrahim et al., 2015; Le Moigne et al., 2018) or incorporating it with synthetic fibers (Ahmed & Vijayarangan, 2008; Faruk et al., 2012).

2.2.2 Cellulose-based Natural Fibers

The chemical constituent of plant fibers depends on several factors: origin, locality, age, growth development, and variety. Several researchers have presented the composition of different natural fibers comprehensively (Bledzki & Gassan, 1999; Jawaid & Abdul Khalil, 2011; Namvar et al., 2014; Peças et al., 2018). Table 2.1 summarizes the chemical composition of some plant fibers.

Cellulosic fiber can be derived from several parts of the plant, for instance, from the trunk (oil palm trunk [OPT]), bast (hemp, flax, jute, kenaf, ramie), leaf (pineapple, banana, oil palm frond [OPF], sisal), and seed (rice husk, cotton, oil palm EFBs, coir). In plant fibers, cellulose is a major substance, which mainly comprises cellulose fibrils embedded in a lignin matrix that contribute to the fiber strength (Abdul Khalil et al., 2010). Cellulose is long, consists of linear chain polymers of hundreds of glucose molecules, and is a hydrophilic glucan polymer consisting of hydroxyl groups, crystalline, and amorphous regions. These hydroxyl groups form intermolecular hydrogen bonds inside the macromolecule itself and intermolecular hydrogen bonds among other cellulose macromolecules as well as with hydroxyl groups from the air. Thus, all plant fibers are hydrophilic. Strong intermolecular hydrogen bonds with large molecules are formed by the crystallite cellulose (Hazrol et al., 2020; Kabir et al., 2012).

Moreover, different plant fibers contain different hemicellulose components. From Table 2.1, OPF fiber contains the highest composition of hemicelluloses compared to other fibers. Meanwhile, lignin content is the highest in OPT as lignified cellulose fibers that retain their strength better than delignified fibers, acting as a bonding or cementing material. Lignin is also responsible for the tough and stiff

TABLE 2.1
Chemical Composition of Some Plant Fibers

Fiber	Cellulose	Hemicellulose	Lignin	Ash Content	Extractive
		Chemical Composition (%)			
Jute[1, 2]	64.4	12.0	11.8	1.0	0.7
Ramie[1]	68.6	13.1	0.6	<1	1.9–2.2
Flax[1]	64.1	16.7	2.0	<1	1.5–3.3
Kenaf[3]	55	31.8	14.7	5.4	5.5
Bamboo[4]	73.83	12.49	10.15	—	3.16
Banana[5, 6]	60–65	19.0	5–10	1.5	4.6
Pineapple[6, 7]	81.5	—	12.7	2.0	—
EFB[8]	43–65	17–33	17.2	0.7	—
OPT[9]	29–37	12–17	22.6	1.6	—
OPF[10, 11]	40–50	34–38	16.4	4.2	4.4

Source: [1]Gassan and Bledzki (1996), [2]Del Río et al. (2012), [3]Abdul Khalil et al. (2010), [4]Wang et al. (2010), [5]Reddy and Yang (2005), [6]Abdul Khalil et al. (2006), [7]Devi et al. (1997), [8]Peh et al. (1976), [9]Khoo and Lee (1985), [10]Jalaluddin (1993), [11]Abdul Khalil and Rozman (2004).

TABLE 2.2
Fiber Properties of Some Plant Fibers and Synthetic Fibers

Fiber	Fiber Properties			
	Density (g/cm³)	Tensile Strength (MPa)	Young's Modulus (GPa)	Elongation at Break (%)
Jute	1.3–1.49	393–800	13–26.5	1.16–1.5
Ramie	1.55	400–938	61.4–128	1.2–3.8
Flax	1.50	345–1,500	27.6	2.7–3.2
Kenaf	1.2	930	53	1.6
Hemp	1.48	550–900	70	1.6
Banana	1.35	500	12	5.9
Pineapple	1.5	170–1,627	60–82	2.4
EFB	0.7–1.55	248	3.2	2.5
Glass fiber	2.55	3400	73	2.5
Carbon fiber	1.78	3,400–4,800	240–425	1.4–1.8
Kevlar	1.44	3,000	60	2.5–3.7

Source: Abdul Khalil et al., 2010; Bismarck et al., 2005; Cheung et al., 2009; Hattali et al., 2002; Hoareau et al., 2004; Jawaid and Abdul Khalil, 2011.

properties of the fiber and affects plant structure as well as morphology (Jawaid & Abdul Khalil, 2011).

Generally, plant fibers are suitable to be used as a reinforcement agent due to their mechanical properties, specifically their low density but relatively high strength and stiffness. Table 2.2 compares the mechanical properties of some plant and synthetic fibers. Plant fibers display elongation at break comparable to synthetic fibers, which is advantageous as it confers better damage tolerance in composites (Jawaid & Abdul Khalil, 2011).

The mechanical properties of fiber such as tensile strength, flexural strength, and rigidity depend on the alignment of cellulose fibrils and internal structure, which are generally arranged along the fiber length (John & Thomas, 2008). Dufresne (2008) stated that the mechanical properties of fiber are strongly influenced by chemical composition, fiber structure, and microfibril angle. These properties differ among different parts of plants as well as among different plants. The tensile strength and Young's modulus of plant fibers increase with increasing cellulose content (Aji et al., 2009). Flax, ramie, and jute are among the plant fibers that have high-tensile strength.

2.2.3 BIOCOMPOSITES

Biocomposites are materials or products that contain natural fibers as a reinforcement agent and use a matrix as a binding agent. For the past few years, biocomposites have become popular among scientists due to their multiple usage and high-end products. Several research and development efforts have been carried out to utilize natural fibers, focusing mainly on the production of polymer-based composites (Abdullah et al., 2006; Abral & Ariksa et al., 2019a, 2019b, 2020; Abral & Atmajaya et al., 2020; Abral & Basri et al., 2019b; Atikah et al., 2019; Halimatul et al., 2019a, 2019b; Ilyas et al., 2018; 2019; Ilyas et al., 2020; Ilyas, Sapuan, Ibrahim, et al., 2019b; Ilyas, Sapuan, Ishak, & Zainudin, 2018a, 2018b; Mishra et al., 2002; Nayak et al., 2010; Paiva Júnior et al., 2004; Rozman et al., 2001), pulp and paper (Ashori, 2006; Astimar et al., 2002; Tanaka et al., 2002), particleboards, and fiberboards (Juliana et al., 2012; Ridzuan et al., 2002).

2.3 BIOMATERIALS IN BIOMEDICINE

In the medical sector, materials that make a function of living materials are called biomaterials. Biomaterials have been used to replace body parts such as joints, legs, knees, hips, and other organs in the human body. Tathe et al. (2010) define biomaterial as a material that comprises the whole or part of a living structure or biomedical device, which performs, augments, or replaces a natural function. The biomaterial can be made from natural or synthetic materials. A comprehensive definition of biomaterial in the biomedical sector has been given by several authors and institutions (Helmus et al., 2008; Patel & Gohil, 2012; Ratner & Hoffman, 2004).

2.3.1 BIOMATERIAL CLASSIFICATIONS

Generally, materials used in biomedical applications are classified into four main groups: metals, ceramics, polymers, and composites, and are produced individually or by a mixture of these groups. The material selection is based on several factors, but the key factor is biocompatibility, which specifies the chemical, biological, and physical appropriateness of the materials, especially in terms of mechanical properties (Namvar et al., 2014; Ramakrishna et al., 2001). Other than biocompatibility, biological response and flexibility are also the main factors that must be considered (Kutz, 2003).

Among the materials, metal is the strongest with high-tensile strength (600–1,085 MPa) and elastic modulus (120–210 GPa), depending on its type (Ramakrishna et al., 2001). Several types of metal are used for medical applications such as amalgam, tungsten, tantalum, titanium alloys, stainless steels, and aluminum alloys. Metal has been applied for use in the medical sector to make dental materials, prosthetic devices, artificial knee joints, artificial hip joints, clips, and bone devices. The selection of metal in medical implants is due to its mechanical properties: high resistance to load bearing, high wear resistance, and good fatigue limit. These properties are suitable in artificial joints, which are subjected to repeated loading and unloading; they also can hold good mechanical loads. Other than that, metal-based materials have good corrosion resistance and low cost (Niinomi, 2002). Stainless steel is a common metal that forms the backbone of medical device manufacturing, in which 316L stainless steel is the most typically used metal in all implants. Tantalum is used as a material that encourages bone growth and also used in dentistry as well as tissue regeneration (Mohandas et al., 2014).

Ceramics are mostly used in orthopedics and dentistry, as they are chemically similar to natural skeletal and dental materials. However, in terms of mechanical properties, ceramics have some limitations when compared to metal. Ceramics are widely used as prosthetic hips; prosthetic knees; replacement for tendons, ligaments, and jawbones; bone grafts; dental implants; ear implants; heart valves; and parts of the skeletal system. Applications and types of ceramic-based materials in biomedicine have been explained by Sarkar and Banerjee (2010).

Polymer in biomedical applications is widely used for replacement and production of human tissues due to its good biocompatibility and degradability, as well as high strength, toughness, and ductility (Pertici, 2016). Two types of polymers are used in biomedical applications: natural and synthetic. Polysaccharides, polypropylene, poly(glycolic acid), PU, polyvinyl acetate, and polyethylene are examples of common polymers used, with poly(lactic acid) being the most widely used synthetic polymer. Wei et al. (2018) and Ibrahim et al. (2017) have reported on the polymers used for biomedical applications.

A composite is a combination of two or more materials that physically and chemically differ (matrix and reinforcement) and produce material with desired properties and strength. The use of composite materials in biomedicine includes orthopedic applications, dental application, prosthetic devices, and tissue or skin regeneration. Intensive reviews on the use of these type of materials for biomedical applications have been presented by Mehboob and Chang (2014), Iftekhar (2004), and Salernitano and Migliaresi (2003).

2.3.2 APPLICATION OF NATURAL FIBER BIOCOMPOSITE IN BIOMEDICINE

Several researchers have explored the application of biocomposites in the biomedical sector. The use of natural fiber has gained more interest among the researchers due to the increasing price of synthetic and petroleum-based products, together with the environmental concern regarding biomedical devices. In addition, natural fiber is suitable for biomedical applications. Reported works on biocomposites are shown in Table 2.3.

Much research has been conducted to develop composite materials reinforced with natural fiber for biomedical applications. For instance, in bone repair, stainless steel and titanium are common materials used as bone plates because of their biocompatibility. Several researchers have studied and developed biocomposites for bone repair. Bagheri et al. (2013), Bagheri et al. (2014), and Bagheri et al. (2015) developed a hybrid composite material for long bone fracture plate application to replace the usage of metal material. They studied the mechanical, fatigue, wetting, as well as cytotoxic and osteogenetic properties of carbon fiber–epoxy composite reinforced with flax fiber. The hybridization of the flax fibers with carbon fibers and epoxy as a matrix increased the tensile, bending, and hardness properties of the sandwich composite compared to those in previous studies (Chłopek & Kmita, 2003; Fiore et al., 2012; Khanam et al., 2010) and compared to the properties of human cortical bone (Brydone et al., 2010). The tensile strength and Young's modulus of the biocomposite were 41.7 GPa and 399.8 MPa, respectively, compared to those of cortical bone of 7–25 GPa and 50–150 MPa (Brydone et al., 2010). The composite

TABLE 2.3

Reported Works on Biocomposites in Biomedical Applications

Type of Fiber	Matrix Polymer	Medical Application	References
Sisal, hemp, and jute fiber	Epoxy	Femur bone prosthesis	Gouda et al. (2014)
Corn fiber	Polymethylmethacrylate (PMMA) and epoxy	Femur bone fracture plates	Hashim et al. (2016)
Flax/Carbon fiber	Epoxy	Long bone fracture plate	Bagheri et al. (2013), Bagheri et al. (2014), and Bagheri et al. (2015)
Flax/Glass fiber	Epoxy	Bone fracture plate	Manteghi et al. (2017) and Manteghi et al. (2019)
Sisal, banana, and roselle fiber		Internal and external bone fixation	Chandramohan and Marimuthu (2011)
Siwak and bamboo fiber	PMMA	Denture applications	Oleiwi et al. (2017)
Pineapple skin, coconut shell, coconut husk, corn cob, rambutan, mangosteen, and banana skin fiber	Polyurethane (PU)	Blood bag, tissue engineering, drug delivery, and scaffold	Ahad et al. (2018a) and Ahad et al. (2018b)
Luffa fiber	Polylactic acid (PLA)	External bone plate application	Kakar et al. (2015)
Ramie fiber	Epoxy and polyester	Socket prosthesis	Irawan et al. (2009) and Irawan et al. (2011)
Bamboo fiber	Epoxy	Orthotic and prosthetic application	Kramer et al. (2015)
Kenaf fiber		Below-knee prosthesis socket	Nurhanisah et al. (2017), Nurhanisah et al. (2018a), and Nurhanisah et al. (2018b)

showed good fatigue properties, indicating its potential use in orthopedic applications, in this case for long bone fracture plate where cyclic loading is required in daily activities.

The mechanical and moisture absorption properties of flax/glass fiber-reinforced epoxy composites for bone fracture plates were studied by Manteghi et al. (2017). The hybrid with flax fiber has improved mechanical properties, where the hybrid composite has ultimate flexural properties compared to conventional metallic plates due to the flexibility of flax fiber as a core material in the laminated composite. In addition, the composite also showed ideal stiffness and strength in terms of axial, shear, and bending, thus making it suitable for designing long bone fracture fixation devices such as plates as it can offer specific stress protection. Manteghi et al. (2019) further studied the static and fatigue compressive properties of this hybrid composite. It showed comparatively constant static and fatigue compressive properties over the cyclic loading routine.

Chandramohan and Marimuthu (2011) reported that flexural strength, moisture absorption, and elongation properties of three types of natural fibers, namely sisal, banana, and roselle composite improved when treated with NaOH. The treatment enhanced bonding properties between the fibers and the matrix, thus creating better stress transfer, resulting in an increase of flexural strength. The best fiber was sisal fiber, although roselle fiber can also be applied for internal and external bone fixation. The hybrid composite was coated with calcium phosphate to make it suitable for internal and external replacement of ruptured bones. Tensile, compression, and bending strength of sisal, jute, and hemp composites with different fiber weight fractions for femur bone prosthesis devices were studied by Gouda et al. (2014). Hybrid composites with 36% natural fiber showed optimally improved mechanical properties and higher density compared to 12% and 14%. The strength properties were also significantly higher for all fiber types when compared to those of the femur (Havaldar et al., 2013).

The tensile and compression strength in a model hybrid of corn fiber with polymethylmethacrylate (PMMA) and epoxy for femur bone plates have been investigated (Hashim et al., 2016). Better tensile strength and modulus of elasticity were observed in corn fiber-reinforced PMMA composites than in corn fiber-reinforced epoxy composites, where alkali treatment by 1% NaOH prior to composite preparation enhanced the composite strength. The authors also conducted the analysis of femur bone plates by finite element method (FEM) using software packages and compared with stainless steel and titanium alloy femur bone plates. The bone plates from stainless steel and titanium alloy can withstand maximum equivalent stresses at femur bone plates better than corn composites.

Oleiwi et al. (2017) developed biocomposite material by adding siwak and bamboo fibers at different fiber volume fractions (3, 6, and 9 wt%) and fiber lengths (2, 6, and 12 mm) with PMMA acrylic resin as a matrix. The composite properties were strongly influenced by weight fraction and fiber length of the reinforcing fibers. The best overall tensile strength and Young's modulus were 72.4 MPa and 5.208 GPa, respectively, and obtained at weight fraction of 9% and fiber length of 12 mm for the bamboo composite. Bamboo composite showed better values compared to pure PMMA, which had lower values of tensile strength and elasticity, 47.6 MPa and 1.7 GPa, respectively. However, the elongation properties of both biocomposites decreased with increasing weight fraction and fiber length.

Ahad et al. (2018a) and Ahad et al. (2018b) studied oil and water absorption behavior of fibers from pineapple skin, coconut shell, coconut husk, corn cob, rambutan, mangosteen, and banana skin as a filler and reinforced with PU resin. Composites were prepared by varying the types and percentage of natural fibers (5%, 10%, 15%, and 20%) using the melt mixing method. The absorption properties of composites with high filler percentage were higher than those with low filler percentage. In addition, different types of natural fibers had different absorption rates of oil and water due to different chemical constituents, specifically cellulose content, as well as fiber structures. Composites with fiber from rambutan, pineapple, and banana at 20% of filler percentage had higher water absorption rate, and pineapple and rambutan composites had higher oil absorption rate. The authors concluded that this kind of composite may be utilized for biomedical applications.

Kakar et al. (2015) studied the tensile properties and interface adhesion of luffa fibers reinforced with polylactic acid (PLA) composite in relation to heat treatment and volume percentage of luffa fibers by utilizing compression molding techniques. Luffa–PLA composite performed better in tensile strength and had better interfacial adhesion after heat treatment due to the removal of hemicellulose, which resulted in better fiber–matrix bonding strength. Luffa–PLA composite at 15% volume of treated fibers produced the highest tensile strength (40.20 MPa) compared to other types of composites. This kind of composite is suitable to be used as external bone plate application for certain hard and soft tissues in the orthopedic field.

Researchers have investigated the possibilities to produce prosthetic devices for replacement components of body parts from natural fibers reinforced with polymer materials. The usage of natural fibers is supposed to cut down on manufacturing cost, as it partially substitutes socket prosthesis with fiberglass polyester composites, thus enabling less financially capable wearers to get affordable prosthetic devices (Jensen & Raab, 2007). Irawan et al. (2009) and Irawan et al. (2011) developed socket prosthesis by using ramie fibers reinforced with epoxy resin. They designed the socket prosthesis by using a computer-based simulation of FEM by varying the angle orientation subjected to torsional load and compressive load of 180°, 45°, and −45°. The findings from this simulation were used as a guideline for the composite production. The tensile and flexural strength of the ramie polyester and epoxy composites were then investigated. The ramie epoxy composite had the highest values of tensile and flexural strength compared to ramie polyester composite and fiberglass polyester composites. The researchers also studied the comfort of socket prosthesis and found that the socket from ramie epoxy composite was more comfortable to be used by the wearer due to better elasticity.

Kramer et al. (2015) studied the tensile strength of bamboo fibers in the form of plain fabric, satin fabric, and knit fabric and produced a composite by utilizing epoxy resin. Bamboo epoxy composites were tested and compared with cotton, nylon, and carbon fiber-based epoxy composite. The bamboo epoxy composites showed comparable strength, good stiffness, and strain properties compared to cotton epoxy composite. Among bamboo types, plain fabric bamboo epoxy composite had higher strength values than satin and knit fabric. In addition, the bamboo and cotton composites showed higher ultimate tensile strength and plastic deformation; however, nylon composite performed better in strain properties. Under different circumstances, carbon fiber composite showed high values of flexural and tensile strength, but catastrophic failure happened due to excess stress. These findings make bamboo fabric preferable for orthotic and prosthetic devices to replace carbon fiber and nylon composites that are presently being used.

The mechanical properties of kenaf fabric and fiberglass polyester composites for below-knee prosthesis sockets were studied by laminated process with different numbers of kenaf fabric layers (Nurhanisah et al., 2017; Nurhanisah et al., 2018a). The hybrid composite consisting of two kenaf fabric layers showed better flexural and impact properties than single kenaf fabric layer-based composites. The prosthetic socket ability was also evaluated and it was satisfactory in terms of comfort and easiness to remove or refit. A further study on thermal management for prosthetic socket from kenaf fibers was conducted. Holes in the socket control the temperature inside the socket, which is important for air circulation and thus avoiding thermal discomfort for the wearer. Thus, kenaf fabric is a potential material for prosthetic applications because of its strength, biodegradability, low cost, functionality, and heat removal ability (Nurhanisah et al., 2018b).

2.4 CONCLUSION

The development of natural fiber usage in the medical field is one of the noteworthy approaches to producing products that are cost efficient, biodegradable, and having comparable properties with existing materials. Researchers must pay attention to several considerations and issues, because the material for medical applications requires specific properties, particularly biocompatibility, in the sense that it is not harmful and is safe for the human body. The properties of natural fibers, specific application, and interaction must be considered in the selection and development of biocomposites in the medical field. Furthermore, in certain applications, the strength of biocomposites should be sufficient for medical applications.

REFERENCES

Abdul Khalil, H. P. S., & Rozman, H. D. (2004). Gentian dan komposit lignoselulosik. Pulau Pinang: Penerbit Universiti Sains Malaysia.

Abdul Khalil, H. P. S., Siti Alwani, M., & Mohd Omar, A. K. (2006). Chemical composition, anatomy, lignin distribution, and cell wall structure of Malaysian plant waste fibers. *BioResources*, 1(2), 220–232.

Abdul Khalil, H. P. S., Yusra, A. I., Bhat, A. H., & Jawaid, M. (2010). Cell wall ultrastructure, anatomy, lignin distribution, and chemical composition of Malaysian cultivated kenaf fiber. *Industrial Crops and Products*, 31(1), 113–121.

Abdullah, A. K., Abedin, M. Z., Beg, M. D. H., Pickering, K. L., & Khan, M. A. (2006). Study on the mechanical properties of jute/glass fiber-reinforced unsaturated polyester hybrid composites: Effect of surface modification by ultraviolet radiation. *Journal of Reinforced Plastics and Composites*, 25(6), 575–588.

Abral, H., Ariksa, J., Mahardika, M., Handayani, D., Aminah, I., Sandrawati, N., Pratama, A. B., Fajri, N., Sapuan, S. M., & Ilyas, R. A. (2020). Transparent and antimicrobial cellulose film from ginger nanofiber. *Food Hydrocolloids*, 98, 105266.

Abral, H., Ariksa, J., Mahardika, M., Handayani, D., Aminah, I., Sandrawati, N., Sapuan, S. M., & Ilyas, R. A. (2019a). Highly transparent and antimicrobial PVA based bionanocomposites reinforced by ginger nanofiber. *Polymer Testing*, 106186.

Abral, H., Atmajaya, A., Mahardika, M., Hafizulhaq, F., Kadriadi, Handayani, D., Sapuan, S. M., & Ilyas, R. A. (2020). Effect of ultrasonication duration of polyvinyl alcohol (PVA) gel on characterizations of PVA film. *Journal of Materials Research and Technology*, 9(2), 2477–2486.

Abral, H., Basri, A., Muhammad, F., Fernando, Y., Hafizulhaq, F., Mahardika, M., Sugiarti, E., Sapuan, S. M., Ilyas, R. A., & Stephane, I. (2019b). A simple method for improving the properties of the sago starch films prepared by using ultrasonication treatment. *Food Hydrocolloids*, 93, 276–283.

Ahad, N. A., Rozali, F. Z., Hanif, N. I. H., & Rosli, N. H. (2018b). Oil and water absorption behavior of natural fibers filled TPU composites for biomedical applications. *Journal of Engineering Research and Education*, 10, 11–16.

Ahad, N. A., Rozali, F. Z., Rosli, N. H., Hanif, N. I. H., & Parimin, N. (2018a). Oil and water absorption behavior of TPU/natural fibers composites. In Solid state phenomena (Vol. 280, pp. 374–381). Trans Tech Publications.

Ahmed, K. S., & Vijayarangan, S. (2008). Tensile, flexural and interlaminar shear properties of woven jute and jute-glass fabric reinforced polyester composites. *Journal of Materials Processing Technology*, 207(1-3), 330–335.

Aisyah, H. A., Paridah, M. T., Sapuan, S. M., Khalina, A., Berkalp, O. B., Lee, S. H., Lee, C. H., Nurazzi, N. M., Ramli, N., Wahab, M. S., & Ilyas, R. A. (2019). Thermal properties of woven kenaf/carbon fibre-reinforced epoxy hybrid composite panels. *International Journal of Polymer Science*, 2019 (December), 1–8.

Aji, I. S., Sapuan, S. M., Zainudin, E. S., & Abdan, K. (2009). Kenaf fibres as reinforcement for polymeric composites: A review. *International Journal of Mechanical and Materials Engineering*, 4(3), 239–248.

Ali, A., Shaker, K., Nawab, Y., Jabbar, M., Hussain, T., Militky, J., & Baheti, V. (2018). Hydrophobic treatment of natural fibers and their composites—A review. *Journal of Industrial Textiles*, 47(8), 2153–2183.

Ashori, A. (2006). Pulp and paper from kenaf bast fibers. *Fibers and Polymers*, 7(1), 26–29.

Astimar, A. A., Mohamad, H., & Anis, M. (2002). Preparation of cellulose from oil palm empty fruit bunches via ethanol digestion: Effect of acid and alkali catalysts. *Journal of Oil Palm Research*, 14(1), 9–14.

Asyraf, M. R. M., Ishak, M. R., Sapuan, S. M., Yidris, N., & Ilyas, R. A. (2020). Woods and composites cantilever beam: A comprehensive review of experimental and numerical creep methodologies. *Journal of Materials Research and Technology*, January.

Atikah, M. S. N., Ilyas, R. A., Sapuan, S. M., Ishak, M. R., Zainudin, E. S., Ibrahim, R., Atiqah, A., Ansari, M. N. M., & Jumaidin, R. (2019). Degradation and physical properties of sugar palm starch/sugar palm nanofibrillated cellulose bionanocomposite. *Polimery*, 64(10), 27–36.

Atiqah, A., Jawaid, M., Sapuan, S. M., Ishak, M. R., Ansari, M. N. M., & Ilyas, R. A. (2019). Physical and thermal properties of treated sugar palm/glass fibre reinforced thermoplastic polyurethane hybrid composites. *Journal of Materials Research and Technology*, July.

Azammi, A. M. N., Ilyas, R. A., Sapuan, S. M., Ibrahim, R., Atikah, M. S. N., Asrofi, M., & Atiqah, A. (2020). Characterization studies of biopolymeric matrix and cellulose fibres based composites related to functionalized fibre-matrix interface. In Interfaces in particle and fibre reinforced composites (1st ed., Issue November, pp. 29–93). Elsevier.

Bagheri, Z. S., El Sawi, I., Bougherara, H., & Zdero, R. (2014). Biomechanical fatigue analysis of an advanced new carbon fiber/flax/epoxy plate for bone fracture repair using conventional fatigue tests and thermography. *Journal of the Mechanical Behavior of Biomedical Materials*, 35, 27–38.

Bagheri, Z. S., El Sawi, I., Schemitsch, E. H., Zdero, R., & Bougherara, H. (2013). Biomechanical properties of an advanced new carbon/flax/epoxy composite material for bone plate applications. *Journal of the Mechanical Behavior of Biomedical Materials*, 20, 398–406.

Bagheri, Z. S., Giles, E., El Sawi, I., Amleh, A., Schemitsch, E. H., Zdero, R., & Bougherara, H. (2015). Osteogenesis and cytotoxicity of a new carbon fiber/flax/epoxy composite material for bone fracture plate applications. *Materials Science and Engineering: C*, 46, 435–442.

Bismarck, A., Mishra, S., & Lampke, T. (2005). Plant fibers as reinforcement for green composites. *Natural Fibers, Biopolymers and Biocomposites*, 2, 37–108.

Bledzki, A. K., & Gassan, J. (1999). Composites reinforced with cellulose based fibres. *Progress in Polymer Science*, 24(2), 221–274.

Brydone, A. S., Meek, D., & Maclaine, S. (2010). Bone grafting, orthopaedic biomaterials, and the clinical need for bone engineering. *Proceedings of the Institution of Mechanical Engineers, Part H: Journal of Engineering in Medicine*, 224(12), 1329–1343.

Chandramohan, D., & Marimuthu, K. (2011). Characterization of natural fibers and their application in bone grafting substitutes. *Acta of Bioengineering & Biomechanics*, 13(1), 77–84.

Cheung, H. Y., Ho, M. P., Lau, K. T., Cardona, F., & Hui, D. (2009). Natural fibre-reinforced composites for bioengineering and environmental engineering applications. *Composites Part B: Engineering*, 40(7), 655–663.

Chłopek, J., & Kmita, G. (2003). Non-metallic composite materials for bone surgery. *Engineering Transactions*, 51(2-3), 307–323.

Del Río, J. C., Prinsen, P., Rencoret, J., Nieto, L., Jiménez-Barbero, J., Ralph, J., Martínez, Á. T., & Gutiérrez, A. (2012). Structural characterization of the lignin in the cortex and pith of elephant grass (*Pennisetum purpureum*) stems. *Journal of Agricultural and Food Chemistry*, 60, 3619–3634.

Devi, L. U., Bhagawan, S. S., & Thomas, S. (1997). Mechanical properties of pineapple leaf fiber-reinforced polyester composites. *Journal of Applied Polymer Science*, 64(9), 1739–1748.

Dufresne, A. (2008). Cellulose-based composites and nanocomposites. In Monomers, polymers and composites from renewable resources, edited by P. Gabbott. Oxford: Blackwell Publishing Ltd, UK.

Faruk, O., Bledzki, A. K., Fink, H. P., & Sain, M. (2012). Biocomposites reinforced with natural fibers: 2000–2010. *Progress in Polymer Science*, 37(11), 1552–1596.

Fiore, V., Valenza, A., & Bella, G. D. (2012). Mechanical behavior of carbon/flax hybrid composites for structural applications. *Composite Materials*, 46(17), 2089–2096.

Gassan, J., & Bledzki, A. K. (1996). Modification methods on nature fibers and their influence on the properties of the composites. *Journal of Engineering and Applied Science*, 2, 2552–2557.

Gouda, T. A., Jagadish, S. P., Dinesh, K. R., Gouda, H., & Prashanth, N. (2014). Characterization and investigation of mechanical properties of hybrid natural fiber polymer composite materials used as orthopaedic implants for femur bone prosthesis. *IOSR Journal of Mechanical and Civil Engineering*, 11(4), 40–52.

Halimatul, M. J., Sapuan, S. M., Jawaid, M., Ishak, M. R., & Ilyas, R. A. (2019a). Effect of sago starch and plasticizer content on the properties of thermoplastic films: Mechanical testing and cyclic soaking-drying. *Polimery*, 64(6), 32–41.

Halimatul, M. J., Sapuan, S. M., Jawaid, M., Ishak, M. R., & Ilyas, R. A. (2019b). Water absorption and water solubility properties of sago starch biopolymer composite films filled with sugar palm particles. *Polimery*, 64(9), 27–35.

Hashim, A. M., Tanner, K., & Ulaiwe, J. K. (2016). Tensile and fracture properties of coir fiber green composites bone plate fixation. *ZANCO Journal of Pure and Applied Sciences*, 28.

Hattali, S., Benaboura, A., Ham-Pichavant, F., Nourmamode, A., & Castellan, A. (2002). Adding value to Alfa grass (Stipa tenacissima L.) soda lignin as phenolic resins, lignin characterization. *Polymer Degradation and Stability*, 75, 259–264.

Havaldar, R., Pilli, S., & Putti, B. (2013). Effects of magnesium on mechanical properties of human bone. *IOSR Journal of Pharmacy and Biological Science*, 7(3), 8–14.

Hazrol, M. D., Sapuan, S. M., Ilyas, R. A., Othman, M. L., & Sherwani, S. F. K. (2020). Electrical properties of sugar palm nanocrystalline cellulose, reinforced sugar palm starch nanocomposites. *Polimery*, 55(5), 33–40.

Helmus, M. N., Gibbons, D. F., & Cebon, D. (2008). Biocompatibility: Meeting a key functional requirement of next-generation medical devices. *Toxicol Pathol*, 36(1), 70–80.

Hoareau, W., Trindade, W. G., Siegmund, B., Castellan, A., & Frollini, E. (2004). Sugar cane bagasse and curaua lignins oxidized by chlorine dioxide and reacted with furfuryl alcohol: Characterization and stability. *Polymer Degradation and Stability*, 86, 567–576.

Ibrahim, I., Sadiku, E., Jamiru, T., Hamam, A., & Kupolati, W. K. (2017). Applications of polymers in the biomedical field. *Current Trends in Biomedical Engineering & Biosciences*, 4, 9–11.

Ibrahim, Z., Aziz, A. A., Ramli, R., Jusoff, K., Ahmad, M., & Jamaludin, M. A. (2015). Effect of treatment on the oil content and surface morphology of oil palm (Elaeis guineensis) empty fruit bunches (EFB) fibres. *Wood Research*, 60(1), 157–166.

Iftekhar, A. (2004). Biomedical composites. Standard handbook of biomedical engineering and design. McGraw-Hill Companies.

Ilyas, R. A., & Sapuan, S. M. (2020a). *Biopolymers and Biocomposites: Chemistry and Technology*, 16, 1–4.

Ilyas, R. A., & Sapuan, S. M. (2020b). The preparation methods and processing of natural fibre bio-polymer composites. *Current Organic Synthesis*, 16(8), 1068–1070.

Ilyas, R. A., Sapuan, S. M., & Ishak, M. R. (2018). Isolation and characterization of nanocrystalline cellulose from sugar palm fibres (Arenga pinnata). *Carbohydrate Polymers*, 181, 1038–1051.

Ilyas, R. A., Sapuan, S. M., Atiqah, A., Ibrahim, R., Abral, H., Ishak, M. R., Zainudin, E. S., Nurazzi, N. M., Atikah, M. S. N., Ansari, M. N. M., Asyraf, M. R. M., Supian, A. B. M., & Ya, H. (2019). Sugar palm (Arenga pinnata [Wurmb.] Merr) starch films containing sugar palm nanofibrillated cellulose as reinforcement: Water barrier properties. *Polymer Composites*, 1–9.

Ilyas, R. A., Sapuan, S. M., Ibrahim, R., Abral, H., Ishak, M. R., Zainudin, E. S., Atikah, M. S. N., Mohd Nurazzi, N., Atiqah, A., Ansari, M. N. M., Syafri, E., Asrofi, M., Sari, N. H., & Jumaidin, R. (2019a). Effect of sugar palm nanofibrillated cellulose concentrations on morphological, mechanical and physical properties of biodegradable films based on agro-waste sugar palm (Arenga pinnata (Wurmb.) Merr) starch. *Journal of Materials Research and Technology*, 8(5), 4819–4830.

Ilyas, R. A., Sapuan, S. M., Ibrahim, R., Abral, H., Ishak, M. R., Zainudin, E. S., Atikah, M. S. N., Mohd Nurazzi, N., Atiqah, A., Ansari, M. N. M., Syafri, E., Asrofi, M., Sari, N. H., & Jumaidin, R. (2019b). Effect of sugar palm nanofibrillated cellulose concentrations on morphological, mechanical and physical properties of biodegradable films based on agro-waste sugar palm (Arenga pinnata (Wurmb.) Merr) starch. *Journal of Materials Research and Technology*, 8(5), 4819–4830.

Ilyas, R. A., Sapuan, S. M., Ibrahim, R., Abral, H., Ishak, M. R., Zainudin, E. S., Atiqah, A., Atikah, M. S. N., Syafri, E., Asrofi, M., & Jumaidin, R. (2020). Thermal, biodegradability and water barrier properties of bio-nanocomposites based on plasticised sugar palm starch and nanofibrillated celluloses from sugar palm fibres. *Journal of Biobased Materials and Bioenergy*, 14(2), 234–248.

Ilyas, R. A., Sapuan, S. M., Ibrahim, R., Abral, H., Ishak, M. R., Zainudin, E. S., Asrofi, M., Atikah, M. S. N., Huzaifah, M. R. M., Radzi, A. M., Azammi, A. M. N., Shaharuzaman, M. A., Nurazzi, N. M., Syafri, E., Sari, N. H., Norrrahim, M. N. F., & Jumaidin, R. (2019).

Sugar palm (Arenga pinnata (Wurmb.) Merr) cellulosic fibre hierarchy: A comprehensive approach from macro to nano scale. *Journal of Materials Research and Technology*, 8(3), 2753–2766.

Ilyas, R. A., Sapuan, S. M., Ibrahim, R., Atikah, M. S. N., Atiqah, A., Ansari, M. N. M., & Norrrahim, M. N. F. (2019). Production, processes and modification of nanocrystalline cellulose from agro-waste: A review. In Nanocrystalline materials (pp. 3–32). IntechOpen.

Ilyas, R. A., Sapuan, S. M., Ishak, M. R., & Zainudin, E. S. (2017). Effect of delignification on the physical, thermal, chemical, and structural properties of sugar palm fibre. *BioResources*, 12(4), 8734–8754.

Ilyas, R. A., Sapuan, S. M., Ishak, M. R., & Zainudin, E. S. (2018). Water transport properties of bio-nanocomposites reinforced by sugar palm (Arenga Pinnata) nanofibrillated cellulose. *Journal of Advanced Research in Fluid Mechanics and Thermal Sciences Journal*, 51(2), 234–246.

Ilyas, R. A., Sapuan, S. M., Ishak, M. R., & Zainudin, E. S. (2018a). Development and characterization of sugar palm nanocrystalline cellulose reinforced sugar palm starch bionanocomposites. *Carbohydrate Polymers*, 202, 186–202.

Ilyas, R. A., Sapuan, S. M., Ishak, M. R., & Zainudin, E. S. (2018b). Sugar palm nanocrystalline cellulose reinforced sugar palm starch composite: Degradation and water-barrier properties. *IOP Conference Series: Materials Science and Engineering*, 368(1).

Ilyas, R. A., Sapuan, S. M., Ishak, M. R., & Zainudin, E. S. (2019). Sugar palm nanofibrillated cellulose (Arenga pinnata (Wurmb.) Merr): Effect of cycles on their yield, physic-chemical, morphological and thermal behavior. *International Journal of Biological Macromolecules*, 123, 379–388.

Ilyas, R. A., Sapuan, S. M., Ishak, M. R., Zainudin, E. S., & Atikah, M. S. N. (2018). Characterization of sugar palm nanocellulose and its potential for reinforcement with a starch-based composite. In *Sugar palm biofibers, biopolymers, and biocomposites* (pp. 189–220). CRC Press.

Ilyas, R. A., Sapuan, S. M., Sanyang, M. L., Ishak, M. R., & Zainudin, E. S. (2018). Nanocrystalline cellulose as reinforcement for polymeric matrix nanocomposites and its potential applications: A review. *Current Analytical Chemistry*, 14(3), 203–225.

Irawan, A. P., Soemardi, T. P., Kusumaningsih, W., & Reksoprodjo, A. H. (2009). *Computer Based Loading Simulation on Socket Prosthesis Design*. In Proceedings of the 11th International Conference on Electronic Materials and Packaging (EMAP 2009), Penang Malaysia, EP09-19-06.

Irawan, A. P., Soemardi, T. P., Widjajalaksmi, K., & Reksoprodjo, A. H. S. (2011). Tensile and flexural strength of ramie fiber reinforced epoxy composites for socket prosthesis application. *International Journal of Mechanical and Materials Engineering*, 6(1), 46–50.

Jalaluddin, H. B. (1993). Chemical modification of oil palm fibres and its application in composite board (Doctoral dissertation, University of Wales, Bangor).

Jawaid, M. H. P. S., & Khalil, H. A. (2011). Cellulosic/synthetic fibre reinforced polymer hybrid composites: A review. *Carbohydrate Polymers*, 86(1), 1–18.

Jensen, J. S., & Raab, W. (2007). Clinical field testing of vulcanized Jaipur rubber feet for trans-tibial amputees in low-income countries. *Prosthetics and Orthotics International*, 31(1), 105–115.

John, M. J., & Thomas, S. (2008). Biofibres and biocomposites. *Carbohydrate Polymers*, 71(3), 343–364.

Juliana, A. H., Paridah, M. T., Rahim, S., Azowa, I. N., & Anwar, U. M. K. (2012). Properties of particleboard made from kenaf (Hibiscus cannabinus L.) as function of particle geometry. *Materials & Design*, 34, 406–411.

Jumaidin, R., Ilyas, R. A., Saiful, M., Hussin, F., & Mastura, M. T. (2019a). Water transport and physical properties of sugarcane bagasse fibre reinforced thermoplastic potato starch biocomposite. *Journal of Advanced Research in Fluid Mechanics and Thermal Sciences*, 61(2), 273–281.

Jumaidin, R., Khiruddin, M. A. A., Asyul Sutan Saidi, Z., Salit, M. S., & Ilyas, R. A. (2019b). Effect of cogon grass fibre on the thermal, mechanical and biodegradation properties of thermoplastic cassava starch biocomposite. *International Journal of Biological Macromolecules,* 146, 746–755.

Jumaidin, R., Saidi, Z. A. S., Ilyas, R. A., Ahmad, M. N., Wahid, M. K., Yaakob, M. Y., Maidin, N. A., Rahman, M. H. A., & Osman, M. H. (2019c). Characteristics of cogon grass fibre reinforced thermoplastic cassava starch biocomposite: Water absorption and physical properties. *Journal of Advanced Research in Fluid Mechanics and Thermal Sciences*, 62(1), 43–52.

Kabir, M. M., Wang, H., Lau, K.T., & Cardona, F. (2012). Chemical treatments on plant-based natural fibre reinforced polymer composites: An overview. *Composite Part B*, 43(7), 2883–2892.

Kakar, A., Jayamani, E., Kok Heng, S., Bakri, M. K. B., & Sinin, H. (2015). Optimization of hot press compression molding and fabrication of poly lactic acid (PLA) luffa biocomposites for biomedical applications. *Australian Journal of Basic and Applied Sciences*, 9(8), 105–112

Khanam, P. N., Khalil, H. A., Jawaid, M., Reddy, G. R., Narayana, C. S., & Naidu, S. V. (2010). Sisal/carbon fibre reinforced hybrid composites: Tensile, flexural and chemical resistance properties. *Journal of Polymers and the Environment*, 18(4), 727–733.

Khoo, K. C., & Lee, T. W. Sulphate pulping of the oil palm trunk. Proceedings of the National Symposium on Oil Palm by-products for agro-based industries, Kuala Lumpur, PORIM (1985).

Kramer, A., Sardo, K., & Slocumb, W. (2015). Analysis of bamboo reinforced composites for use in orthotic and prosthetic application. Washington: American Academy of Orthotists & Prosthetists.

Kutz, M. (2003). Standard handbook of biomedical engineering and design. McGraw-Hill.

Le Moigne, N., Otazaghine, B., Corn, S., Angellier-Coussy, H., & Bergeret, A. (2018). Modification of the interface/interphase in natural fibre reinforced composites: Treatments and processes. In Surfaces and interfaces in natural fibre reinforced Composites (pp. 35–70). Cham: Springer.

Manteghi, S., Mahboob, Z., Fawaz, Z., & Bougherara, H. (2017). Investigation of the mechanical properties and failure modes of hybrid natural fiber composites for potential bone fracture fixation plates. *Journal of the Mechanical Behavior of Biomedical Materials*, 65, 306–316.

Manteghi, S., Sarwar, A., Fawaz, Z., Zdero, R., & Bougherara, H. (2019). Mechanical characterization of the static and fatigue compressive properties of a new glass/flax/epoxy composite material using digital image correlation, thermographic stress analysis, and conventional mechanical testing. *Materials Science and Engineering: C*, 99, 940–950.

Mazani, N., Sapuan, S. M., Sanyang, M. L., Atiqah, A., & Ilyas, R. A. (2019). Design and fabrication of a shoe shelf from kenaf fiber reinforced unsaturated polyester composites. In Lignocellulose for future bioeconomy (Issue 2000, pp. 315–332). Elsevier.

Mehboob, H., & Chang, S. H. (2014). Application of composites to orthopedic prostheses for effective bone healing: A review. *Composite Structures*, 118, 328–341.

Mishra, S., Tripathy, S. S., Misra, M., Mohanty, A. K., & Nayak, S. K. (2002). Novel eco-friendly biocomposites: Biofiber reinforced biodegradable polyester amide composites—fabrication and properties evaluation. *Journal of Reinforced Plastics and Composites*, 21(1), 55–70.

Mohandas, G., Oskolkov, N., McMahon, M. T., Walczak, P., & Janowski, M. (2014). Porous tantalum and tantalum oxide nanoparticles for regenerative medicine. *Acta Neurobiolgiae Experimentalis (Wars)*, 74(2), 188–196.

Namvar, F., Jawaid, M., Md Tahir, P., Mohamad, R., Azizi, S., Khodavandi, A., Rahman, H. S., & Nayeri, M. D. (2014). Potential use of plant fibres and their composites for biomedical applications. *BioResources*, 9, 3, 5688–5706.

Nayak, S. K., Mohanty, S., & Samal, S. K. (2010). Hybridization effect of glass fibre on mechanical, morphological and thermal properties of polypropylene–bamboo/glass fibre hybrid composites. *Polymers and Polymer Composites*, 18(4), 205–218.

Niinomi M. (2002). Recent metallic materials for biomedical applications. *Metal Mater Trans A*, 33 (3), 477–486.

Norizan, M. N., Abdan, K., Ilyas, R. A., & Biofibers, S. P. (2020). Effect of fiber orientation and fiber loading on the mechanical and thermal properties of sugar palm yarn fiber reinforced unsaturated polyester resin composites. *Polimery*, 65(2), 34–43.

Nurazzi, N. M., Khalina, A., Sapuan, S. M., Ilyas, R. A., Rafiqah, S. A., & Hanafee, Z. M. (2019). Thermal properties of treated sugar palm yarn/glass fiber reinforced unsaturated polyester hybrid composites. *Journal of Materials Research and Technology*, December.

Nurhanisah, M. H., Hashemi, F., Paridah, M. T., Jawaid, M., & Naveen, J. (2018a). Mechanical properties of laminated kenaf woven fabric composites for below-knee prosthesis socket application. *In IOP Conference Series: Materials Science and Engineering*, 368(1), 012050.

Nurhanisah, M. H., Jawaid, M., Ahmad Azmeer, R., & Paridah, M. T. (2018b). The AirCirc: Design and development of a thermal management prototype device for below-knee prosthesis leg socket. *Disability and Rehabilitation: Assistive Technology*, 1–8.

Nurhanisah, M. H., Saba, N., Jawaid, M., & Paridah, M. T. (2017) Design of prosthetic leg socket from kenaf fibre based composites. In *Green Biocomposites: Green Energy and Technology,* M. Jawaid, M. Salit, O. Alothman (Eds.), Cham: Springer.

Oleiwi, J. K., Salih, S. I., & Hwazen, S. F. (2017). Effect of siwak and bamboo fibers on tensile properties of self-cure acrylic resin used for denture applications. *Journal of Material Sciences and Engineering*, 6(5), 1–6.

Paiva Júnior, C. Z., De Carvalho, L. H., Fonseca, V. M., Monteiro, S. N., & D'Almeida, J.R. M. (2004). Analysis of the tensile strength of polyester/hybrid ramie–cotton fabric composites. *Polymer Testing*, 23(2), 131–135.

Patel, N. R., & Gohil, P. P. (2012). A review on biomaterials: Scope, applications & human anatomy significance. *International Journal of Emerging Technology and Advanced Engineering*, 2(4), 91–101.

Peças, P., Carvalho, H., Salman, H., & Leite, M. (2018). Natural fibre composites and their applications: A review. *Journal of Composites Science*, 2(4), 66.

Peh, T. B., Khoo, K. C., & Lee, T. W. (1976). Pulping studies on empty fruit bunches of oil palm (Elaeis guineensis Jacq.). *The Malaysian Forester*, 39(1), 23–37.

Pertici, G. (2016) Introduction to bioresorbable polymers for biomedical, Bioresorbable Polymers for Biomedical Applications: From Fundamentals to Translational Medicine p. 1.

Ramakrishna, S., Mayer, J., Wintermantel, E., & Leong, K. W. (2001). Biomedical applications of polymer-composite materials: A review. *Composites Science and Technology*, 61(9), 1189–1224.

Ratner, B. D., Hoffman, A. S., Schoen, F. J., & Lemons, J. E. (2004). Biomaterials science: An introduction to materials in medicine. Elsevier.

Reddy, N., & Yang, Y. (2005). Biofibers from agricultural byproducts for industrial applications. *Trends in Biotechnology*, 23(1), 22–27.

Ridzuan, R., Shaler, S., & Jamaludin, M.A. (2002). Properties of medium density fibreboard from oil palm empty fruit bunch fibre. *Journal of Oil Palm Research*, 14(2), 34–40.

Romanzini, D., Ornaghi Jr, H. L., Amico, S. C., & Zattera, A. J. (2012). Influence of fiber hybridization on the dynamic mechanical properties of glass/ramie fiber-reinforced polyester composites. *Journal of Reinforced Plastics and Composites*, 31(23), 1652–1661.

Rozman, H. D., Tay, G. S., Kumar, R. N., Abusamah, A., Ismail, H., & Mohd. Ishak, Z. A. (2001). Polypropylene–oil palm empty fruit bunch–glass fibre hybrid composites: A preliminary study on the flexural and tensile properties. *European Polymer Journal*, 37(6), 1283–1291.

Salernitano, E., & Migliaresi, C. (2003). Composite materials for biomedical applications: A review. *Journal of Applied Biomaterials and Biomechanics*, 1(1), 3–18.

Sarkar, R., & Banerjee, G. (2010). Ceramic based bio-medical implants. *Interceram*, 59(2), 98–102.

Sreekumar, P. A. (2008). Matrices for natural-fibre reinforced composites. In Properties and performance of natural-fibre composite, edited by K. L. Pickering (p. 541). UK: Birmingham, Woodhead Publication Limited.

Syafri, E., Sudirman, Mashadi, Yulianti, E., Deswita, Asrofi, M., Abral, H., Sapuan, S. M., Ilyas, R. A., & Fudholi, A. (2019). Effect of sonication time on the thermal stability, moisture absorption, and biodegradation of water hyacinth *(Eichhornia crassipes)* nanocellulose-filled bengkuang *(Pachyrhizus erosus)* starch biocomposites. *Journal of Materials Research and Technology*, 8(6), 6223–6231.

Tanaka, R., Peng, L. C., & Wan Rosli, W. D. (2002). Preparation of cellulose pulp from oil palm EFB by processes including pre-hydrolysis and ozone bleaching. Proceeding of the USM-JIRCAS Joint International Symposium-Lignocellulose Material of the Millennium: Technology and Application, March 20-21, Penang, Malaysia, 33–38.

Tathe, A., Ghodke, M., & Nikalje, A. P. (2010). A brief review: Biomaterials and their application. *International Journal of Pharmacy and Pharmaceutical Sciences*, 2(4), 19–23.

Ticoalu, A., Aravinthan, T., & Cardona, F. (2010). A review of current development in natural fiber composites for structural and infrastructure applications. In Proceedings of the Southern Region Engineering Conference (SREC 2010), Engineers Australia, 113–117.

Wang, Y., Wang, G., Cheng, H., Tian, G., Liu, Z., & Xiao Qun, F., (2010). Structures of bamboo fiber for textiles. *Textile Research Journal*, 80(4), 334–343.

Wei, Q., Deng, N. N., Guo, J., & Deng, J. (2018). Synthetic polymers for biomedical applications. *International Journal of Biomaterials*, 2018.

3 Biocomposites in Advanced Biomedical and Electronic Systems Applications

Faris M. AL-Oqla
Department of Mechanical Engineering,
The Hashemite University, Zarqa, Jordan

CONTENTS

3.1   INTRODUCTION

Newly emerging technologies and techniques have made it possible to invent and fabricate new biocomposite materials that are able to serve fundamental roles in diverse applications. Such newly invented biopolymers are being used not only for the enhancement of the performance, properties, and characteristics of the devices, but also to provide designers with new possibilities for satisfying new attractive features required by a specific application. Transparency, flexibility, and degradability are examples of such new requirements to be satisfied (AL-Oqla & El-Shekeil, 2019; AL-Oqla & Salit, 2017b; Alaaeddin et al., 2019a, c; Alaaeddin et al., 2019d).

Bio-based composites are increasingly replacing conventional composites in a wide array of applications for sustainable industries. However, enormous efforts are still required to better exploit these composites and to expand their applications to include high-tech ones in several fields such as electronics, biomedical, drugs, civil, and sports. Also, the need to improve their qualities for reliable performance is still demanding (AL-Oqla, 2017; AL-Oqla et al., 2019). To achieve these goals, establishing efficient and robust evaluation and selection techniques for the composites' constituents is the most important step. In these techniques, the various desired fibers as well as polymer properties such as physical, mechanical, economic, environmental, and thermal have to be considered together and evaluated to discover the best type of both fibers and polymers for a certain application (AL-Oqla et al., 2014a; AL-Oqla et al., 2014b; AL-Oqla et al., 2015b; AL-Oqla et al., 2017; Aridi et al., 2016a, b, 2017; Sapuan et al., 2013; Sapuan et al., 2016). Besides, potentials of new materials with new levels of performance would be revealed, which in turn, expand the applications for the biomaterials.

Recently, new biomaterials derived from natural plants and agronomic residues called natural cellulosic fibers (NCF) have been utilized as a reinforcing filler with starch-based materials. NCF have various characteristics, such as availability, affordability, low density, biodegradability, recyclability, and secure handling (AL-Oqla & Omar, 2015; AL-Oqla & Sapuan, 2015, 2018a). Remains from agricultural fields, such as maize stover, sugar palm residues, cassava bagasse, and pineapple wastes, are generated yearly in bulk and cost less; these are the most abundant sources of natural fiber. In addition to its availability, cellulose plant fiber is characterized by flexibility, affordability, durability, high extendibility, and low density, making it an attractive study area. However, the addition of cellulosic fibers to reinforce the starch-based materials has provided significant enhancement, specifically in terms of tensile performance and thermal stability, as well as water barrier characteristics. This enhancement is associated with the high compatibility and compositional similarity between fiber and starch because both originate from biological sources. In fact, the proper selection of biomaterial constituents for a certain performance in a particular application is a multi-criteria decision-making problem where suitable fields of optimizations and decision-making (e.g., the analytical hierarchy process and TOPSIS method) were implemented to enhance the performance of biomaterials as well as others in various applications (AL-Oqla & Hayajneh, 2007; AL-Oqla & Omar, 2012; AL-Oqla & Omar, 2015; AL-Oqla & Salit, 2017a; Al-Widyan & AL-Oqla, 2011, 2014).

3.2 BIOPOLYMER PROCESSING AND ITS DEVELOPMENT

3.2.1 EXTRUSION

Extrusion is a plastic industry manufacturing process with a high-volume rate. It is practically a combination of screw conveyor and compressor with other accessories to cause raw heated plastics to melt and form under a continuous cycle as seen

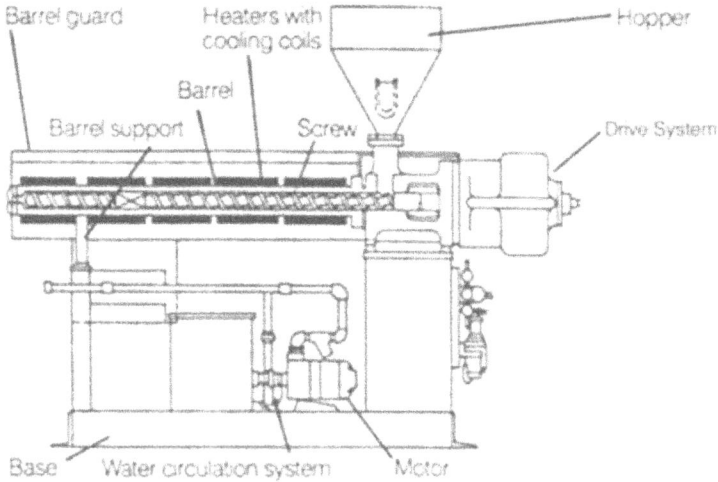

FIGURE 3.1   Extrusion process device.

in Figure 3.1. A heated helical screw (either single- or twin-screw) compresses the raw plastics with other additives like chain-extenders, stabilizers, plasticizers, and/or colorants through the machine hopper to make a homogeneous melt plastic by forcing the air out of the barrel without causing damage to the molten.

Various benefits can be achieved using single-screw or twin-screw based upon the polymer being processed and the products to be fabricated. A comparison between single- and twin-screw is illustrated in Figure 3.2. The ratio of the screw length to its diameter (L/D) is responsible for homogeneous mixing and varies for various polymer types. Generally speaking, screws with large L/D ratios can give higher shear heating, as well as better blends and longer melt residence times in the extruder (AL-Oqla et al., 2015c; AL-Oqla & Sapuan, 2014a).

In polylactic acid (PLA) polymer extrusion for instance, hydrolytic degradation is the main drawback. Thus, it should be dried to 0.025% w/w moisture content before extrusion. Accordingly, to avoid reduction of molecular weight, higher working temperature (about 240°C) must maintain its moisture content below 0.005% w/w. However, drying of PLA resin is an exciting issue as it degrades at elevated temperatures and high moisture content environments. PLA resins are formed by means of conventional extruder with a general-purpose screw (L/D ratio = 24–30). To make sure that all PLA crystalline are melted as well as attaining an optimal melt viscosity for processing, the heater set point is usually between 200–210°C.

A PLA/sisal biocomposite has been recycled for eight times via the extrusion process, to illustrate its recyclability (Chaitanya et al., 2019). The specimen's performance has significantly dropped after the third recycle processing as a result of severe damage of fibers and PLA degradation. A screw speed of 60 RPM combined with temperature profile of 150–185°C have caused thermal degradation to the composite. An increase in surface roughness also appeared as a result of the hydrolytic

Criteria	Single-Screw	Twin-Screw
Flow type	Drag	Near positive
Residence time and distribution	Medium and wide	Low and narrow
Effect of back pressure on output	Reduces output	Moderate effect on output
Shear in channel	High (useful for stable polymer)	Low
Maximum screw speed	High (limited by melting and stability of polymer)	Medium
Thrust capacity	High	Low
Mechanical construction	Robust, simple	Complicated
Initial costing	Moderate	High

FIGURE 3.2 Comparison between single and twin screw.

degradation that reduced the PLA molecular weight and increased its melt flow rate, which led to improper fiber wetting as illustrated in the morphological images in Figure 3.3.

3.2.2 INJECTION MOLDING

Injection molding is one of the most promising methods to fabricate plastic products as well as thermoplastic starch. Figure 3.4 demonstrates a typical injection machine. This method allows the sample to be fabricated into the desired shape; hence, no further machining process is required to prepare the sample for testing (such as the dumbbell shape). The injection unit consists of a reciprocal screw that dissolves polymer with high temperature and conveys it to a mold that clamps in the clamping unit. Then the molten is allowed to cool and solidify in the mold before ejected.

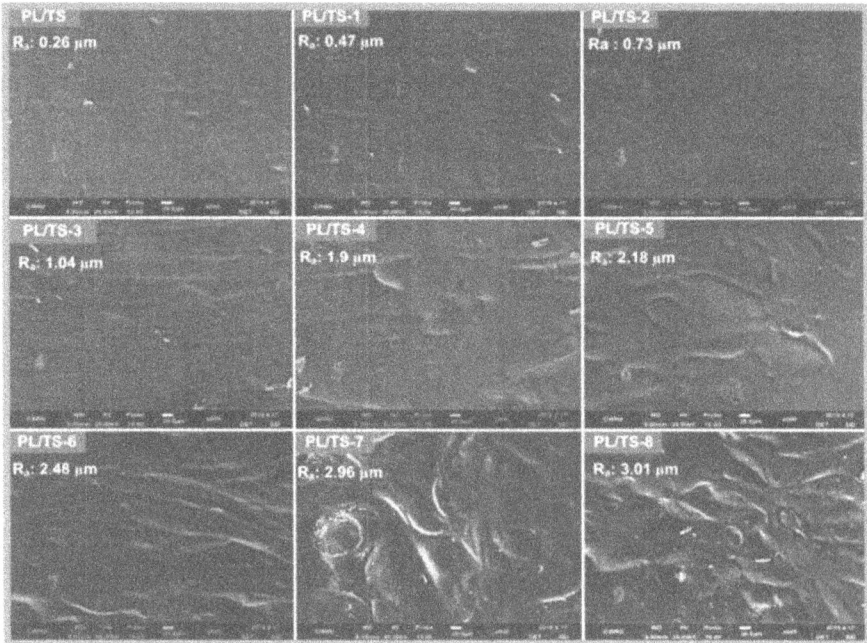

FIGURE 3.3 Morphological surface of recycled PLA/sisal biocomposites during eight extrudes. (Adapted from Chaitanya et al., 2019.)

In hot pressing, the samples faced the difficulty of melting uniformly when heated in a small mold; that is, dumbbell shape. However, cutting the specimen into a dumbbell shape is not a good option because it will introduce cut edges or cracks that may become a stress concentration point during tensile testing. Injection molding is the most common process used for the production of conventional plastics; hence, this

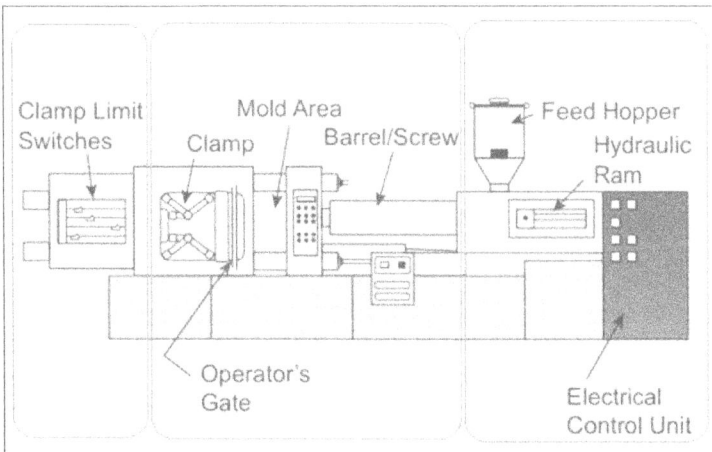

FIGURE 3.4 A typical injection machine.

method will provide better opportunities for the commercialization of thermoplastic starch in the near future. However, there are some limitations to this process as compared to others. In terms of cost, injection molding requires high-end equipment to carry out the process. In addition, huge amounts of materials are required to produce samples; this is not entirely suitable for studies that employ nanoscale materials as fillers or for certain types of starch where the amount is very limited. Injection molding also requires a series of parameters that are much more complex than solution casting and hot press. Therefore, a higher number of trials and replications are necessary to obtain optimum parameters for the desired thermoplastic starch conditions. Again, this process will require huge amounts of materials to be used, which can be difficult depending on the material involved.

Injected PLA products are usually brittle due to rapid physical aging as its glass transition temperature, T_g, is higher than the ambient temperature. It was reported that the lower the aging temperature from T_g, the slower the aging rate. However, no sign of physical aging is taking place for aging temperature higher than T_g of PLA. Thus, injected products aged at room temperature might become very brittle. But PLA polymer with high crystallinity could minimize the effect of physical aging as such type of aging occurring below T_g usually is governed by its amorphous phase. Further, cooling temperature and rate, as well as the mold temperature and backpressures, are key parameters that may dramatically influence PLA aging properties.

Other manufacturing processes are in fact available and accessible for biopolymers and their composites including blow molding, thermoforming, and 3D printing. Three kinds of blow molding systems are available, injection blow molding, extrusion blow molding, and stretch blow molding. All plastic bottles are made from thermoplastic materials by using blow molding processing. For injection blow molding process, the material is injected. The hot injected plastic is then blown into a bottle at a blow molding station and allowed to cool (1–4 days) then ejected at the next station. It is often used to produce irregular shapes of bottles, such as bore necks or wide neck bottles. Alternatively, stretch blow molding is similar to that of injection blow molding, but it is further stretched throughout. On the other hand, thermoforming is a process of heating a thermoplastic sheet to its glass transition temperature to stretch and then allowing it to cool into final products. However, elastomers and thermosets are prohibited from the thermoforming process because they do not soften when heated. The process of thermoforming is capable of creating several finishing parts on the same product. Hence, time and cost savings are achievable. There are three available types of thermoforming processing: vacuum thermoforming, pressure thermoforming, and mechanical thermoforming. Figure 3.5 shows the effect of some parameters (heating temperature and time) during thermoforming processing on edge, corner, and thickness of various polymer blend products, where PLA reproduced accurately the mold's geometry and kept a regular thickness distribution, while PHBV products observed to have irregular shape with nonhomogeneous thickness throughout the specimens (red boxes in Figure 3.5).

On the other hand, 3D printing creates parts by building up objects one layer at a time. It generally has several advantages over traditional manufacturing techniques. Compared to traditional manufacturing methods, additive manufacture has the main advantage of the speed at which parts can be produced where complex designs

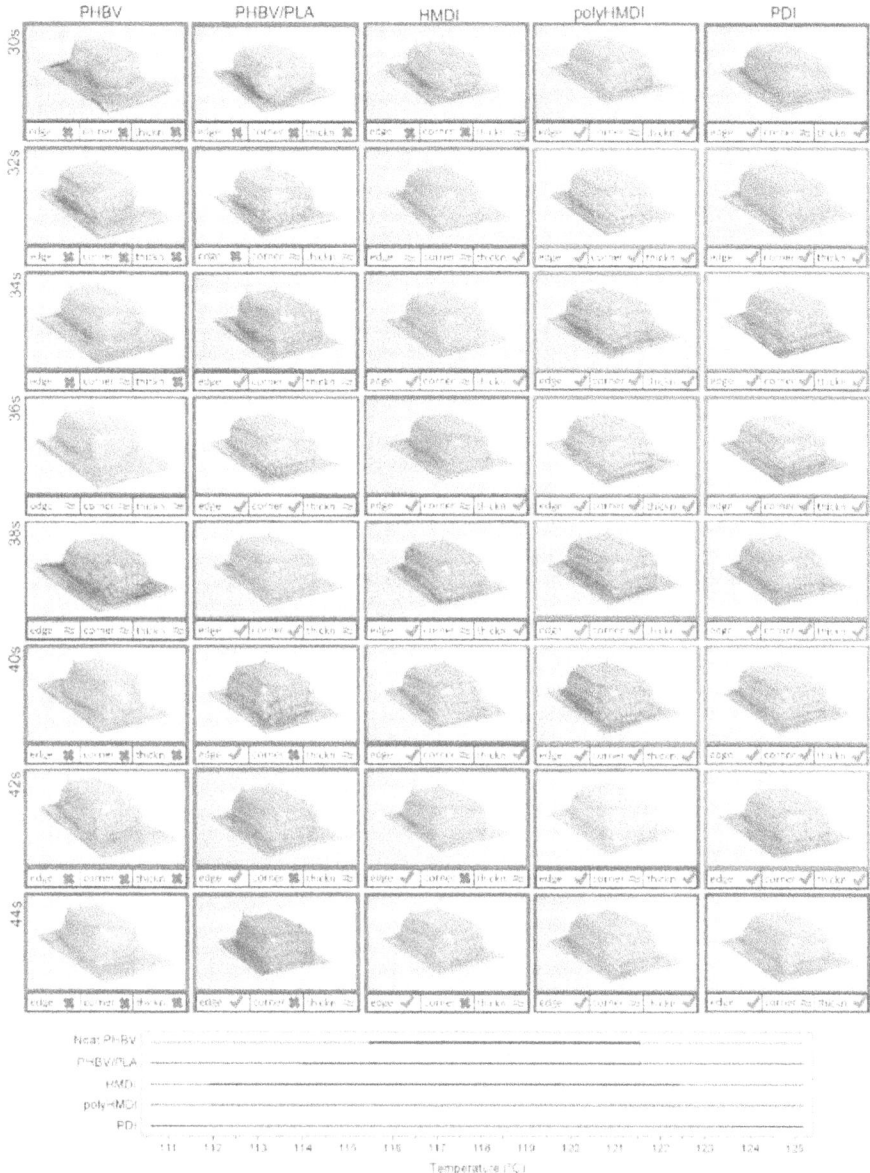

FIGURE 3.5   The parameter inspection (heating temperature and time) on edge, corner, and thickness of PHBV, PHBV/PLA blend, blend with hexamethylene diisocyanate (HMDI), blend with poly(hexamethylene) diisocyanate (polyHMDI), and blend with 1,4-phenylene diisocyanate (PDI) compatibilizers. Red boxes represent an unsatisfied product; blue boxes represent two investigates passed; green boxes show all investigates passed the inspection. (Adapted from González-Ausejo et al., 2017.)

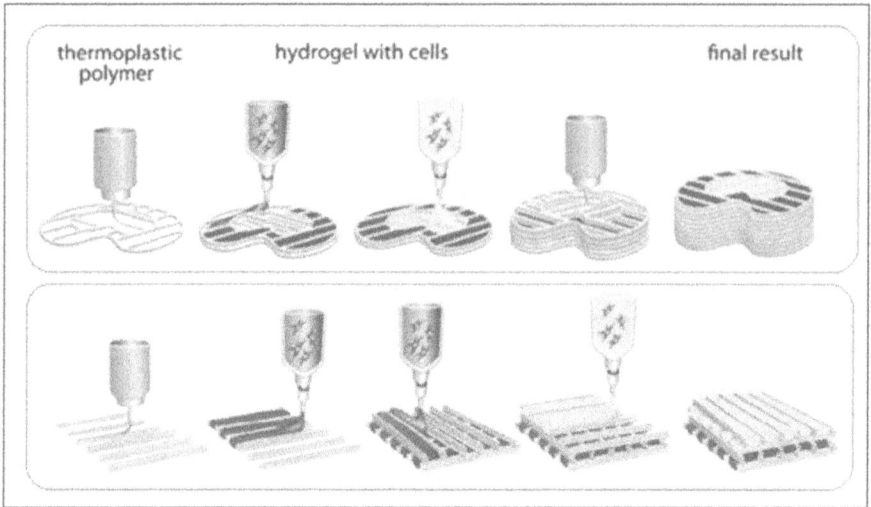

FIGURE 3.6 3D printing schematic for tissue and organ engineering. (Adapted from Chiulan et al., 2018.)

can be uploaded from a CAD system and printed in a few hours. This allows the opportunity to rapidly verify and develop more realistic design ideas.

Utilizing 3D printing technology in tissue engineering offered a more detailed, complex, and versatile scaffold system. Figure 3.6 displays the 3D printing diagram for tissue and organ engineering.

3.3   ELECTRONICS APPLICATIONS OF BIOCOMPOSITES

3.3.1   BIOCOMPOSITE MATERIALS IN THE FIELD OF LEDs

The use of biocomposite materials in high-tech applications including the electronic ones has been growing due to their beneficial integrated characteristics. Moreover, the use of biocomposite materials in the field of LEDs has added many attractive values to the overall performance and characteristics of display systems. Among these are: lower power consumption, higher degree of degradability, renewable, lighter weight, lower costs, and extended color quality (AL-Oqla et al., 2015c; AL-Oqla et al., 2016; AL-Oqla & Salit, 2017a). Many biocomposite materials have been extensively studied and used in the fabrication of light-emitting displays leading to bio-based organic light-emitting diodes (OLEDs). For instance, DNA polymer has been used as an electron blocking layer/hole transporting layer (EBL/HTL) in OLEDs resulting in augmented conversion efficiency and luminescence. On the other hand, and due to its transparent nature and exceptional electrical construction, graphene is becoming widely used as a promising substitute for indium tin oxide (ITO), which has been extensively used as a conductive material. Graphene structure has a zero band gap resulting in extraordinary electronic conductivity. This material enjoys also a very high electronic mobility of around 20×10^3 cm^2/V s with a wide surface area

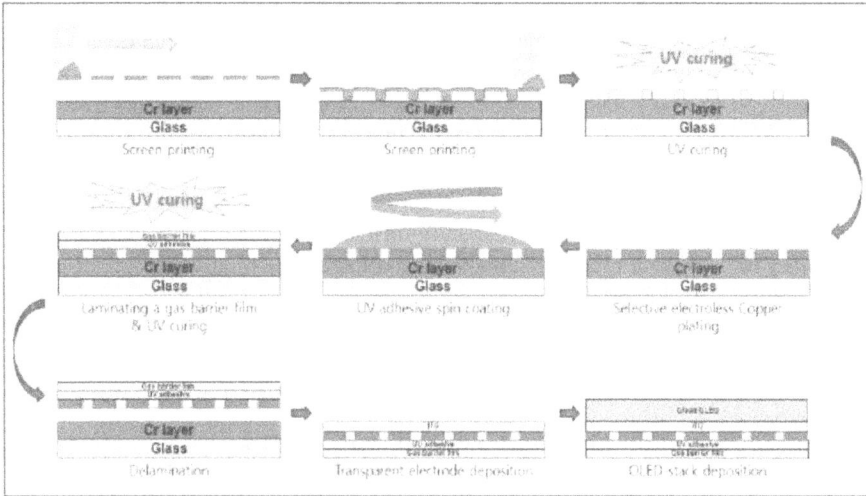

FIGURE 3.7   Schematics illustration for the steps in fabricating flexible OLEDs. (Adapted from Cho et al., 2017.)

that could reach more than 2,600 m^2/g. This is beside the high degree of transparency, which could reach 97%. Graphene shows excellent performance in building the anode parts of OLEDs.

Cellulose-based OLEDs are another advanced and very hopeful flexible OLED family. One advanced method to enhance the characteristics of coefficient of thermal expansion (CTE) in OLEDs, nanocellulose-based composite materials are largely utilized in the production of OLED substrates to give them the desirable flexibility (Cho et al., 2017). Figure 3.7 illustrates the required steps for fabricating flexible OLEDs. On the other hand, starch is considered a cheap, widely available sustainable material for producing renewable biocomposites as it contains a large number of functionalized hydroxyl groups, which are around the surface. Therefore, such structure makes it possible to mix them with carbon nanodots (CDs) resulting in more ecological CD-based and phosphors-based LEDs.

3.3.2   BIOSENSORS AND ACTUATORS

Biosensors are integrated systems that typically contain two main parts combined together. Such parts are the receptors and the transducers, where receptors are responsible for interaction and recognition of the element being sensed. Several kinds of biorecognition elements including enzymes, antibodies, and nucleic acids are utilized for this purpose. However, the basic challenge in this branch of high-tech applications is to enhance the sensitivity as well as the electivity of the receptor elements. Therefore, to achieve such objectives, especially engineered biopolymers are garnering increased attention and investigations worldwide. This is mainly due to the numerous distinguished characteristics and benefits that made them suitable for both medical applications and ecological screening.

Sensors based on organic thin film transistors (OTFTs) found many applications in sensing not only variations in environmental conditions, but also in detecting the presence and concentrations of certain elements. This is mainly because organic materials can be easily inserted to the structure of the organic transistors to work as receptors or even as the active layer. Unlike the organic electrochemical transistors (OECTs), which were being used for bio-sensing applications, OTFTs are solid-state based transistors and thus have the advantage of being easier to integrate with other parts of the complete system. This is in addition to the many merits of OTFTs, such as the ability to operate at relatively low-biasing voltages, the relatively high efficiency in converting ionic signals into electronic signals for amplification and processing, and the lower power consumption levels. They are made of nontoxic materials making them very attractive for biomedical applications. Figure 3.8 demonstrates two sensors utilizing organic-based materials. The first sensor (Figure 3.8A) was planned to sense the concentrations of mercuric ion (Hg^{2+}) (Rullyani et al., 2019), whereas the second sensor (Figure 3.8B) was aimed to be able to display numerous ecological factors at the same time (Surya et al., 2019).

Moreover, the principle of mechanical actuation in electro-active papers (EAPaps) happens mainly as a result of ion migration phenomenon and the piezoelectric effect that occurs within cellulose sheets. When a cellulose sheet is subjected to an electric field, these two mechanisms originate a mechanical displacement, which forces the EAPap to bend (Kim et al., 2018). The piezoelectric sensor fabrication sequence using chitin biopolymers is seen in Figure 3.9. The amount of this mechanical bending depends on many factors including the applied electric field strength and frequency, the type of host paper, and bonding agents used. Generally, EAPaps can produce relatively small bending, which results in poor actuation performance. In order to enhance the actuation behavior, conducting polymers, such as polypyrrole (PPY) and polyaniline (PANI), are used as coating materials for EAPaps.

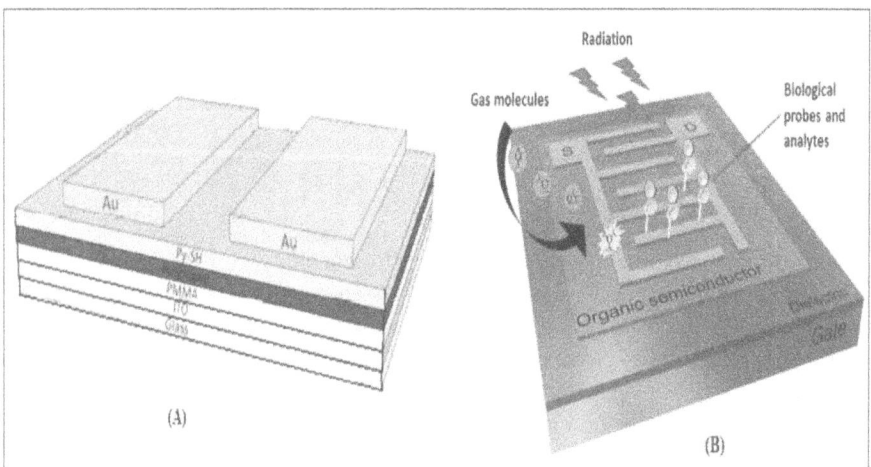

FIGURE 3.8 3D schematic illustrations of sensor made of organic materials (A) for the detection of Hg^{2+} ions and (B) an environmental multipurpose monitoring sensor.

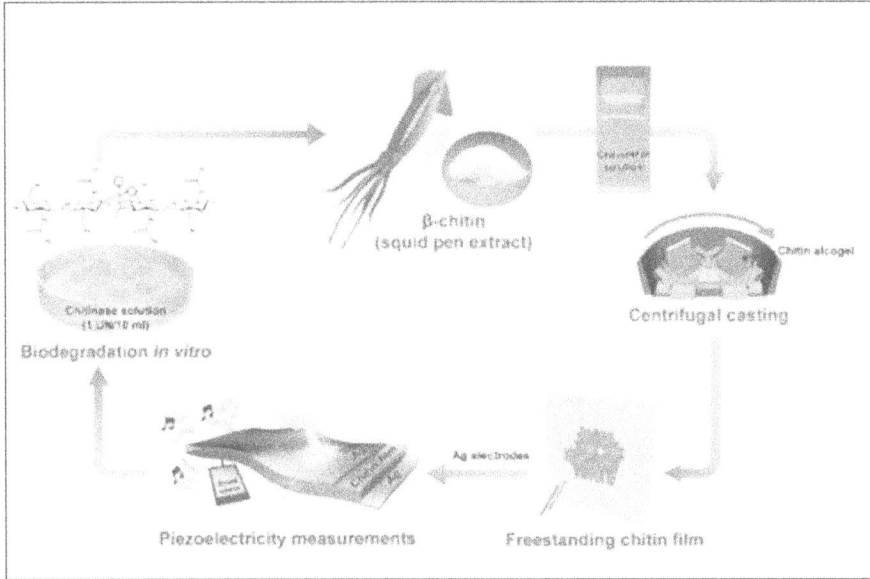

FIGURE 3.9   Piezoelectric sensor fabrication sequence using chitin biopolymers. (Adapted from Kim et al., 2018.)

EAPaps coated with PANI have been reported to have a mechanical displacement of about 10 mm (Kim, 2017).

3.3.3 SUPERCAPACITORS

Supercapacitors can be categorized into two main types; the electrochemical double layer capacitors (EDLL) and the pseudocapacitors. EDLL supercapacitors are very similar in principle to ordinary capacitors as they are formed of two parallel conducting plates with no charge transfer. Pseudocapacitors, on the other hand, depend on charge transfer mechanisms occurring on the electrodes (Ramesh et al., 2018). One main factor affecting the choice of a supercapacitor for a certain application is its life cycle, which is heavily affected and determined by the life cycle of the biopolymer composite used in the fabrication.

3.3.4 PHOTODIODES AND PHOTOVOLTAIC SOLAR CELLS

Biopolymers received lots of attention in various fields of high-tech applications. Optoelectronics is a major field. In such a field, the electronic circuits of the optoelectronic systems operate electronics as well as optics not only to convert electronic signals into light and vice versa, but also for storing, transmitting, and further processing. The most widespread inorganic photodiodes are fabricated from gallium arsenide (GaAS). Photodiodes have many applications in several fields both of civilian and military nature.

In principle, a photovoltaic (PV) solar cell is a silicon-based multi-layer device used to convert light energy into electric energy. The fact that biopolymers like starch, cellulose, collagen, and many other widely available worldwide have repeated carboxyl, hydroxyl, and amino functional groups, gives them the ability to connect and interrelate with their surroundings. To get benefit out of the many advantages biocomposite polymers can offer to the industry of solar cells, recent research has been focusing on introducing new generations of organic solar cells. Biopolymers and nanoparticles are being used not only in the substrate part of the solar cells, but also within all other layers including the photoelectric layer, the dye, and electrolyte. Organic-based PV solar cells can be of a single-layer or a two-layer structure. In a single-layer cell, the two conduction electrodes are separated by the organic photosensitive electronic composite. On the other hand, to construct a two-layer cell, the photosensitive layer is formed out of two layers; one as an electron donor and the other as an electron acceptor. The illustrations presented in Figure 3.10 show the device structure, layout, and photonic efficiency versus bending cycles of the proposed device presented in (La Notte et al., 2018).

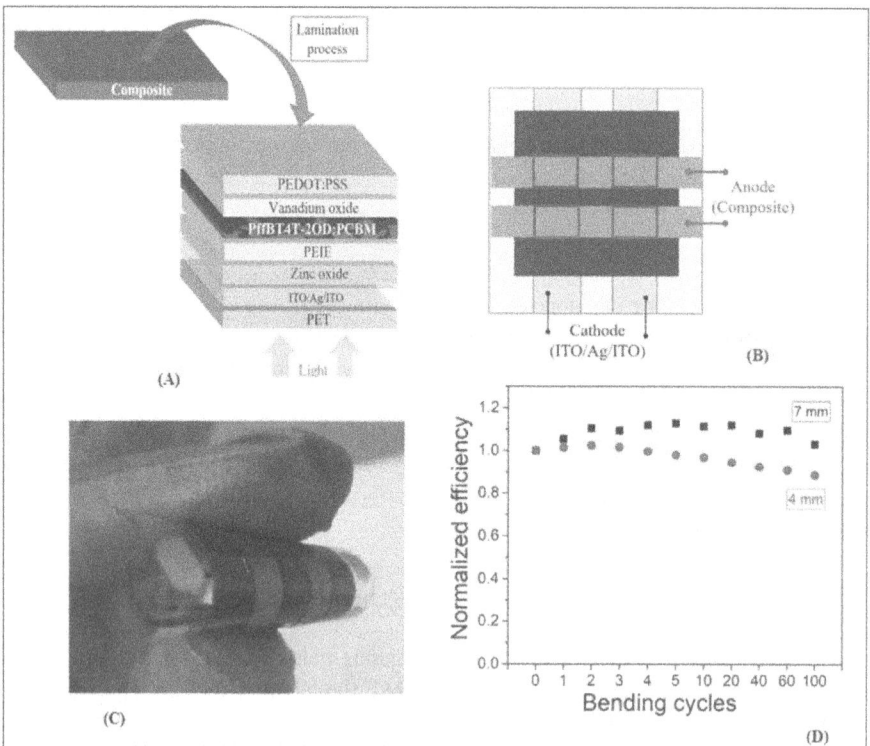

FIGURE 3.10 (A) 3D illustration of the organic photovoltaic. (B) OPV layout. (C) Image of the substrate under mechanical bending. (D) Performance characteristics the OPV under different degrees of binding. (Adapted from La Notte et al., 2018.)

PLA is a thermoplastic polyester, which means that it can be heated to its melting point at around 150–160°C, cooled, and heated again without any degradation. In contrast, the thermoset plastic can be heated only once and it is irreversible. Due to its characteristics, PLA has been the most studied and utilized biodegradable plastic in human history. It is replacing conventional petrochemical-based polymers slowly and becoming the leading biomaterial for medical application as well as in other plastic industries.

Ultraviolet photodetectors are special types of photodiodes that are designed and fabricated to be used to detect and interact with the part of the electromagnetic wave spectrum with wave lengths within the range of 280–400 nm. Such light detectors found very important roles in a wide range of applications including environment monitoring, communications, and astronomy. ZnO-cellulose composite is one of the very attractive biocomposite materials that have been used in the fabrication of ultraviolet (UV) detectors (Mohiuddin et al., 2017). This is mainly because this biocomposite combines the very attractive features of both ZnO and cellulose. In this biocomposite, cellulose is used to form a flexible matrix for the ZnO active element. When exposed to oxygen, zinc oxide becomes sensitive to light. Cellulose fibers, on the other hand, are very absorbent, mechanically strong, and molecular oxygen can easily permeate through cellulose matrix. Photonic sensitivity of this composite is highly dependent on the ratio of ZnO to cellulose contents.

In the face of the various benefits they enjoy, biopolymer-based PV solar cells still have some drawbacks leading to real challenges to its industry. The poor light to electric energy conversion rates, in addition to the instability in the overall performance and comparatively short life cycle are among their weaknesses. However, biopolymer composites, particularly the nanocellulosic materials, are believed to have the capability of improving the characteristics of specific properties of PV solar cells as one of the most promising materials in this regard due to their outstanding mechanical properties, ease of adaptation, elevated aspect ratio, and low density.

3.3.5 OTHER ELECTRICAL APPLICATIONS OF BIOPOLYMERS

Biopolymer-based materials have found significant roles in high-tech components in various modern and advanced applications. This includes but is not limited to electromagnetic interference shielding materials, fuel cell fabrication, piezoelectric and thermoelectric materials, and microwave absorbers.

The key challenge in the fuel cell industry is to have stable materials at a wide range of operating temperatures with high efficiency as well as having extended lifetimes, while being eco-friendly, sustainable, and commercially available (AL-Oqla et al., 2014c; AL-Oqla & Sapuan, 2014b; AL-Oqla & Sapuan, 2018b). Chitosan and starch, cellulose, and chitin as well as other biomaterials are being considered as good candidates to be used in more economic and eco-friendly fuel cells.

Energy harvesting systems employing the piezoelectric principle as well as thermoelectric principle are imperative fields for the use of the notable properties of biopolymers (AL-Oqla et al., 2018). Biopolymers' excellent flexibility makes them reasonable candidates as energy harvesting materials. They have the ability to change any applied mechanical stress into free electric charges regardless of the direction of the stress. Polyamides, liquid crystal polymers, and Parylene C are reported for their piezoelectricity. Examples of thermoelectric materials are PANI, PPY, and polythiophene (PTh) and its derivatives.

3.4 BIOPOLYMERS IN MEDICAL APPLICATIONS

Several biopolymers are utilized in fabricating biomaterials. Some are more desirable than others because they are more compatible with the human body. Proteins (like albumen), silk, and collagen are examples of these biopolymers (Butcher et al., 2014). Moreover, some scaffolds, which are utilized as tissue engineering and controlled drug delivery devices, consist of biopolymers that are used *in vitro* and *in vivo* applications. Such biopolymers include films, fibers, micro- and nanoparticles, hydrogels, as well as 2D and 3D structure scaffolds.

Regardless of the well-known benefits of biomaterials, there are some limitations that bound the usage of biopolymers in biomedical applications (AL-Oqla et al., 2018; Alaaeddin et al., 2018, 2019a, b, c). For instance, materials made from biopolymers have limited mechanical characteristics, and their stability in physiological and aqueous circumstances essential for medical usages is low (Jiang et al., 2010). Biomaterials are typically taken from several biological species like cellulose, silk, chitosan, starch, collagen, and poly(lactic acid). Such biopolymers have numerous returns such as cytocompatibility and the ability to be dissolved in the human body devoid of any toxic or harmful materials. Moreover, several biomaterials are used in bone fracture treatment; an essential role is to be a substitute option for bone grafts, as the number of donors decreases with time (AL-Oqla et al., 2015a; AL-Oqla & Omari, 2017).

3.4.1 Biopolymer Uses in Bone Tissue Engineering (BTE)

Biopolymers are similar to biological macromolecules. They are produced by normal procedures and are extracted from animals, bacteria, and plants (Okamoto & John, 2013). Numerous biopolymer manufacturing processes have been accepted by the Food and Drug Administration (FDA). Natural biopolymers are allocated into three classes based upon their chemical structure, polysaccharides, proteins, and polyesters. Table 3.1 illustrates several types of popular biopolymers applied in BTE, and presents their treating, antigenicity (i.e., the ability of a chemical assembly to be precisely adhesive with a set of certain products with adaptive immunity), and crosslinking agent, which helps create crosslinking bonds that grow the material's mechanical possession and provide stability. Also, the table contains the biopolymer material's ability for *in vitro* forming.

3.4.2 Scaffolds Including Calcium Phosphate (CaP)

Extracellular matrix of the bone is fundamentally built of fibrous collagen and hydroxyapatite (HAp) (Li et al., 2014). One of the common approaches utilized to create scaffolds for BTE is to imitate the organic–inorganic composite, by means of a combination of calcium phosphate (CaP) into polymer matrices (Li et al., 2014).

TABLE 3.1
Biopolymers Commonly Used in BTE (Adapted from Li et al., 2014)

Biopolymer		Origin	Major Antigenicity	Crosslinking Agent	Dominating Nucleation Sites
Protein	Collagen	Animal tissue	Pathogen, terminal telopeptides	Transglutaminase, dehydrothermal carbodiimide, treatment, UV light; aldehydes; riboflavin	Charged amino acid
	Silk fibroin	Silkworm, bombyx mori	Sericin	Glutaraldehyde, carbodiimide	Electronegative aminoacidic sequences
Polysaccharides	Chitosan	Crab shells, shrimps, fungal fermentation	Low toxicity	Glutaraldehyde, genipin, epoxy compound, sodium tripolyphosphate	Cationic amine groups
	Starch	Plant	Nontoxicity	Sodium trimetaphosphate, malonic acid, formaldehyde, anhydride	OH – groups
	Cellulose	Plant; bacteria	Low toxicity	Aldehydes, carbodiimides, carboxylic acids, irradiation	OH – groups

The combination practice will improve bone bonding performance and will increase cell adhesion. Some methods used in scaffold combinations to make natural bio-polymer-based CaP with some examples of the biopolymer's composite produced are listed below:

- Physical mixture: collagen/hydroxyapatite (HAp), Chitosen/nHAp, Silk fibron/HAp with low crystallinity, and Alginate/HAp partied.
- Chemical deposition: Collagen/HAp, Chitosen/nHAp, and Silk fibron/Calcium-deficient HAp.
- Biomimetic mineralization: Collagen/CaP, Chitosen/HAp, and bacterial cellulose/Calcium-deficient HAp.

To produce a "living" scaffold in tissue engineering, scaffolds are typically combined with stem cells in one method, while in other approaches they are joined with bioactive molecules (Li et al., 2014). In all cases, the scaffold offers an appropriate situation for the cell's growth and a conveyance issue for cell-based therapies (Li et al., 2014).

3.4.3 STRUCTURE AND ORGANIZATION OF PROTEIN BIOPOLYMERS

Amino acids are the chief component of forming protein. The arrangements of the amino acid control the complication of the biomolecule. Collagen is the most investigated type of protein biopolymer. Collagen offers stiffness and strength to the most human tissues (Gautieri et al., 2011). In fact, the human body is rich in collagen type I. It contains three polypeptides called α chain and a sequence of any amino acid. Then three α chains carry together a triple helix tertiary structure. If it became stable by a hydrogen bond, it would be assembled into fibrils then into a supramolecular set of buildings (Gautieri et al., 2011). Collagen fibril diameter is usually small (20 nm) as in corneas and can be large (500 nm) as in tendons. Figure 3.11 displays the main stages of polymer assembly and the self-assembly of biopolymers and the formulation of supramolecular polymers.

3.4.4 POLYMERIC BIOMATERIALS IN OPHTHALMOLOGY

Generally, ophthalmology is the science of the eye and the treatment of its disorders and diseases. Biomaterials have been utilized in ophthalmology science since the

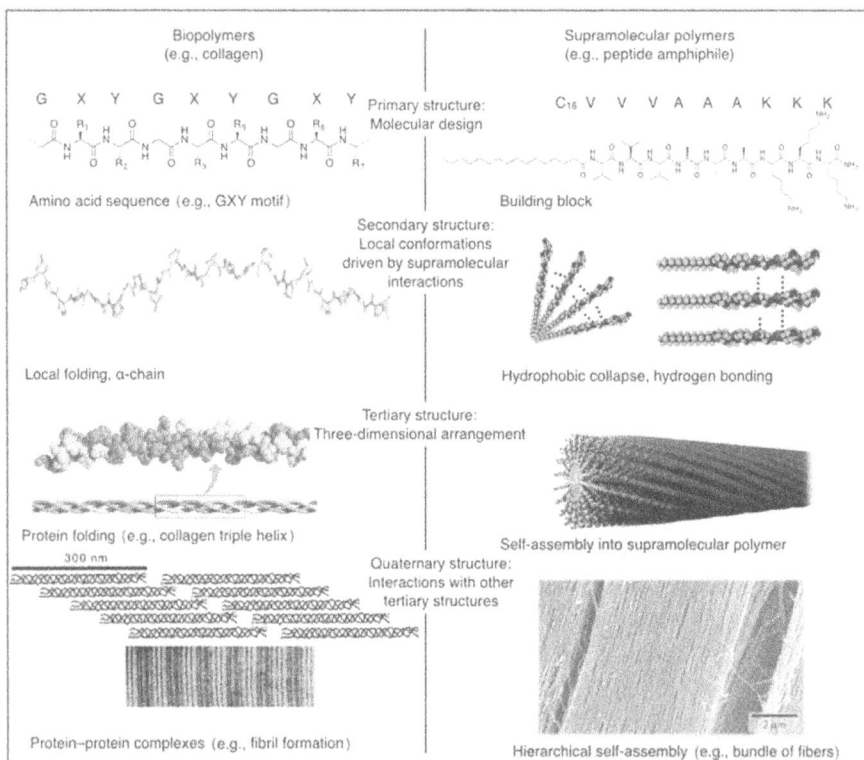

FIGURE 3.11 Self-assembly of biopolymers (left) and supramolecular polymers (right). (Adapted from Freeman et al., 2015.)

middle of the nineteenth century. In the meantime, a wide variety of biomaterials used in ophthalmological applications are now recognized. Some common uses of biomaterials in ophthalmology are the following (Modjarrad & Ebnesajjad, 2013):

a. Contact lenses
b. Intraocular lenses (IOLs)
c. Artificial orbital walls
d. Artificial corneas
e. Artificial lacrimal ducts
f. Glaucoma filtration implants
g. Viscoelastic replacements
h. Drug delivery systems
i. Scleral buckles
j. Retinal tacks and adhesives
k. Ocular endotamponades

3.4.5   POLYMERIC BIOMATERIALS FOR CARDIOVASCULAR DISEASE THERAPY

Biomaterials have an imperative role in treatment procedures of cardiovascular diseases; some examples of these applications are:

1. Heart valve prostheses
2. Vascular grafts and stents
3. Indwelling catheters
4. Ventricular assist devices
5. Total implantable artificial hearts
6. Pacemakers
7. Automatic internal cardioverter defibrillators
8. Intra-aortic balloon pumps

Biomedical uses of biopolymers are increasingly achieved via innovations in creating biopolymers with desirable characteristics to be natural functionalized materials. Such advanced and high-tech medical applications of biopolymers including ophthalmologic applications such as artificial corneas, dental applications such as endosteal root form dental implants, cardiovascular applications such as heart valves, orthopedic applications such as artificial knees, and other applications such as organs (e.g., artificial skin), organ implantations (e.g., breast), synthetic kidneys, drug delivery, and development of synthetic heart lung machines.

3.5   CONCLUSIONS

Biopolymers are green biodegradable functionalized materials that have massive success in high-tech applications due to their promising performances. Processing methods of biodegradable polymers are similar to conventional thermoplastics but with some modifications on the processing parameters or incorporations of compatibilizers. Alterations in terms of viscosity, melt flow rate, and melt strength have made the existing processing factors inappropriate for biopolymers. The use of extruder is the most appropriate procedure when processing biopolymers. In addition, 3D printing offers many advantages over old-style manufacturing methods. Employing of 3D printing technology has been applied in numerous fields particularly for medical applications. Biopolymers have numerous vital accepted chemical, electrical, and mechanical characteristics that enable them to be utilized in modern advanced biomedical and electrical systems. The invention of cellulosic electro-active paper (EAPap) was an innovation in the industry of bio-sensing and actuating. Cellulose biopolymers have advanced the fabrication of OLEDs to achieve point sources as well as panel displays that are biodegradable, eco-friendly, flexible, and economic. Starch has also been effectively developed to make piezoelectric and thermoelectric materials with excellent conversion efficiencies. Chitin is another biopolymer that is widely available in nature. It is used as fillers in batteries and electrolyte fabrication. Biopolymer-based materials have found significant roles in high-tech components in various modern and advanced applications. This includes but is not limited to electromagnetic interference shielding materials, fuel cell fabrication, piezoelectric and thermoelectric materials, and microwave absorbers.

REFERENCES

AL-Oqla, F.M., 2017. Investigating the mechanical performance deterioration of Mediterranean cellulosic cypress and pine/polyethylene composites. *Cellulose*, 24, 2523–2530.
AL-Oqla, F.M., Almagableh, A., Omari, M.A., 2017. Design and Fabrication of Green Biocomposites, *Green Biocomposites*, Springer, Cham, Switzerland, pp. 45–67.
AL-Oqla, F.M., Alothman, O.Y., Jawaid, M., Sapuan, S.M., Es-Saheb, M., 2014a. Processing and Properties of Date Palm Fibers and Its Composites, *Biomass Bioenergy*, Springer, Cham, Switzerland, pp. 1–25.
AL-Oqla, F.M., El-Shekeil, Y., 2019. Investigating and predicting the performance deteriorations and trends of polyurethane bio-composites for more realistic sustainable design possibilities. *Journal of Cleaner Production*, 222, 865–870.
AL-Oqla, F.M., Hayajneh, M.T., 2007. A design decision-making support model for selecting suitable product color to increase probability, Design Challenge Conference: Managing Creativity, Innovation, and Entrepreneurship. Yarmouk University, Amman, Jordan.
AL-Oqla, F.M., Hayajneh, M.T., Fares, O., 2019. Investigating the mechanical thermal and polymer interfacial characteristics of Jordanian lignocellulosic fibers to demonstrate their capabilities for sustainable green materials. *Journal of Cleaner Production*, 241, 118256.
Al-Oqla, F.M., Omar, A.A., 2012. A decision-making model for selecting the GSM mobile phone antenna in the design phase to increase overall performance. *Progress in Electromagnetics Research C*, 25, 249–269.
Al-Oqla, F.M., Omar, A.A., 2015. An expert-based model for selecting the most suitable substrate material type for antenna circuits. *International Journal of Electronics*, 102, 1044–1055.
AL-Oqla, F.M., Omar, A.A., Fares, O., 2018. Evaluating sustainable energy harvesting systems for human implantable sensors. *International Journal of Electronics*, 105, 504–517.

AL-Oqla, F.M., Omari, M.A., 2017. Sustainable Biocomposites: Challenges, Potential and Barriers for Development, In: Jawaid, M., Sapuan, Salit Mohd, Alothman, Othman Y (Ed.), *Green Biocomposites: Manufacturing and Properties*, Springer International Publishing (Verlag), Cham, Switzerland, pp. 13–29.

AL-Oqla, F.M., Salit, M.S., 2017a. Material Selection of Natural Fiber Composites Using the Analytical Hierarchy Process. *Materials Selection for Natural Fiber Composites*, Woodhead Publishing, Elsevier Cambridge, USA, pp. 169–234.

AL-Oqla, F.M., Salit, M.S., 2017b. *Materials Selection for Natural Fiber Composites*, Woodhead Publishing, Elsevier, Cambridge, USA.

AL-Oqla, F.M., Sapuan, S.M., 2014a. Date Palm Fibers and Natural Composites, Postgraduate Symposium on Composites Science and Technology 2014 & 4th Postgraduate Seminar on Natural Fibre Composites 2014, 28/01/2014, Putrajaya, Selangor, Malaysia.

AL-Oqla, F.M., Sapuan, S.M., 2014b. Enhancement selecting proper natural fiber composites for industrial applications, Postgraduate Symposium on Composites Science and Technology 2014 & 4th Postgraduate Seminar on Natural Fibre Composites 2014, 28/01/2014, Putrajaya, Selangor, Malaysia.

AL-Oqla, F.M., Sapuan, S.M., 2015. Polymer selection approach for commonly and uncommonly used natural fibers under uncertainty environments. *Journal of the Minerals, Metals and Materials Society*, 67, 2450–2463.

AL-Oqla, F.M., Sapuan, S.M., 2018a. Investigating the inherent characteristic/performance deterioration interactions of natural fibers in bio-composites for better utilization of resources. *Journal of Polymers and the Environment*, 26, 1290–1296.

AL-Oqla, F.M., Sapuan, S.M., 2018b. Natural Fiber Composites. Kenaf Fibers and Composites.

AL-Oqla, F.M., Sapuan, S.M., Ishak, M.R., Aziz, N.A., 2014b. Combined multi-criteria evaluation stage technique as an agro waste evaluation indicator for polymeric composites: Date palm fibers as a case study. *BioResources*, 9, 4608–4621.

AL-Oqla, F.M., Sapuan, S.M., Ishak, M.R., Aziz., N.A., 2015a. Selecting Natural Fibers for Industrial Applications, Postgraduate Symposium on Biocomposite Technology Serdang, Malaysia.

AL-Oqla, F.M., Sapuan, S.M., Ishak, M.R., Nuraini, A.A., 2014c. A novel evaluation tool for enhancing the selection of natural fibers for polymeric composites based on fiber moisture content criterion. *BioResources*, 10, 299–312.

AL-Oqla, F.M., Sapuan, S.M., Ishak, M.R., Nuraini, A.A., 2015b. Decision making model for optimal reinforcement condition of natural fiber composites. *Fibers and Polymers*, 16, 153–163.

AL-Oqla, F.M., Sapuan, S.M., Ishak, M.R., Nuraini, A.A., 2015c. A model for evaluating and determining the most appropriate polymer matrix type for natural fiber composites. *International Journal of Polymer Analysis and Characterization*, 20, 191–205.

AL-Oqla, F.M., Sapuan, S.M., Jawaid, M., 2016. Integrated mechanical-economic–environmental quality of performance for natural fibers for polymeric-based composite materials. *Journal of Natural Fibers*, 13, 651–659.

Al-Widyan, M.I., AL-Oqla, F.M., 2011. Utilization of supplementary energy sources for cooling in hot arid regions via decision-making model. *International Journal of Engineering Research and Applications*, 1, 1610–1622.

Al-Widyan, M.I., AL-Oqla, F.M., 2014. Selecting the most appropriate corrective actions for energy saving in existing buildings A/C in hot arid regions. *Building Simulation*, 7, 537–545.

Alaaeddin, M., Sapuan, S.M., Zuhri, M., Zainudin, E., AL-Oqla, F.M., 2018. Properties and Common Industrial Applications of Polyvinyl fluoride (PVF) and Polyvinylidene fluoride (PVDF), IOP Conference Series: Materials Science and Engineering. IOP Publishing, p. 012021.

Alaaeddin, M., Sapuan, S.M., Zuhri, M., Zainudin, E., AL-Oqla, F.M., 2019a. Photovoltaic applications: Status and manufacturing prospects. *Renewable and Sustainable Energy Reviews*, 102, 318–332.

Alaaeddin, M., Sapuan, S.M., Zuhri, M., Zainudin, E., AL-Oqla, F.M., 2019b. Polymer matrix materials selection for short sugar palm composites using integrated multi criteria evaluation method. *Composites Part B: Engineering*, 107342.

Alaaeddin, M., Sapuan, S.M., Zuhri, M., Zainudin, E., AL-Oqla, F.M., 2019c. Polyvinyl fluoride (PVF); Its Properties, Applications, and Manufacturing Prospects, IOP Conference Series: Materials Science and Engineering. IOP Publishing, p. 012010.

Alaaeddin, M., Sapuan, S.M., Zuhri, M., Zainudin, E., AL-Oqla, F.M., 2019d. Lightweight and durable PVDF–SSPF composites for photovoltaics backsheet applications: Thermal, optical and technical properties. *Materials*, 12, 2104.

Aridi, N., Sapuan, S.M., Zainudin, E., AL-Oqla, F.M., 2016a. Investigating morphological and performance deterioration of injection-molded rice husk–polypropylene composites due to various liquid uptakes. International Journal of Polymer Analysis and Characterization, 21, 675–685.

Aridi, N., Sapuan, S.M., Zainudin, E., AL-Oqla, F.M., 2016b. Mechanical and morphological properties of injection-molded rice husk polypropylene composites. International Journal of Polymer Analysis and Characterization, 21, 305–313.

Aridi, N., Sapuan, S.M., Zainudin, E., AL-Oqla, F.M., 2017. A review of rice husk bio-based composites. *Current Organic Synthesis*, 14, 263–271.

Butcher, A.L., Offeddu, G.S., Oyen, M.L., 2014. Nanofibrous hydrogel composites as mechanically robust tissue engineering scaffolds. *Trends Biotechnol*, 32, 564–570.

Chaitanya, S., Singh, I., Song, J.I., 2019. Recyclability analysis of PLA/Sisal fiber biocomposites. *Composites Part B: Engineering*, 173, 106895.

Chiulan, I., Frone, A.N., Brandabur, C., Panaitescu, D.M., 2018. Recent advances in 3D printing of aliphatic polyesters. *Bioengineering*, 5, 2.

Cho, D.-H., Kwon, O.E., Park, Y.-S., Yu, B.G., Lee, J., Moon, J., Cho, H., Lee, H., Cho, N.S., 2017. Flexible integrated OLED substrates prepared by printing and plating process. Organic *Electronics*, 50, 170–176.

Freeman, R., Boekhoven, J., Dickerson, M.B., Naik, R.R., Stupp, S.I., 2015. Biopolymers and supramolecular polymers as biomaterials for biomedical applications. *MRS Bulletin*, 40, 1089–1101.

Gautieri, A., Vesentini, S., Redaelli, A., Buehler, M.J., 2011. Hierarchical structure and nanomechanics of collagen microfibrils from the atomistic scale up. *Nano Letters*, 11, 757–766.

González-Ausejo, J., Sanchez-Safont, E., Lagaron, J.M., Olsson, R.T., Gamez-Perez, J., Cabedo, L., 2017. Assessing the thermoformability of poly(3-hydroxybutyrate-co-3-hydroxyvalerate)/poly(acid lactic) blends compatibilized with diisocyanates. *Polymer Testing*, 62, 235–245.

Jiang, Q., Reddy, N., Yang, Y., 2010. Cytocompatible cross-linking of electrospun zein fibers for the development of water-stable tissue engineering scaffolds. *Acta Biomater*, 6, 4042–4051.

Kim, J., 2017. Multifunctional Smart Biopolymer Composites as Actuators. *Biopolymer Composites in Electronics*. Elsevier, United Kingdom, pp. 311–331.

Kim, K., Ha, M., Choi, B., Joo, S.H., Kang, H.S., Park, J.H., Gu, B., Park, C., Park, C., Kim, J., 2018. Biodegradable, electro-active chitin nanofiber films for flexible piezoelectric transducers. Nano Energy 48, 275–283.

La Notte, L., Cataldi, P., Ceseracciu, L., Bayer, I.S., Athanassiou, A., Marras, S., Villari, E., Brunetti, F., Reale, A., 2018. Fully-sprayed flexible polymer solar cells with a cellulose-graphene electrode. Materials Today Energy 7, 105–112.

Li, J., Baker, B.A., Mou, X., Ren, N., Qiu, J., Boughton, R.I., Liu, H., 2014. Biopolymer/calcium phosphate scaffolds for bone tissue engineering. Advanced Healthcare Materials 3, 469–484.

Modjarrad, K., Ebnesajjad, S., 2013. *Handbook of Polymer Applications in Medicine and Medical Devices*. Elsevier, United Kingdom.

Mohiuddin, M., Kumar, B., Haque, S., 2017. Biopolymer Composites in Photovoltaics and Photodetectors. *Biopolymer Composites in Electronics*. Elsevier, United Kingdom, pp. 459–486.

Okamoto, M., John, B., 2013. Synthetic biopolymer nanocomposites for tissue engineering scaffolds. *Progress in Polymer Science*, 38, 1487–1503.

Ramesh, S., Khandelwal, S., Rhee, K.Y., Hui, D., 2018. Synergistic effect of reduced graphene oxide, CNT and metal oxides on cellulose matrix for supercapacitor applications. *Composites Part B: Engineering*, 138, 45–54.

Rullyani, C., Shellaiah, M., Ramesh, M., Lin, H.-C., Chu, C.-W., 2019. Pyrene-SH functionalized OTFT for detection of Hg2+ ions in aquatic environments. *Organic Electronics*, 69, 275–280.

Sapuan, S.M., Haniffah, W., AL-Oqla, F.M., 2016. Effects of reinforcing elements on the performance of laser transmission welding process in polymer composites: A systematic review. *International Journal of Performability Engineering*, 12, 553.

Sapuan, S.M., Pua, F., El-Shekeil, Y., AL-Oqla, F.M., 2013. Mechanical properties of soil buried kenaf fibre reinforced thermoplastic polyurethane composites. *Materials and Design*, 50, 467–470.

Surya, S.G., Raval, H.N., Ahmad, R., Sonar, P., Salama, K.N., Rao, V.R., 2019. Organic field effect transistors (OFETs) in environmental sensing and health monitoring: A review. *TrAC Trends in Analytical Chemistry*, 111, 27–36.

4 Resin-Based Composites in Dentistry—A Review

Z. Radzi, R. A. Diab, N. A. Yahya, and M. A. G. Gonzalez
University of Malaya, Kuala Lumpur, Malaysia

CONTENTS

4.1 INTRODUCTION: RESIN-BASED COMPOSITES (RBC)

Dental composite is one of the most commonly used dental materials in restorative dentistry, pediatric dentistry, and orthodontics. The widespread use of dental composite is heavily accredited to its versatility, wide biocompatibility, and superiority in mechanical, physical, and chemical properties. Additionally, over a period of time, the advancement of technology has made significant improvements to the composite and its manufacturing process.

A composite is a material that consists of at least two distinct phases, normally formed by blending components of different structures and properties (McCabe & Walls, 2008). Resin-based composites (RBCs) are defined as "a highly cross-linked polymeric material reinforced by a dispersion of amorphous silica, glass, crystalline, or organic resin filler particles and/or short fibers bonded to the matrix by a coupling agent" (The Glossary of Prosthodontic Terms, 2017). RBCs are used for a variety of applications in dentistry, including but not limited to restorative materials, cavity liners, pit and fissure sealants, cores and build-ups, inlays, onlays, crowns, provisional restorations, cements for single- or multiple-tooth prostheses, and for orthodontic devices, endodontic sealers, and root canal posts. It was first introduced in the 1950s (Ferracane, 2011) and has been continuously improved with the incorporation of different types and sizes of fillers, polymeric matrices, and photoinitiators. RBC composition defines its chemical and physical properties depending on filler weight percent (wt%) and volume percent (vol%), which has led to different formulations tailored for their particular application.

Despite the different indications for RBCs, their main components comprise of a blend of hard inorganic particles dispersed in a soft resin matrix (Figure 4.1).

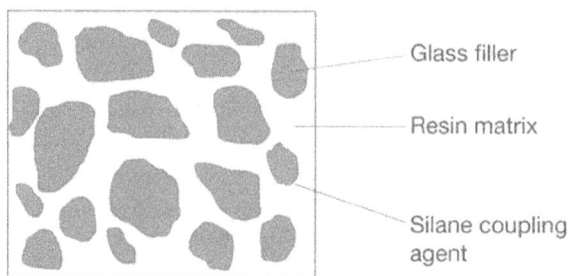

FIGURE 4.1 Structure of resin-based composite material.

Generally, RBCs encompass three main components: First is the resin matrix comprising a monomer system, an initiator system for free radical polymerization, and stabilizers. Second, the inorganic filler consists of particles such as glass, quartz, and/or fused silica. Third is the coupling agent, which is usually an organosilane that chemically bonds the reinforcing filler to the resin matrix, for example, (methacryloxy)propyltrimethoxysilane (MPTMS) (Peutzfeldt, 1997).

4.1.1 MATRIX IN RESIN-BASED COMPOSITE

Dental resins are mainly based on methyl methacrylate monomers. In the late 1930s, poly (methyl methacrylate) was introduced to be used as denture base resin. A few years later, indirect filling resins were introduced, which were then followed by direct resins. However, the methyl methacrylate resin was associated with significant defects, including large polymerization shrinkage, high coefficient of thermal expansion, severe discoloration, severe pulp damage, and high incidence of secondary caries (Peutzfeldt, 1997). Epoxy resins were then developed. This is a synthetic resin that hardens at room temperature with little shrinkage, producing an insoluble polymer. It had better properties than methyl methacrylate resin. It was deemed to be a promising material used in the first dental RBC: an epoxy resin with aggregates of fused quartz or porcelain particles (Bowen, 1956). However, the use of epoxy resin was abandoned due to its slow hardening, which prevented its use as a direct filling material.

In 1956, the era of dental RBCs started when Bowen synthesized a new monomer, 2,2-bis[4-(2-hydroxy-3-methacrylyloxypropoxy)phenyl]propane, which is also known as Bis-GMA. It resembles an epoxy resin; however, the epoxy groups are replaced by methacrylate groups. It is prepared from bisphenol A and glycidyl methacrylate or diglycidyl ether of bisphenol A and methacrylic acid; therefore, it is a dimethacrylate (Bowen, 1959). Polymerization of the monomer occurs through carbon-carbon double bonds (C=C) of the two methacrylate groups. Bis-GMA is superior to methyl methacrylate because of its large molecular size and chemical structure, providing lower volatility, lower polymerization shrinkage, more rapid hardening, and production of stronger and stiffer resins.

Due to the high viscosity of Bis-GMA (Figure 4.2), it is sometimes admixed with lower molecular weight dimethacrylates such as ethylene glycol dimethacrylate

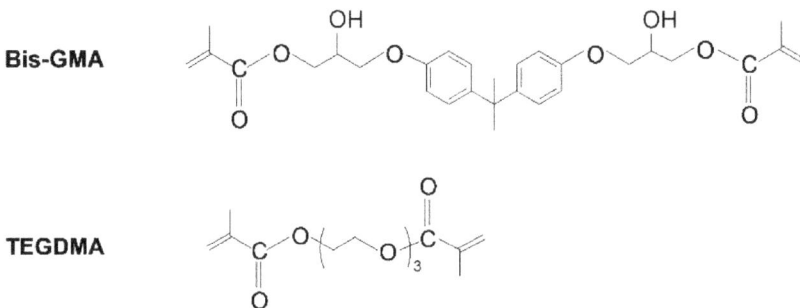

FIGURE 4.2 Chemical structures of Bis-GMA and TEGDMA.

(EGDMA), urethane dimethacrylate (UDMA), or triethylene glycol dimethacrylate (TEGDMA) (Asmussen, 1975; Ruyter & Oysaed, 1987; Vankerckhoven, Lambrechts, van Beylen, & Vanherle, 1981). Lower viscosity monomer allows incorporating fillers in the material (Bowen, 1962 & 1963). The lower the viscosity of the monomer mixture, the more filler may be incorporated. It has been shown that the maximum polymerization rate of pure Bis-GMA occurs at less than 5% conversion due to its very high viscosity, and the final degree of conversion (DC) is limited to about 30%. In contrast, for pure TEGDMA (Figure 4.2), which is far less viscous, the maximum rate was observed at around 22% conversion, with a final DC of over 60%. Mixing the different comonomers together resulted in intermediate values between these two extremes (Lovell, Newman, & Bowman, 1999).

While Bis-GMA is the most commonly used monomer in dental restorative materials, non-dimethacrylate monomers have been incorporated in some dental RBCs. Among these novel monomer technologies are those based on ring-opening epoxy chemistry. Additionally, a high-weight and low-viscosity monomer, Procrylat(2,2-bis-4-(3-hydroxy-propoxy-phenyl)propane dimethacrylate, was also introduced in some contemporary flowable RBCs. Procrylat resin is characterized by a lower viscosity and lack of pendant hydroxyl groups. However, contemporary dimethacrylate-based RBCs still represent the vast majority of commercially available materials for direct restorations (Leprince, Palin, Hadis, Devaux, & Leloup, 2013).

4.1.2 FILLERS IN RESIN-BASED COMPOSITE

The resinous matrix of dental RBC is incorporated with fillers that considerably influence the properties of the material. Fillers have several roles, which include enhancing the modulus of elasticity, providing radiopacity, altering thermal expansion behavior, and reducing polymerization shrinkage by reducing the resin fraction. Fillers in RBC restorations can be classified as macrofillers, microfillers (pyrogenic silica), or microfiller-based complexes (Lutz & Phillips, 1983). Additionally, nanofillers are among the fillers that are incorporated in dental RBCs (Moszner & Klapdohr, 2004).

Macrofillers were the first to be used in dental RBCs. They are mechanically prepared from larger pieces of the material by grinding and/or crushing. The inorganic fillers are usually splinter shaped, and are made of quartz, glass, borosilicate, or a ceramic. Heavy metal glasses that provide adequate radiopacity are also available. Initially, manufactured macrofiller sizes ranged between 0.1 and 100 µm with an average particle size between 5 and 30 µm. However, recent macrofiller sizes are ranging between 0.2 and 5 µm (Craig, 1981). Traditional or macrofilled RBCs allow the packing of the organic matrix providing higher inorganic filler loading. The use of these materials has been limited due to filler dislodgement leaving a rough surface restoration (Lambrechts & Van Herle, 1982).

Microfillers were then introduced to produce different RBCs. Derived chemically by hydrolysis and precipitation, microfillers consist of finely dispersed radiolucent glass spheres. Originally, the primary particles commonly used had an average size of 0.04 µm. A tendency to use larger average particle sizes of approximately 0.05–0.1 µm was observed (Lutz & Phillips, 1983). In an attempt to avoid the problems encountered with macrofilled RBCs, inorganic macrofillers and microfillers were combined to produce hybrid RBCs. They constitute a wide range of filler sizes within the range of 10–50 μm and 0.01–0.05 μm (Ferracane, 2011). The use of different filler sizes allowed intermediate aesthetic properties and good mechanical performance. It also led to efficient packing of the restoration and allowed the possibility of high filler loadings as high as 90 wt%. Acceptable surface smoothness can be achieved in these hybrid RBCs; however, wear can result due to the presence of traditional macrofillers (Heuer, Garman, Sherrer, & Williams, 1982). Therefore, homogeneous microfilled RBCs were developed where the fillers are purely inorganic microfillers, which are smaller than the wavelength of visible light. They have a high degree of surface polishability. Even though the microfillers can be dislodged because of wear, the polished surface retains its enamel-like luster because the induced surface irregularities cannot be detected optically. Pyrogenic silica is an important representative of the microfillers used. It has a strong thickening effect if added directly to a liquid mixture. The viscosity of the material might be increased due to the large specific surface area of the microfillers. Therefore, alternative ways of admixing microfillers to an organic matrix were developed to increase the filler load without jeopardizing the handling properties (Lutz & Phillips, 1983).

Microfiller-based complexes were developed. They are divided into three different types depending on the manufacturing process: splintered prepolymerized microfilled complexes, spherical polymer-based microfilled complexes, and agglomerated microfiller complexes (Schmitt, Purrmann, Jochum, & Hubner, 1981). Splintered prepolymerized microfilled complexes are made by incorporating pyrogenic silica into a resin matrix and then heat-curing the mixture. After it sets, it is milled into particles with a size range between 1 and 200 μm. As for the spherical polymer-based microfilled complexes, they are manufactured by incorporating pyrogenic silica into partially cured polymer spheres that have an average diameter of 20–30 μm. These types of fillers are commonly named spherical prepolymerized particles. The final microfiller-based complex is the agglomerated microfiller, one which consists of artificially agglomerated microfillers in the range between 1 and 25 μm. This type is purely inorganic unlike the above-mentioned types of complexes. Heterogeneous microfilled RBCs were developed where the filler content is admixed microfillers and microfiller-based complexes. These RBCs have increased filler loading; however, it resulted in an overall lower inorganic glass content, around 50 wt% in comparison to traditional macrofilled RBCs (Anusavice, Shen, & Rawls, 2012).

Improvements in RBC technology have led to the introduction of new fillers with sizes ranging from around 5 to 100 nm (Moszner & Klapdohr, 2004). Besides that, considering the diversity of nano-based dental materials described herein, resin-based dental materials containing nanofiller or nanocomposites currently have the highest viability for clinical use. The use of nanomaterials allows the addition of larger amounts of fillers into dental resin composites compared to those containing macro- and microfillers (Figure 4.3). Nanofiller particles allow a homogeneous distribution in the resin matrix and lead to a larger interface area between the fillers and resin matrix. Nanofillers are predominantly made of crystalline silica and zirconia (Ilie, Rencz, & Hickel, 2013a). Nanofilled RBCs were introduced in the early 2000s. These contain discrete colloidal silica spherical nanoparticles or nanomers (5–100 nm).

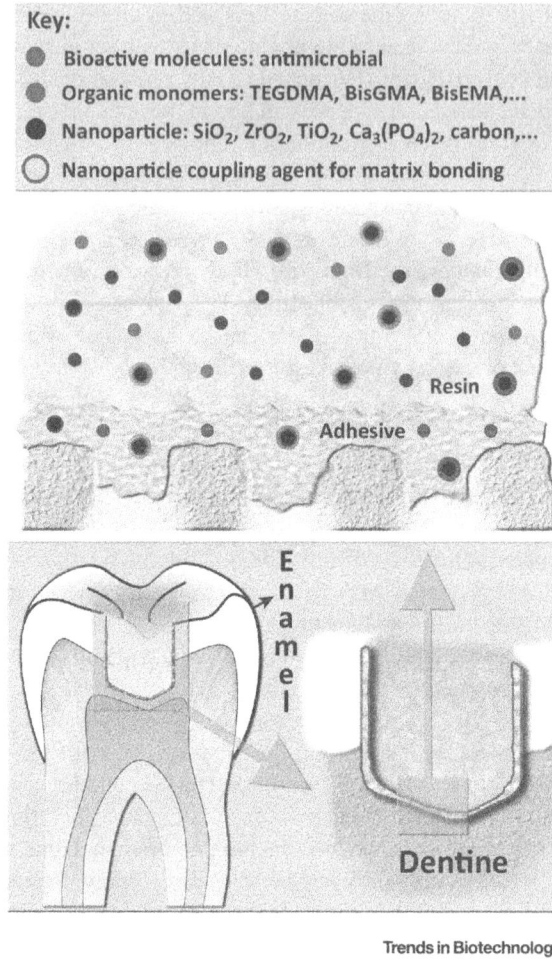

FIGURE 4.3 Application of resin-based composites (RBCs) incorporated with nanoparticles for dental restorations. (Adapted, with permission, from Padovani et al., 2015.)

They also contain agglomerations of particles described as "nanoclusters", in which the cluster size may be significantly above 100 nm, with a resulting improvement in filler loading (Cramer, Stansbury, & Bowman, 2011). The milling procedures are unable to reduce the filler particle size below 100 nm. Therefore, these nanoparticles are synthesized using synthetic chemical processes to produce building blocks on a molecular scale. These materials are then assembled into progressively larger structures and transformed into nanosized fillers suitable for a dental RBC (Mitra, Wu, & Holmes, 2003). The main advantages of nanofill RBCs are their superior resistance to wear (Yap, Tan, & Chung, 2004) and their high degree of polishability (Mitra et al., 2003). The nanoclusters provide a distinct reinforcing mechanism with enhanced entanglement between the nanofillers and the resin matrix compared with either microfill or hybrid systems, resulting in significant improvements in their strength and reliability (Beun, Glorieux, Devaux, Vreven, & Leloup, 2007; Curtis, Palin, Fleming, Shortall, & Marquis, 2009). Due to the high filler loading that can be obtained, polymerization shrinkage is also reduced (Moszner & Salz, 2001). The major drawback of nanofill composites is that they tend to be thick and sticky and to slump during clinical application (Al-Ahdal, Silikas, & Watts, 2014).

Also introduced were nanohybrid RBCs that contain milled glass fillers of diameter within the range 0.6–3.0 µm. These RBCs are claimed to combine the wear characteristics provided by the small size of the particles in the agglomerated nanoclusters with the improved handling characteristics and aesthetics of the microhybrid RBCs. However, a study assessing a number of commercially available materials marketed as "nanohybrids" has shown that nanohybrid materials exhibited inferior diametral tensile strength compared to nanofilled and microhybrid RBCs, and similar or slightly better wear resistance compared to microhybrids in terms of retention of surface smoothness (de Moraes et al., 2009).

RBC classification based on filler size, such as the macrofilled, microfilled, and hybrid materials, has shown useful correlation with the aesthetic, surface, and mechanical performance with early types of RBC. However, the usefulness of this classification may be compromised nowadays because most current composites are either microhybrids or nanohybrids. Also, there is an increasing diversity of filler shapes, materials, distribution, and organic matrix components of current RBC, which may affect the performance of the materials; thus, the final properties are more material-dependent rather than being dependent on filler size alone.

4.1.3 ADDITIVES IN RESIN-BASED COMPOSITE

In general, dental RBCs contain a mixture of at least two different fillers. A number of additives are further incorporated within the material. Some elements such as Yttrium fluoride (YbF_3), which serves as a radioopacifier, fluorosilicate glasses, or sparing amounts of soluble fluoride salts are added to the RBCs. Inhibitors such as phenols, for example, 2,6-di-tert-butyl-methylphenol (BHT) and hydroquinone monomethylether (MEHQ), are also added to the resin formulation. Their amounts range between 200 and 1,000 ppm preventing premature polymerization during storage of the RBC and avoiding uncontrolled photopolymerization by normal room light during their use. Ultraviolet (UV) photostabilizers, such as 2-hydroxy-benzophenones or 3-(2-hydroxyphenyl)-benzotriazoles, are also added to protect the RBCs against photodegradation of the organic matrix, which may cause color changes of the material. Finally, color pigments are added in order to meet the aesthetic demands by incorporating a mixture of different inorganic pigments (yellow, red, white, and black) to imitate the color of the natural teeth (Klapdohr & Moszner, 2004).

4.2 POLYMERIZATION OF RESIN-BASED COMPOSITES

Early RBCs were supplied as power/liquid, self-curing direct, acrylic-based restorative materials. Dual-paste, self-cure RBC systems were then introduced and eventually the light-cured single-paste system was developed (Craig, 1981).

This continuous development was the result of attempts to improve the materials' properties. Nowadays, light-activated RBCs are predominantly used, some of which might be dual-cured containing a chemically cured component. Photopolymerization is defined as "a chemical reaction wherein photons activate an initiator which will react in the presence of an aliphatic amine with the urethane dimethacrylate oligomer and an acrylic copolymer" (The Glossary of Prosthodontic Terms, 2017). In other words, photopolymerization is the conversion of monomers into polymers when RBC material is exposed to light. This process usually starts at the surface where light is applied leading to a rapid increase in the DC and crosslinking density.

4.2.1 KINETICS OF PHOTOPOLYMERIZATION REACTION

Previous studies (Dionisio, Mahabadi, O'Driscoll, Abuin, & Lissi, 1979) have been carried out to describe the kinetic behavior of photopolymerization and to clarify the onset of RBC network formation. Generally, photopolymerization is divided into three main steps. The first is the initiation process, which is characterized by free radical formation leading to the start of the reaction. Second is the propagation process where chain growth takes place. This stage is further subdivided into: the quasi-static process where the number of high molecular weight chains increases resulting in viscosity increases; the gel phase or the Trommsdorff-Norrish effect (Dionisio et al., 1979) where the reaction rate increases and viscosity is further increased due to the slow termination reactions; and finally, the glass phase, which is a gelatinous phase and is accompanied by a decrease in the reaction rate. The third step of the photopolymerization process is termination, which occurs when two free radicals join, or when one molecule abstracts a hydrogen atom from another, forming a C=C.

During photopolymerization of dental RBCs, the DC and the crosslinking density increase rapidly. This results in a rapid increase in viscosity that reaches the gelation state, where the polymer matrix becomes rigid. Both propagation and termination reactions are diffusion-controlled (Anseth, Wang, & Bowman, 1994). The propagation process involves the reaction of polymeric radical and methacrylate monomer. This does not become diffusion controlled until the polymer reaches the glassy phase in a process known as "vitrification". As the polymer vitrifies, the propagation reaction slows and the polymerization ceases, that is, auto-deceleration occurs. This process is particularly important in dental RBCs, where auto-deceleration results in residual, unreacted methacrylates that remain in the RBC restoration. In contrast to the propagation reaction, in the termination reaction, the radical concentration and polymerization rate increase significantly, a phenomenon referred to as "auto-acceleration". This process is important for dental RBCs since it results in rapid photopolymerization within a clinically acceptable time.

In addition to the complex polymerization kinetics, the polymer structure formed has numerous complexities. Two critical macroscopic demarcations occur during polymerization. The first is the gel point conversion, which represents the point at which an infinitely cross-linked polymer network first appears, accompanied by an increase in viscosity and poorer flow (Leprince et al., 2013). The DC of dimethacrylates at the gel point is expected to be less than 1–5% (Stansbury et al., 2005). The second macroscale demarcation is the vitrification point, which represents the conversion at which the polymer becomes glassy, accompanied by a significant increase in modulus and viscosity. When the vitrification point is reached, the reactive dimethacrylate groups become increasingly less able to migrate to the reaction sites than in the gel stage.

It should be noted that polymer networks are extremely heterogeneous due to two main facts. The first is due to the formation of microgels near the initiation sites and the second is due to the presence of pendant methacrylate groups in Bis-GMA, TEGDMA monomers, or other analogous systems, which might be more reactive than their monomeric counterparts (Elliott, Lovell, & Bowman, 2001). A pendant is a small chain that hangs off of the main chain, which is the backbone of the polymer. This heterogenicity has significant implications on RBC restorations affecting its post-cure behavior (Truffier-Boutry et al., 2006) and the refractive index variation within the polymer matrix enhancing the RBC translucency (Howard, Wilson, Newman, Pfeifer, & Stansbury, 2010).

4.2.2 OXYGEN INHIBITED LAYER

The polymerization reaction of dental RBCs is based on the presence of free radicals that induce the reaction. Oxygen is considered as a free-radical scavenger (Xia & Cook, 2003). It diffuses from the atmosphere into RBCs causing the oxidation of radicals into stable species known as hydroperoxides (Andrzejewska, Lindén, & Rabek, 1998; Schulze & Vogel, 1998), which have low reactivity toward monomers. If the concentration of these hydroperoxides is sufficient, they can alter the properties of the polymer. Diffusion of oxygen into the exposed resin or RBC surface, as polymerization proceeds, results in quenching of both initiator and polymer-based radical species and is responsible for the poorly polymerized, air-inhibited surface layer (Finger, Lee, & Podszun, 1996). The inhibitory effect of oxygen was found to affect resin polymerization to a depth of 53 µm (Rueggeberg & Margeson, 1990).

4.2.3 POST-POLYMERIZATION REACTION

After the photocuring of RBCs, it has been observed that the polymerization reaction is not complete. An additional cure was reported to range between 5 minutes and 24 hours representing 19–26% of the final DC depending upon the material (Halvorson, Erickson, & Davidson, 2002). However, another study noted that DC had a very minimal increase of more or less 2% during the first 24 hours, which was not considered a significant increase (Truffier-Boutry et al., 2006). Extrapolating these findings clinically, it was found that small amounts of unreacted monomers from polymerized material are eluted (Lagocka et al., 2018) into either the oral cavity or through dentinal diffusion (Gerzina & Hume, 1996) into the pulpal tissues, hence leaching into the blood circulation (Bouillaguet, Wataha, Hanks, Ciucchi, & Holz, 1996). Unreacted monomers from dental RBCs were found to be cytotoxic (Lee, Kim, Kwon, Lee, & Kim, 2017) and might cause local adverse effects (Nocca et al., 2007). However, not all residual monomers are eluted into an aqueous solution (Ferracane, 1994). Pulp studies have shown a lack of significant pulpal irritation after the placement of properly sealed RBC restorations (Cox, Keall, Keall,

Ostro, & Bergenholtz, 1987). It was reported that 75% of the elutable molecules are extracted within a few hours and that 95% are extracted within 24 hours (Ferracane & Condon, 1990). Therefore, RBCs do not provide a chronic source of unreacted monomer to the pulp or other oral tissues (Ferracane & Condon, 1990). However, the unreacted monomers and/or functional groups within the polymer can act as plasticizers and, therefore, might have a negative impact on the mechanical properties due to the lower cross-linking of the polymer resulting in lower DC (Durner, Obermaier, Draenert, & Ilie, 2012).

4.3 PHOTOINITIATORS

A photoinitiator is a molecule that can absorb light and, as a result, either directly or indirectly, generate reactive radicals that can initiate polymerization (Fouassier, 1995). Hence, the photoinitiator is a compound that can transform the physical energy of light into suitable chemical energy in the form of reactive intermediates (Figure 4.4). Photoinitiators can be classified into Norrish Type I and II initiators. Norrish Type I initiators are typically compounds containing benzoyl groups, which undergo cleavage when exposed to visible light to generate two free radicals where at least one the free radicals reacts with monomers to initiate polymerization. They do not require a co-initiator to produce free radicals. However, Norrish Type II initiators absorb visible light to form excited molecules, which abstract a hydrogen atom from a donor molecule (synergist). The donor then reacts with a monomer to initiate polymerization. Photoinitiators are incorporated in dental RBCs to start the polymerization reaction.

4.3.1 CAMPHORQUINONE (CQ)

The CQ (1,7,7-trimethylbicyclo [2.2.1] heptane-2,3-dione) Type II initiator system is the most widely used photoinitiator in dental RBCs (Stansbury, 2000). It is a diketone molecule that creates free radicals to initiate the photopolymerization process. A co-initiator is required for an efficient polymerization process to occur. In case of dental RBCs restorative containing CQ, tertiary amine photoreductant is the co-initiator. Both the CQ and the tertiary amine molecule react together to provide reactive radicals that begin polymerization of the methacrylate resin system.

A CQ molecule has two carbonyl groups in its structure. Homolytic cleavage of the C=C in these two carbonyl groups results in the formation of two carbonyl radicals. These radicals escape forming photo-decomposed products, which result in a decrease in the concentration of the active CQ molecules. These pair of radicals remain structurally connected to each other and might recombine again to reform the CQ molecules. Recombination can be greatly reduced by the addition of aliphatic or aromatic amine co-initiators, which may also serve to increase the rate of photo-decomposition and polymerization efficiencies (Sun & Chae, 2000).

The maximum absorbance peak of CQ is at 468 nm, but it could be activated in a range between 400 and 500 nm (Neumann, Schmitt, Ferreira, & Correa, 2006). It was also found to be activated by LCUs of wavelengths below 320 nm (Jakubiak et al., 2003).

FIGURE 4.4 The polymerization of RBCs is directly influenced by the quality of light produced by the dentist's light curing units (LCUs), including the strength of irradiation, peak wavelength emission, and the interaction between that and the constituents of the individual RBC. (Adapted, with permission, from Santini, Gallegos, & Felix, 2013.)

4.3.2 RECENT PHOTOINITIATORS

CQ color is intense yellow (Janda, Roulet, Kaminsky, Steffin, & Latta, 2004) and only a portion of CQ is utilized during photopolymerization (Jakubiak, Sionkowska, Lindén, & Rabek, 2001). Therefore, alternative lighter-colored initiators that completely bleach out after photopolymerization have been recently used. These include acyl phosphine oxide (APO), phenyl propanedione (PPD), and Ivocerin®. All these photoinitiators have different spectral absorbance ranges of activity, and also differ greatly in their ability to absorb light (Neumann et al., 2006).

4.3.2.1 Acyl Phosphine Oxide (APO)

APO is a Type I initiator that has high absorbency and efficient quantum yields (Neumann et al., 2006). The sensitivity peak of APO is approximately 370 nm, which is considerably lower than that of CQ. Two variations of APO initiators are used in dental restorative materials: monoacylphosphine oxide (MAPO) and bisacyl-phosphine oxide (BAPO). Lucirin® TPO (2,4,6-trimethylbenzoyldiphenylphosphine oxide) is a representative of MAPO initiator. The absorption spectrum of TPO is seen mainly in the UV range. It was incorporated in dental RBCs as the sole initiator, but currently, TPO is combined with CQ (and other photoinitiators) to provide enhanced resin photopolymerization and decrease the yellowing of the restoration (de Oliveira et al., 2016; Leprince et al., 2011). Irgacure 819 is a representative of BAPO. It was found to be very sensitive to light below 440 nm (Neumann et al., 2006).

4.3.2.2   Phenyl Propanedione (PPD)

Phenyl propanedione (1-phenyl-1,2-propanedione) is another Type II initiator that has a more broad-banded absorption spectrum, which extends from the UV wavelength range to approximately 490 nm. It is usually combined with CQ resulting in a syner-gistic effect, leading to enhanced resin polymerization and also slowing the overall rate of the reaction. It also contributes to the reduction of the residual yellow color of the restorative material (Schneider, Cavalcante, Consani, & Ferracane, 2009).

4.3.2.3   Germanium-Based Photoinitiator

Germanium-based photoinitiator has a broader spectrum and absorbs light at a higher wavelength range than APO and has a sensitivity peak of about 420 nm. It is characterized by high-quantum efficiency, high-absorption capacity, and very good bleaching properties. It is claimed to produce highly reactive polymerization in very small amounts; therefore, it is considered a polymerization booster.

4.4   BULK-FILL RESIN-BASED COMPOSITES

Polymerization shrinkage is considered the most significant problem with current RBC restorations. It is an intrinsic property of conventional resin-based materials caused by the approximation of monomers during polymerization, that is, the dis-tance between monomers is reduced due to the conversion of the weak van der Waals forces into covalent bonds. Polymerization shrinkage is a primary contributor to premature failure due to its contribution to internal and marginal gaps, microleak-age, and microcracking of tooth structure due to cuspal deflection (Ferracane & Mitchem, 2003; Goncalves, Azevedo, Ferracane, & Braga, 2011). Polymerization stress occurs in most of the clinically relevant cavity configurations and is dictated by multiple factors including resin viscosity, volume shrinkage, polymerization rate, DC, modulus development, and network structural evolution. The C-factor is among the determinants of polymerization shrinkage. It is defined as the ratio of the bonded to the unbonded surface area of the restoration. The degree of resin flow is determined by the material and the C-factor (Davidson, 1986). An increased rate of shrinkage stress development with a high C-factor leads to a decrease in flow capacity, which is not sufficient to preserve adhesion to dentine by dentine bonding agents (Feilzer, De Gee, & Davidson, 1987). Thus, the utilization of modified resin monomers, photoinitiators, and filler has increasingly contributed to less shrinkage stress development (Figure 4.5).

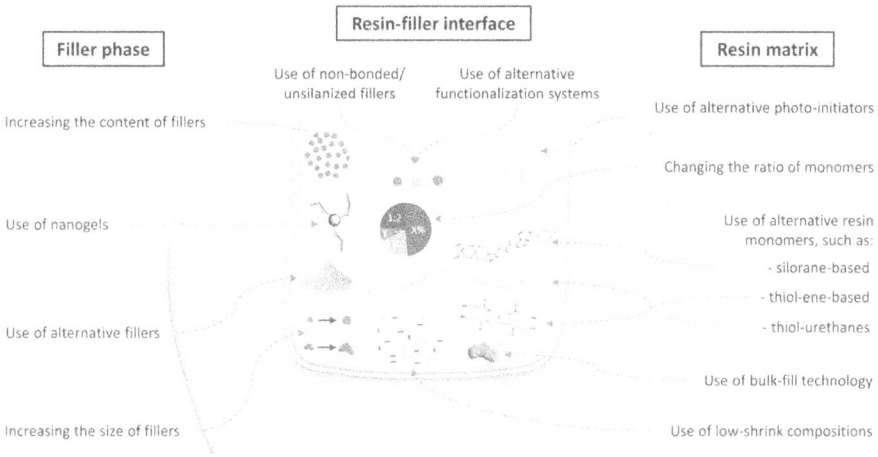

FIGURE 4.5 The use of alternative or modified resin monomers and photoinitiators, as well as improved filler particles, has increasingly contributed to less shrinkage stress development. (Adapted, with permission, from Meereis, Münchow, de Oliveira da Rosa, da Silva, & Piva, 2018.)

4.4.1 Techniques for Resin-Based Composite Application

4.4.1.1 Incremental Technique

Multiple studies were carried out to determine how to reduce the overall polymerization shrinkage and internal stresses to decrease the clinical issues of RBC restorations (Figure 4.6). Insertion of RBC in increments of 2 mm or less was among the proposed techniques to reduce the overall volume of the resin, and therefore, the

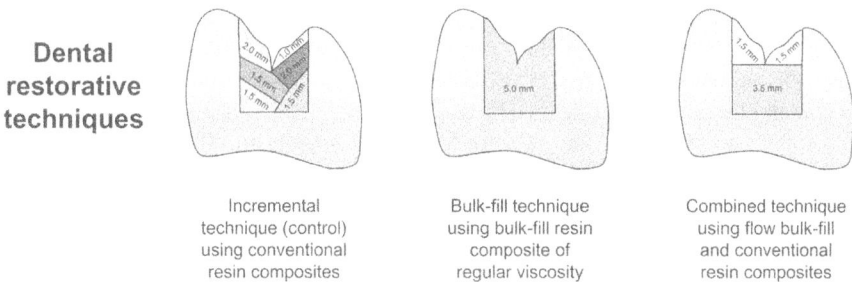

	Incremental technique (control) using conventional resin composites	Bulk-fill technique using bulk-fill resin composite of regular viscosity	Combined technique using flow bulk-fill and conventional resin composites
Load-to-fracture (N)	1206 (309)	1411 (332)	1176 (206)
Work of fracture (J/m²)	363 (128)	508 (180)	405 (85)
Weibull modulus (95% CI)	5.1 (2.8 – 8.0)	4.9 (2.6 – 7.7)	6.9 (3.7 – 10.8)
Characteristic strength (MPa; 95% CI)	246 (219 – 311)	352 (275 – 357)	300 (275 – 357)

FIGURE 4.6 Several dental restorative techniques have been used to reduce the overall polymerization shrinkage and internal stresses to decrease the clinical issues of RBC restorations. (Adapted, with permission, from Rosa de Lacerda et al., 2019.)

polymerization shrinkage. This was shown to reduce the stresses generated at the cavity walls (Lutz, Krejci, & Barbakow, 1991).

Several techniques were suggested for the application of these 2 mm increments to lower the C-factor, such as horizontal incremental, oblique incremental (Lutz, Krejci, & Oldenburg, 1986), centripetal incremental (Bichacho, 1994), and split horizontal incremental techniques (Hassan & Khier, 2005). However, these techniques are time-consuming, where each layer has to be incorporated in a certain way and photopolymerized before proceeding to the following layer. This has led to more time spent to complete the chairside procedure. Another problem encountered was the presence of air bubbles incorporated between layers. These porosities enhance the propagation of cracks within the resin, reducing the fatigue strength and increasing the wear rate of the material (McCabe & Ogden, 1987). The incorporation of air into the resin matrix may, therefore, be detrimental to the durability of the material and the need to avoid procedures or reduce the number of procedures that are likely to influence the levels of porosity within the material.

4.4.1.2 Bulk-Fill Technique

In an attempt to solve the shortcomings of conventional RBCs, bulk-fill materials were introduced where they could be applied in 4–5 mm layers according to the manufacturer instruction (Ilie & Hickel, 2011). Although they are advertised as a new material category, they do not essentially differ in their chemical composition from regular microhybrid and nanohybrid RBCs. The matrix resin is mainly based on monomers of Bis-GMA, UDMA, TEGDMA, and EBPDMA. However, some bulk-fill RBCs have modified Bis-GMA resulting in a monomer of lower viscosity. The fillers in bulk-fill RBCs are lower in percentage by volume, but higher by weight in comparison to conventional microhybrid and nanohybrid RBCs; in other words, bulk-fill RBCs contain a lower filler content and an enlarged filler size.

Alternative resin composites have been introduced in dentistry in order to solve the undesirable limitations of conventional materials. For instance, bulk-fill resin composites can be applied using much thicker increments of ~4.0–5.0 mm (El-Damanhoury and Platt, 2014), which is only possible due to their increased polymerization depth, allowing a thorough photoactivation of each increment.

4.4.2   BULK-FILL RESIN-BASED COMPOSITES

In 2009, the first bulk-fill RBC with flowable consistency was introduced to the market. It was indicated for use as a base in Class I and Class II cavity preparations, requiring an additional layer of 2 mm of conventional RBC on the occlusal surface. The stress decreasing resin (SDR) allows greater molecule flexibility, thus avoiding the stress generated during photocuring (Ilie & Hickel, 2011).

Later, bulk-fill RBCs with similar consistency and application methods were developed for clinical use. For Filtek™ BulkFill Flowable (3M ESPE), the manufacturer claims that it is based on four monomers: BisGMA, UDMA, Procrylat, and BisEMA, which have high molecular weight, to reduce the development of polymerization shrinkage. Procrylat monomer is added to allow greater fluidity and thus lower the polymerization stress (Ilie & Hickel, 2011).

For all the above RBCs, the manufacturers have not announced major changes to the polymerization system. Subsequently, a different bulk-fill RBC variation (Tetric N Ceram® Bulk Fill, Ivoclar Vivadent and x-tra fil, VOCO) was introduced, which had conventional consistency and was indicated to be used in increments of up to 4 mm without the need for an extra occlusal layer. In the case of flowable bulk-fill RBCs, the inorganic filler ranges between 64–75 wt% and 38–61 vol%. On the other hand, in most of the high-viscosity bulk-fill RBCs, this component was increased up to 79–86 wt% and 61–81 vol%.

Tetric N Ceram® Bulk Fill (Ivoclar Vivadent) has a new polymerization initiation booster called Ivocerin®, which is a germanium-based photoinitiator of greater reactivity than that of CQ, due to its higher absorption of visible light (400–450 nm) (Moszner, Fischer, Ganster, Liska, & Rheinberger, 2008). It was also reported that it has a filter for light pollution that ensures proper clinical work time.

A third variation of bulk-fill was introduced (SonicFill™, Kerr), which depends on a special sonic vibration handpiece and it is used in 5 mm increments. It is activated by means of sound vibration, producing a momentary drop in viscosity during application. This resin is also indicated for Class I and II preparations with no occlusal layer.

4.4.3 PROPERTIES OF BULK-FILL COMPOSITES

Generally, bulk-fill RBCs have a lower proportion of fillers with an increase in their size, which was claimed to decrease problems encountered with conventional RBCs. Multiple *in vitro* studies evaluated different properties for the newly introduced bulk-fill RBCs. One of the main concerns was the depth of cure of 4 mm increment, which was recommended by the manufacturers. In conventional photopolymerized RBCs, limited curing depth was found and the possibility of insufficient monomer conversion in the bottom of the cavity preparation was speculated (Lindberg, Peutzfeldt, & van Dijken, 2005). Some studies found an improved depth of cure of bulk-fill RBCs in comparison to conventional ones, while others found no improvement. Several reasons have been proposed to explain the improved depth of cure. The first suggested reason was the inclusion of more efficient initiation systems in some resins, which was claimed to improve the light penetration in the RBC (Alrahlah, Silikas, & Watts, 2014). A second suggested reason is that bulk-fill RBCs have higher translucency, which would allow deeper light penetration (Bucuta & Ilie, 2014). Another reason is the decreased matrix-filler surface interface, which reduces light refraction and improves light penetration (Ilie, Bucuta, & Draenert, 2013c). Reduction in refractive index differences between resin and filler improved the DC (Fujita, Nishiyama, Nemoto, Okada, & Ikemi, 2005), and increased depth of cure as well as color shade matching (Shortall, Palin, & Burtscher, 2008).

4.4.3.1 Polymerization Shrinkage and Stress

It was found that some bulk-fill flowable RBCs were effectively cured in 4 mm bulk, but shrank more than the conventional non-flowable RBCs (Jang, Park, & Hwang, 2015). However, bulk-fill non-flowable RBC showed comparable shrinkage to conventional non-flowable RBC, but it was not sufficiently cured in 4 mm bulk

(Jang, Park, & Hwang, 2015). When evaluating the polymerization stress of bulk-fill RBCs, it was found to be lower than conventional RBCs and conventional flowable RBCs (El-Damanhoury & Platt, 2014; Ilie & Hickel, 2011). However, in another study, it was found that polymerization shrinkage of bulk-fill RBCs might be smaller, similar, or larger than conventional flowable RBCs (Garcia, Yaman, Dennison, & Neiva, 2014). This has shown the relationship between the proportion of filler and polymerization shrinkage, where the resins with the least amount of filler loading, and therefore with the greater proportion of resinous matrix, experienced higher levels of shrinkage by polymerization, and vice versa.

4.4.3.2 Cuspal Flexure

Cuspal flexure was evaluated in premolars with Class II restorations restored with bulk-fill RBCs applied in one increment; it was significantly lower than those observed in conventional RBCs applied in incremental layers (Moorthy et al., 2012). However, it is difficult to determine if the lower cusp flexure observed by using bulk-fill RBC was due to a smaller contraction of the resin or due to changes in the mode of application of the material. A study has determined the existence of a smaller change in the inter-cusp distance when RBC was placed as a single increment in comparison to the incremental method. It was reasoned that each increment would cause deformation of the cavity's walls with the downward movement of the walls and toward the inside. This movement decreases the total volume of the cavity. In contrast, for RBC placed using the single-increment technique, the cavity volume remains relatively the same (Versluis, Douglas, Cross, & Sakaguchi, 1996).

4.4.3.3 Marginal Adaptability

RBC marginal adaptability is closely related to the development of polymerization shrinkage and stress. Studies found no significant differences in marginal integrity when using bulk-fill RBCs in comparison to conventional RBCs (Roggendorf, Kramer, Appelt, Naumann, & Frankenberger, 2011). Several studies evaluated the viscosity of bulk-fill RBCs in relation to different properties. It was found that the use of flowable RBCs results in better adaptation compared to packable RBCs (Opdam, Roeters, Joosten, & Veeke, 2002).

4.5   LIGHT CURING UNITS

A diversity of dental light curing units (LCU) are available in the market. These include quartz-tungsten-halogen light (QTH), plasma-arc light (PAC), argon-ion laser (AL), and light-emitting diode (LED).

4.5.1   QUARTZ-TUNGSTEN-HALOGEN LIGHT (QTH)

Quartz-Tungsten-Halogen Light (QTH) was the first visible light used to cure RBCs. Its bulb consists of a tungsten filament enclosed in a clear, crystalline quartz casing, filled with a halogen-based gas. As electricity flows through the filament, heat develops because of the wire resistance. The heat developed is sufficient to cause tungsten atoms to vaporize from the wire surface. When this happens, tremendous amounts of electromagnetic energy are released, mostly occurring in the infrared (IR) spectral region. Thus, these types of light units typically require filtering to remove heat, as well as excess visible light not required for photopolymerization.

4.5.2 Plasma-Arc Light (PAC)

Plasma-Arc Light (PAC) was introduced to photopolymerize RBCs. These units utilize two tungsten rods, held at a specified distance, encased in a high-pressure envelope of xenon gas, with a sapphire window through which emitted radiation escapes. When high voltage is applied across the electrodes, a spark forms, which produces a tremendous amount of electromagnetic radiation over a wide spectral range: from IR to short wavelength UV. Because of the massive amount of radiation emitted falling outside of the narrow limits needed for dental photopolymerization, a substantial amount of filtering is required in this light. Additionally a special liquid helps to further reduce unwanted IR, UV, and visible light (Rueggeberg, 2011).

4.5.3 Argon-Ion Laser (AL)

Argon-Ion Laser (AL) was used to enhance vital tooth bleaching before being used for intraoral photopolymerization of dental RBCs. The initial delivery of power from the laser to the tooth was directly through the end of a fiber optic cable. The radiation coming out of the fiber optic cable had a divergent nature, and therefore other methods were developed in an attempt to make a collimated beam of coherent energy, whose target power was not related to the curing light guide-to-tooth distance. However, these units were large, heavy, and expensive. Over time, the unit became much smaller, and could easily fit into a clinic, yet, it needed to be on a cart when moved from room-to-room due to its weight. Additionally, after the use of the device, room temperature would be elevated. Due to the above reasons, this curing system became outdated in a short time (Rueggeberg, 2011).

4.5.4 Light-Emitting Diode (LED)

The Light-Emitting Diode (LED) technology has been proven to be an efficient, cost-effective lighting source. It has been widely used shortly after the blue LEDs became available using indium-gallium-nitride (InGaN) substrates (Rueggeberg, 1999). These devices rely on the forward-biased energy difference (band gap) between two dissimilar semiconductor substrates (n-type conduction band and p-type valence band), to determine the wavelength of emitted light (Rueggeberg, 1999). Electrons are forced to traverse from one side of a semiconductor material (the "N" material, having an excess of electrons) to a substrate having an electron deficiency (the "P" material). When electrons travel through this potential energy "gap", they emit light with wavelengths depending on the composition of each semiconductor substrate (Nakamura, 2015). The spectral emission from such units could successfully photoactivate CQ-based products.

4.5.5   Development in Light-Emitting Diode

LED technology was borrowed from other industries and was incorporated in the dental field in the 1990s. The concept that the spectral emission from such units could successfully photoactivate CQ-based products has been proven. LED technology requires low power, no filament, no optical filter, and therefore provides much greater photon-generating efficiency than any competitive light source. In addition, these units can be powered by a battery. LED sources are claimed to last for thousands of hours without needing replacement (Rueggeberg, 1999).

4.5.5.1   First-Generation LED

The first introduced blue LED curing lights were mainly experimental. They were built to test whether they could generate light at the correct wavelength and deliver a sufficient number of photons needed to successfully photopolymerize dental RBCs. The individual LED available at that time had a very low output power, which necessitated the different arrangement of multiple LED elements into a physical array, and a turbo-tip was used, resulting in sufficient output, which was enough for photopolymerization of CQ-based RBCs.

These closely packed arrays generated a significant amount of thermal energy, and therefore heat dissipation was needed. Heat sinking technology was incorporated in some units to draw heat away from the LED chip. Other units have used metal body castings that provided a large area for the thermal dissipation as well as structural durability.

The first-generation LED did not produce adequate output for curing RBCs resulting in insufficient depths of cure of dental RBCs (Ernst et al., 2004). However, if the curing unit was used for extended exposures, the output was found to be comparable with QTH source (Ernst et al., 2004). The irradiance value of this LED generation varied greatly between different units. Nickel-cadmium (NiCAD) battery was used, but careful recharging routines had to be followed or the useful lifetime of these power sources was significantly reduced. The spectral emission of this light was effective for activation of CQ and PPD, but it was not suitable with TPO (Rueggeberg, 1999).

4.5.5.2   Second-Generation LED

In the illumination industry, high emission area LEDs became available and were adapted in dentistry. One-Watt chips were available, all on one body, consisting of four main areas of illumination, each consisting of four, bar-shaped emitting surfaces: a total of sixteen emission areas. Incorporation of these chip types greatly boosted irradiant output and allowed blue LEDs to be able to accomplish effective photopolymerization in a much shorter time. A higher-power LED became available shortly thereafter containing 5-Watt devices and rapid advances have led to chips reaching 10 and 15 Watts, which were incorporated in dental LEDs. These resulted in a greater photon density increasing irradiance values and allowed lower exposure time to achieve optimal photopolymerization of CQ-containing restorative materials (Uhl, Sigusch, & Jandt, 2004).

Concurrently, battery technology also advanced, allowing incorporation of the longer-lasting nickel metal hydride (NiMH) units. Higher-rated LED chips produced a lot of heat, and therefore advanced methods such as internal fans and large metal heat sinks were used to dissipate thermal power from the LED arrays (Rueggeberg, 1999).

As for the emission spectra of this generation, it is much greater than the first one. However, the peak emission is located at a shorter wavelength than previously seen leading to an increased overlap with PPD while still providing activation of CQ. No interaction with APO was possible.

4.5.5.3   Third-Generation LED

The need to provide radiant energy to activate TPO as well as Ivocerin® drove manufacturers to incorporate more than one color into the LED chipset. These LCUs are known as polywave as they emit irradiation with different peaks of wavelengths. Different arrays were used to provide a simultaneous combination of violet and blue wavelengths. Among the chip arrangements was an array with centrally positioned, high-wattage blue LED, surrounded by four lower-powered, converging violet LEDs. Another array incorporated two blue LEDs and one violet LED. This array arrangement is seen in Bluephase Style light. A third array was the arrangement of three different color chips into the single array set: two blue (emitting near to 460 nm), a shorter wavelength blue (emitting near to 445 nm), and one violet (emitting close to 400 nm).

With the inclusion of the violet emission near to 407 nm, photons are delivered to a wide bandwidth for all photoinitiators used in dental restorative materials, particularly for TPO, PPD, and CQ. However, in the third generation, the blue emission is reduced compared to that of an all-blue emitting light, which means less potential for CQ activation at RBC depths. In materials that contain Lucirin® TPO or Ivocerin® in addition to CQ, improved curing is possible, even with the lower blue light present. Despite the differences in the amount of violet and blue light emitted by the polywave LED, previous evidence supports the idea that the spectral output of this polywave LED used was enough to efficiently cure bulk-fill RBCs containing either CQ or CQ associated with alternative photoinitiators (Schneider, Pfeifer, Consani, Prahl, & Ferracane, 2008). However, the light beam profile received by the specimen should not be ignored.

The third-generation curing lights are present in two different designs. The first is the traditional gun-style light, which has the chipset inside the gun body and uses fiber optic light guides to transmit emitter photons onto the target area. Another concept is the use of a pencil-style body, which can still use removable fiber optic guides or can have the emitting chipset placed directly at the distal guide end of the unit. This allows the light to directly shine onto the target, without the use of fiber optic light guides. It also has the advantage of greater ease of placement intraorally facilitating light guide position and allowing more direct illumination and maximum transfer of light to the restoration. Battery technology has been advancing, and the use of lithium-ion batteries is common in most LED units of this generation. They are stable and durable allowing long-usage energy storage sources providing a reliable output over extended clinical operation time.

4.6   EFFECTIVENESS OF CURE OF RESIN-BASED COMPOSITES

Effectiveness of cure of RBCs is a crucial parameter that evaluates the extent of material polymerization. An RBC that is cured effectively ensures optimum clinical performance due to an acceptable DC of monomers into polymers, which results in enhanced mechanical properties of the material. The effectiveness of RBC cure may be assessed directly or indirectly. Direct methods using vibrational spectroscopy such as IR spectroscopy and laser Raman spectroscopy assess DC. However, indirect methods include the ISO 4049 method and SH testing.

4.6.1   VIBRATIONAL SPECTROSCOPY

These methods provide a direct approach to determine the effectiveness of cure by measuring the DC (Asmussen, 1982b; Eliades, Vougiouklakis, & Caputo, 1987; Ferracane & Greener, 1984; Rueggeberg & Craig, 1988). This is achieved by measuring both the percentage of carbon-carbon single bonds in the cured material and the percentage of unreacted C=C bonds. Vibrational spectroscopy can be classified into two techniques. The first technique is Fourier transform infrared spectroscopy (FTIR), which is sometimes documented as IR spectroscopy and is based on light absorption. The second technique, Raman spectroscopy is based on light scattering (Figure 4.7). Two devices are popularly used: Fourier transform-Raman spectroscopy (FT-Raman) and micro-Raman spectroscopy (MRS) (De Santis & Baldi, 2004). Other available techniques include differential thermal analysis (DTA), differential scanning calorimetry (DSC), and nuclear magnetic resonance (NMR) (Alshihri, Santini, & Aldossary, 2018).

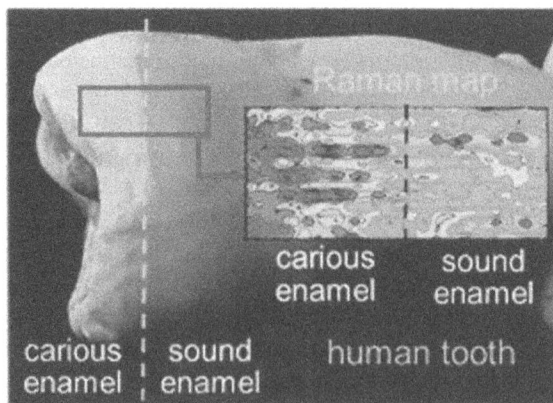

FIGURE 4.7   Raman spectroscopy is based on light scattering. Raman spectroscopy is an effective method in determining the early changes in human enamel caused by artificial caries. (Adapted, with permission, from Buchwald & Buchwald, 2019.)

4.6.2 ISO 4049 METHOD

Manufacturers use the ISO 4049 method primarily to certify that their RBCs will achieve the minimal depth of cure requirement. They also provide recommendations for light curing times relative to RBC increment thickness to adequately polymerize the material. Some studies have also used this method to determine the effectiveness of cure of the tested RBC (Ruyter & Oysaed, 1982). Samples are prepared in a cylindrical stainless steel mold of dimensions 6 mm long × 4 mm diameter. If the manufacturer claims a depth of cure of 3 mm, the mold should be at least 2 mm longer than twice the claimed depth of cure. After polymerization of the specimen, it should be removed from the mold and scraping of uncured material is done with a plastic instrument. A micrometer is used to measure the height of the cured material, where the values are divided by 2 to give the depth of cure of the tested RBC (ISO_4049, 2009). According to ISO 4049-2009, the depth of cure should be no more than 0.5 mm below the value stated by the manufacturer.

The rationale behind ISO 4049 is unspecified and the correlations between the test and clinical performance are lacking. It was concluded that the ISO 4049 method overestimated the depth of cure when it was compared with Knoop hardness profiles (Moore, Platt, Borges, Chu, & Katsilieri, 2008) and Vickers hardness estimations of the DC (Flury, Hayoz, Peutzfeldt, Husler, & Lussi, 2012).

4.6.3 SURFACE HARDNESS (SH)

SH is a well-accepted method and has been used to indirectly probe polymer network conversion; therefore, hardness profiles can be used to measure the depth of cure. Ideally, SH of RBC should be equal or close to equal throughout the restoration and for the life of the restoration. However, the SH of bulk-fill RBC has different values at different depths (Garcia et al., 2014). Therefore, sufficient cure was defined as SH of more than 80% between the bottom and top surfaces (B/T ratio) of RBC samples (Johnston, Leung, & Fan, 1985).

As a measure of the completeness of conversion, the bottom to top Knoop hardness number (B/T-KHN) was approximately 2.5 times more sensitive than the bottom to top DC (B/T-DC) ratio (Bouschlicher, Rueggeberg, & Wilson, 2004). B/T-KHN ratios provide an accurate, simple method of assessing the efficacy of photoinitiation strategies (curing light/exposure duration) instead of using more complex FTIR methods to determine the DC (Bouschlicher et al., 2004).

4.7 THE INFLUENCE OF CURING LIGHT DISTANCE ON THE EFFECTIVENESS OF CURE OF RESIN-BASED COMPOSITES

Among the factors that affect the effectiveness of cure of RBCs is the distance between the surface of the restoration and the LCU guide. Manufacturers recommend positioning the curing light guide as close as possible to the RBC restoration surface. These recommendations are typically based on testing the material in ideal laboratory conditions, commonly at a curing light distance of 0 mm (Shortall, Price, MacKenzie, & Burke, 2016). However, in a clinical setting, this curing light distance is difficult to achieve especially in Class II cavities. One study measured the depths of Class II proximal cavities of 1,146 extracted teeth. They reported an average depth of 4–5 mm for mandibular premolars, 5–6 mm for maxillary premolars, and 5–7 mm for molars (Hansen & Asmussen, 1997). This was in agreement with another study that found that 6.3 mm ($\pm$ 0.7 mm) was the typical Class II cavity depth when the LCU guide was positioned directly on the tooth or the restoration surface (Price, Derand, Sedarous, Andreou, & Loney, 2000). Additionally, more than 15% of the mandibular molars were reported to have a proximal cavity depth of more than 8 mm (Hansen & Asmussen, 1997).

Several studies have investigated the impact of curing light distance on the effectiveness of cure of conventional RBCs. Most of these studies found that the SH of the RBC decreases with increasing the curing light distance (Caldas, de Almeida, Correr-Sobrinho, Sinhoreti, & Consani, 2003; Rode, Kawano, & Turbino, 2007; Thome, Steagall, Tachibana, Braga, & Turbino, 2007). Other studies found that there was no statistical difference of the top SH of RBCs at small curing light distances; however, the bottom SH was statistically significant at the same small curing light distances (Aguiar, Lazzari, Lima, Ambrosano, & Lovadino, 2005; Pires, Cvitko, Denehy, & Swift, 1993). To date, most of the studies assessing the effectiveness of cure of RBCs at different curing light distances were done on conventional RBCs; however, scarce studies were done on bulk-fill RBCs.

4.8 SUMMARY

In general, Resin-Based Composite (RBC) had been significantly improved through innovations in photoinitiator, polymeric matrix, filler type and size. Incremental technique of RBC placement was introduced to reduce polymerization shrinkage and incomplete photopolymerization problems. The technique is, however, time-consuming and if not carried out properly can lead to strength reduction, post-operative sensitivity, and early restoration failure due to microleakage. Subsequently, the bulk-fill RBCs with lower polymerization shrinkage and improved depths of cure were developed. The increased depth of cure was achieved by improving material translucency via incorporating large-size fillers and decreasing the filler load. In some instances, certain novel photoinitiators are included such as germanium derivatives that substantially increase the absorption of visible light resulting in greater depth of cure. Efficient photopolymerization is essential to ensure optimal clinical performance of RBCs. It is influenced by intrinsic (material) and extrinsic factors, including the type of LCUs and curing light distance. Additionally, various modifications have been made to improve the performance of RBCs, which has been the subject of studies by many researchers. The future direction is also toward the development of self-healing RBCs to address the problem of micro-cracks that occur in the polymer matrix. Self-healing RBCs allow for autonomous crack repairs, thus extending their service lives. Numerous studies have provided several approaches to form self-healing RBCs, which include synthesis of microcapsules and the use of urease enzymes. To sum up, the utilization of composites in dentistry is an example of the evolution of technology in the best sense.

REFERENCES

Aguiar, F. H., Lazzari, C. R., Lima, D. A., Ambrosano, G. M., & Lovadino, J. R. (2005). Effect of light curing tip distance and resin shade on microhardness of a hybrid resin composite. *Brazilian Oral Research, 19*(4), 302–306.

Al-Ahdal, K., Silikas, N., & Watts, D. C. (2014). Rheological properties of resin composites according to variations in composition and temperature. *Dental Materials, 30*(5), 517–524.

Alrahlah, A., Silikas, N., & Watts, D. C. (2014). Post-cure depth of cure of bulk fill dental resin-composites. *Dental Materials, 30*(2), 149–154.

Alshihri, A., Santini, A., & Aldossary, M. (2018). The influence of in vitro mold characteristics on the polymerization outcomes of resin based composites: A literature review. *Current Analysis of Dentistry, 1*(2018), 7–11.

Andrzejewska, E., Lindén, L.-Å., & Rabek, J. F. (1998). The role of oxygen in camphorquinone-initiated photopolymerization. *Macromolecular Chemistry and Physics, 199*(3), 441–449.

Anseth, K. S., Wang, C. M., & Bowman, C. N. (1994). Kinetic evidence of reaction-diffusion during the polymerization of multi(meth)acrylate monomers. *Macromolecules, 27*(3), 650–655.

Anusavice, K. J., Shen, C., & Rawls, H. R. (2012). Philips Science of Dental Materials (12th ed.). Elsevier.

Asmussen, E. (1975). NMR-analysis of monomers in restorative resins. *Acta Odontologica Scandinavica, 33*(3), 129–134.

Asmussen, E. (1982b). Restorative resins: Hardness and strength vs. quantity of remaining double bonds. *Scandinavian Journal of Dental Research, 90*(6), 484–489.

Beun, S., Glorieux, T., Devaux, J., Vreven, J., & Leloup, G. (2007). Characterization of nanofilled compared to universal and microfilled composites. *Dental Materials, 23*(1), 51–59.

Bichacho, N. (1994). The centripetal build-up for composite resin posterior restorations. *Practical Periodontics and Aesthetic Dentistry, 6*(3), 17–23; quiz 24.

Bouillaguet, S., Wataha, J. C., Hanks, C. T., Ciucchi, B., & Holz, J. (1996). In vitro cytotoxicity and dentin permeability of HEMA. *Journal of Endodontics, 22*(5), 244–248.

Bouschlicher, M. R., Rueggeberg, F. A., & Wilson, B. M. (2004). Correlation of bottom-to-top surface microhardness and conversion ratios for a variety of resin composite compositions. *Operative Dentistry, 29*(6), 698–704.

Bowen, R. L. (1956). Use of epoxy resins in restorative materials. *Journal of Dental Research, 35*(3), 360–369.

Bowen, R. L. (1959). Method of preparing a monomer having phenoxy and methacrylate groups linked by hydroxy glyceryl groups. United States patent US 3,179,623 A.

Bowen, R. L. (1962). Dental filling material comprising vinyl silane treated fused silica and a binder consisting of the reaction product of bis phenol and glycidyl acrylate. United States patent US 3,066,112 A.

Bowen, R. L. (1963). Properties of a silica-reinforced polymer for dental restorations. *Journal of the American Dental Association, 66*, 57–64.

Buchwald, T., & Buchwald, Z. (2019). Assessment of the Raman spectroscopy effectiveness in determining the early changes in human enamel caused by artificial caries. *The Analyst, 144*(4), 1409–1419.

Bucuta, S., & Ilie, N. (2014). Light transmittance and micro-mechanical properties of bulk fill vs. conventional resin based composites. *Clinical Oral Investigations, 18*(8), 1991–2000.

Caldas, D. B., de Almeida, J. B., Correr-Sobrinho, L., Sinhoreti, M. A., & Consani, S. (2003). Influence of curing tip distance on resin composite Knoop hardness number, using three different light curing units. *Operative Dentistry, 28*(3), 315–320.

Cox, C. F., Keall, C. L., Keall, H. J., Ostro, E., & Bergenholtz, G. (1987). Biocompatibility of surface-sealed dental materials against exposed pulps. *Journal of Prosthetic Dentistry*, *57*(1), 1–8.

Craig, R. G. (1981). Chemistry, composition, and properties of composite resins. *Dental Clinics of North America*, *25*(2), 219–239.

Cramer, N. B., Stansbury, J. W., & Bowman, C. N. (2011). Recent advances and developments in composite dental restorative materials. *Journal of Dental Research*, *90*(4), 402–416.

Curtis, A. R., Palin, W. M., Fleming, G. J., Shortall, A. C., & Marquis, P. M. (2009). The mechanical properties of nanofilled resin-based composites: The impact of dry and wet cyclic pre-loading on bi-axial flexure strength. *Dental Materials*, *25*(2), 188–197.

Davidson, C. L. (1986). Resisting the curing contraction with adhesive composites. *Journal of Prosthetic Dentistry*, *55*(4), 446–447.

de Moraes, R. R., Goncalves Lde, S., Lancellotti, A. C., Consani, S., Correr-Sobrinho, L., & Sinhoreti, M. A. (2009). Nanohybrid resin composites: Nanofiller loaded materials or traditional microhybrid resins? *Operative Dentistry*, *34*(5), 551–557.

de Oliveira, D., Rocha, M. G., Correa, I. C., Correr, A. B., Ferracane, J. L., & Sinhoreti, M. A. C. (2016). The effect of combining photoinitiator systems on the color and curing profile of resin-based composites. *Dental Materials*, *32*(10), 1209–1217.

De Santis, A., & Baldi, M. (2004). Photo-polymerisation of composite resins measured by micro-Raman spectroscopy. *Polymer*, *45*(11), 3797–3804.

Dionisio, J., Mahabadi, H. K., O'Driscoll, K. F., Abuin, E., & Lissi, E. A. (1979). High-conversion polymerization. IV. A definition of the onset of the gel effect. *Journal of Polymer Science: Polymer Chemistry Edition*, *17*(7), 1891–1900.

Durner, J., Obermaier, J., Draenert, M., & Ilie, N. (2012). Correlation of the degree of conversion with the amount of elutable substances in nano-hybrid dental composites. *Dental Materials*, *28*(11), 1146–1153.

El-Damanhoury, H., & Platt, J. (2014). Polymerization shrinkage stress kinetics and related properties of bulk-fill resin composites. *Operative Dentistry*, *39*(4), 374–382.

Eliades, G. C., Vougiouklakis, G. J., & Caputo, A. A. (1987). Degree of double bond conversion in light-cured composites. *Dental Materials*, *3*(1), 19–25.

Elliott, J. E., Lovell, L. G., & Bowman, C. N. (2001). Primary cyclization in the polymerization of bis-GMA and TEGDMA: A modeling approach to understanding the cure of dental resins. *Dental Materials*, *17*(3), 221–229.

Ernst, C. P., Meyer, G. R., Muller, J., Stender, E., Ahlers, M. O., & Willershausern, B. (2004). Depth of cure of LED vs QTH light-curing devices at a distance of 7 mm. *Journal of Adhesive Dentistry*, *6*(2), 141–150.

Feilzer, A. J., De Gee, A. J., & Davidson, C. L. (1987). Setting stress in composite resin in relation to configuration of the restoration. *Journal of Dental Research*, *66*(11), 1636–1639.

Ferracane, J. L. (1994). Elution of leachable components from composites. *Journal of Oral Rehabilitation*, *21*(4), 441–452.

Ferracane, J. L. (2011). Resin composite–state of the art. *Dental Materials*, *27*(1), 29–38.

Ferracane, J. L., & Condon, J. R. (1990). Rate of elution of leachable components from composite. *Dental Materials*, *6*(4), 282–287.

Ferracane, J. L., & Greener, E. H. (1984). Fourier transform infrared analysis of degree of polymerization in unfilled resins–methods comparison. *Journal of Dental Research*, *63*(8), 1093–1095.

Ferracane, J. L., & Mitchem, J. C. (2003). Relationship between composite contraction stress and leakage in Class V cavities. *American Journal of Dentistry*, *16*(4), 239–243.

Finger, W. J., Lee, K. S., & Podszun, W. (1996). Monomers with low oxygen inhibition as enamel/dentin adhesives. *Dental Materials*, *12*(4), 256–261.

Flury, S., Hayoz, S., Peutzfeldt, A., Husler, J., & Lussi, A. (2012). Depth of cure of resin composites: Is the ISO 4049 method suitable for bulk fill materials? *Dental Materials*, *28*(5), 521–528.

Fouassier, J. P. (1995). *Photoinitiation, Photopolymerization and Photocuring: Fundamentals and Applications.* Hanser Publishers, Munich Vienna New York.

Fujita, K., Nishiyama, N., Nemoto, K., Okada, T., & Ikemi, T. (2005). Effect of base monomer's refractive index on curing depth and polymerization conversion of photo-cured resin composites. *Dental Materials Journal*, *24*(3), 403–408.

Garcia, D., Yaman, P., Dennison, J., & Neiva, G. (2014). Polymerization shrinkage and depth of cure of bulk fill flowable composite resins. *Operative Dentistry*, *39*(4), 441–448.

Gerzina, T. M., & Hume, W. R. (1996). Diffusion of monomers from bonding resin-resin composite combinations through dentine in vitro. *Journal of Dentistry*, *24*(1-2), 125–128.

The Glossary of Prosthodontic Terms: Ninth Edition. (2017). *Journal of Prosthetic Dentistry*, *117*(5S), e1–e105.

Goncalves, F., Azevedo, C. L., Ferracane, J. L., & Braga, R. R. (2011). BisGMA/TEGDMA ratio and filler content effects on shrinkage stress. *Dental Materials*, *27*(6), 520–526.

Halvorson, R. H., Erickson, R. L., & Davidson, C. L. (2002). Energy dependent polymerization of resin-based composite. *Dental Materials*, *18*(6), 463–469.

Hansen, E. K., & Asmussen, E. (1997). Visible-light curing units: Correlation between depth of cure and distance between exit window and resin surface. *Acta Odontologica Scandinavica*, *55*(3), 162–166.

Hassan, K., & Khier, S. (2005). Split increment horizontal layering: A simplified placement technique for direct posterior resin restorations. *General Dentistry*, *53*(6), 406–409.

Heuer, G. A., Garman, T. A., Sherrer, J. D., & Williams, H. A. (1982). A clinical comparison of a quartz- and glass-filled composite with a glass-filled composite. *Journal of the American Dental Association*, *105*(2), 246–247.

Howard, B., Wilson, N. D., Newman, S. M., Pfeifer, C. S., & Stansbury, J. W. (2010). Relationships between conversion, temperature and optical properties during composite photopolymerization. *Acta Biomaterialia*, *6*(6), 2053–2059.

Ilie, N., Bucuta, S., & Draenert, M. (2013c). Bulk-fill resin-based composites: An in vitro assessment of their mechanical performance. *Operative Dentistry*, *38*(6), 618–625.

Ilie, N., & Hickel, R. (2011). Investigations on a methacrylate-based flowable composite based on the SDR technology. *Dental Materials*, *27*(4), 348–355.

Ilie, N., Rencz, A., & Hickel, R. (2013a). Investigations towards nano-hybrid resin-based composites. *Clinical Oral Investigations*, *17*(1), 185–193.

Jakubiak, J., Allonas, X., Fouassier, J. P., Sionkowska, A., Andrzejewska, E., Linden, L. Å., & Rabek, J. F. (2003). Camphorquinone–amines photoinitiating systems for the initiation of free radical polymerization. *Polymer*, *44*(18), 5219–5226.

Jakubiak, J., Sionkowska, A., Lindén, L. Å., & Rabek, J. F. (2001). Isothermal photo differential scanning calorimetry. Crosslinking polymerization of multifunctional monomers in presence of visible light photoinitiators. *Journal of Thermal Analysis and Calorimetry*, *65*(2), 435–443.

Janda, R., Roulet, J. F., Kaminsky, M., Steffin, G., & Latta, M. (2004). Color stability of resin matrix restorative materials as a function of the method of light activation. *European Journal of Oral Sciences*, *112*(3), 280–285.

Jang, J. H., Park, S. H., & Hwang, I. N. (2015). Polymerization shrinkage and depth of cure of bulk-fill resin composites and highly filled flowable resin. *Operative Dentistry*, *40*(2), 172–180.

Johnston, W. M., Leung, R. L., & Fan, P. L. (1985). A mathematical model for post-irradiation hardening of photoactivated composite resins. *Dental Materials*, *1*(5), 191–194.

Klapdohr, S., & Moszner, N. (2004). New inorganic components for dental filling composites. *Monatshefte für Chemie - Chemical Monthly*, *136*(1), 21–45.

Lagocka, R., Mazurek-Mochol, M., Jakubowska, K., Bendyk-Szeffer, M., Chlubek, D., & Buczkowska-Radlinska, J. (2018). Analysis of base monomer elution from 3 flowable bulk-fill composite resins using high performance liquid chromatography (HPLC). *Medical Science Monitor, 24,* 4679–4690.

Lambrechts, P., & Van Herle, G. (1982). Observation and comparison of polished composite surfaces with the aid of SEM and profilometer. *Journal of Oral Rehabilitation, 9*(3), 203–216.

Lee, M. J., Kim, M. J., Kwon, J. S., Lee, S. B., & Kim, K. M. (2017). Cytotoxicity of light-cured dental materials according to different sample preparation methods. *Materials (Basel), 10*(3), 288.

Leprince, J. G., Hadis, M., Shortall, A. C., Ferracane, J. L., Devaux, J., Leloup, G., & Palin, W. M. (2011). Photoinitiator type and applicability of exposure reciprocity law in filled and unfilled photoactive resins. *Dental Materials, 27*(2), 157–164.

Leprince, J. G., Palin, W. M., Hadis, M. A., Devaux, J., & Leloup, G. (2013). Progress in dimethacrylate-based dental composite technology and curing efficiency. *Dental Materials, 29*(2), 139–156.

Lindberg, A., Peutzfeldt, A., & van Dijken, J. W. (2005). Effect of power density of curing unit, exposure duration, and light guide distance on composite depth of cure. *Clinical Oral Investigations, 9*(2), 71–76.

Lovell, L. G., Newman, S. M., & Bowman, C. N. (1999). The effects of light intensity, temperature, and comonomer composition on the polymerization behavior of dimethacrylate dental resins. *Journal of Dental Research, 78*(8), 1469–1476.

Lutz, F., Krejci, I., & Barbakow, F. (1991). Quality and durability of marginal adaptation in bonded composite restorations. *Dental Materials, 7*(2), 107–113.

Lutz, E., Krejci, I., & Oldenburg, T. R. (1986). Elimination of polymerization stresses at the margins of posterior composite resin restorations: A new restorative technique. *Quintessence International, 17*(12), 777–784.

Lutz, F., & Phillips, R. W. (1983). A classification and evaluation of composite resin systems. *Journal of Prosthetic Dentistry, 50*(4), 480–488.

McCabe, J. F., & Ogden, A. R. (1987). The relationship between porosity, compressive fatigue limit and wear in composite resin restorative materials. *Dental Materials, 3*(1), 9–12.

McCabe, J. F., & Walls, A. W. G. (2008). *Applied Dental Materials* (9th ed.). Blackwell Publishing Ltd.

Meereis, C. T. W., Münchow, E. A., de Oliveira da Rosa, W. L., da Silva, A. F., & Piva, E. (2018). Polymerization shrinkage stress of resin-based dental materials: A systematic review and meta-analyses of composition strategies. *Journal of the Mechanical Behavior of Biomedical Materials, 82,* 268–281.

Mitra, S. B., Wu, D., & Holmes, B. N. (2003). An application of nanotechnology in advanced dental materials. *Journal of the American Dental Association, 134*(10), 1382–1390.

Moore, B. K., Platt, J. A., Borges, G., Chu, T. M., & Katsilieri, I. (2008). Depth of cure of dental resin composites: ISO 4049 depth and microhardness of types of materials and shades. *Operative Dentistry, 33*(4), 408–412.

Moorthy, A., Hogg, C. H., Dowling, A. H., Grufferty, B. F., Benetti, A. R., & Fleming, G. J. (2012). Cuspal deflection and microleakage in premolar teeth restored with bulk-fill flowable resin-based composite base materials. *Journal of Dentistry, 40*(6), 500–505.

Moszner, N., Fischer, U. K., Ganster, B., Liska, R., & Rheinberger, V. (2008). Benzoyl germanium derivatives as novel visible light photoinitiators for dental materials. *Dental Materials, 24*(7), 901–907.

Nakamura, S., (2015). Background story of the invention of efficient InGaN blue-light-emitting diodes (Nobel Lecture). *Angewandte Chemie 54*(1), 7770–7788

Neumann, M. G., Schmitt, C. C., Ferreira, G. C., & Correa, I. C. (2006). The initiating radical yields and the efficiency of polymerization for various dental photoinitiators excited by different light curing units. *Dental Materials, 22*(6), 576–584.

Nocca, G., De Palma, F., Minucci, A., De Sole, P., Martorana, G. E., Calla, C., ... Lupi, A. (2007). Alterations of energy metabolism and glutathione levels of HL-60 cells induced by methacrylates present in composite resins. *Journal of Dentistry*, *35*(3), 187–194.

Opdam, N. J., Roeters, J. J., Joosten, M., & Veeke, O. (2002). Porosities and voids in Class I restorations placed by six operators using a packable or syringable composite. *Dental Materials*, *18*(1), 58–63.

Padovani, G. C., Feitosa, V. P., Sauro, S., Tay, F. R., Durán, G., Paula, A. J., & Durán, N. (2015). Advances in dental materials through nanotechnology: Facts, perspectives and toxicological aspects. *Trends in Biotechnology*, *33*(11), 621–636.

Peutzfeldt, A. (1997). Resin composites in dentistry: The monomer systems. *European Journal of Oral Sciences*, *105*(2), 97–116.

Pires, J. A., Cvitko, E., Denehy, G. E., & Swift, E. J., Jr. (1993). Effects of curing tip distance on light intensity and composite resin microhardness. *Quintessence International*, *24*(7), 517–521.

Price, R. B., Derand, T., Sedarous, M., Andreou, P., & Loney, R. W. (2000). Effect of distance on the power density from two light guides. *Journal of Esthetic Dentistry*, *12*(6), 320–327.

Rode, K. M., Kawano, Y., & Turbino, M. L. (2007). Evaluation of curing light distance on resin composite microhardness and polymerization. *Operative Dentistry*, *32*(6), 571–578.

Roggendorf, M. J., Kramer, N., Appelt, A., Naumann, M., & Frankenberger, R. (2011). Marginal quality of flowable 4-mm base vs. conventionally layered resin composite. *Journal of Dentistry*, *39*(10), 643–647.

Rosa de Lacerda, L., Bossardi, M., Silveira Mitterhofer, W. J., Galbiatti de Carvalho, F., Carlo, H. L., Piva, E., & Münchow, E. A. (2019). New generation bulk-fill resin composites: Effects on mechanical strength and fracture reliability. *Journal of the Mechanical Behavior of Biomedical Materials*, *96*, 214–218.

Rueggeberg, F. (1999). Contemporary issues in photocuring. *Compendium of Continuing Education in Dentistry Supplement*, (25), S4–15; quiz S73.

Rueggeberg, F. A. (2011). State-of-the-art: Dental photocuring - a review. *Dental Materials*, *27*(1), 39–52.

Ruyter, I. E., & Oysaed, H. (1982). Conversion in different depths of ultraviolet and visible light activated composite materials. *Acta Odontologica Scandinavica*, *40*(3), 179–192.

Ruyter, I. E., & Oysaed, H. (1987). Composites for use in posterior teeth: Composition and conversion. *Journal of Biomedical Materials Research*, *21*(1), 11–23.

Santini, A., Gallegos, I. T., & Felix, C. M. (2013). Photoinitiators in dentistry: A review. *Primary Dental Journal*, *2*(4), 30–33.

Schmitt, W., Purrmann, R., Jochum, P., & Hubner, H. -J. (1981). Use of silicic acid pellets as fillers for dental materials. Patent WO 81/01366-PCT/EP 80/00135.

Schneider, L. F., Cavalcante, L. M., Consani, S., & Ferracane, J. L. (2009). Effect of co-initiator ratio on the polymer properties of experimental resin composites formulated with camphorquinone and phenyl-propanedione. *Dental Materials*, *25*(3), 369–375.

Schneider, L. F., Pfeifer, C. S., Consani, S., Prahl, S. A., & Ferracane, J. L. (2008). Influence of photoinitiator type on the rate of polymerization, degree of conversion, hardness and yellowing of dental resin composites. *Dental Materials*, *24*(9), 1169–1177.

Schulze, S., & Vogel, H. (1998). Aspects of the safe storage of acrylic monomers: Kinetics of the oxygen consumption. *Chemical Engineering & Technology*, *21*(10), 829–837.

Shortall, A. C., Palin, W. M., & Burtscher, P. (2008). Refractive index mismatch and monomer reactivity influence composite curing depth. *Journal of Dental Research*, *87*(1), 84–88.

Shortall, A. C., Price, R. B., MacKenzie, L., & Burke, F. J. (2016). Guidelines for the selection, use, and maintenance of LED light-curing units - Part 1. *British Dental Journal*, *221*(8), 453–460.

Stansbury, J. W. (2000). Curing dental resins and composites by photopolymerization. *Journal of Esthetic Dentistry*, *12*(6), 300–308.

Stansbury, J. W., Trujillo-Lemon, M., Lu, H., Ding, X., Lin, Y., & Ge, J. (2005). Conversion-dependent shrinkage stress and strain in dental resins and composites. *Dental Materials*, *21*(1), 56–67.

Sun, G. J., & Chae, K. H. (2000). Properties of 2,3-butanedione and 1-phenyl-1,2-propanedione as new photosensitizers for visible light cured dental resin composites. *Polymer*, *41*(16), 6205–6212.

Thome, T., Steagall, W., Jr., Tachibana, A., Braga, S. R., & Turbino, M. L. (2007). Influence of the distance of the curing light source and composite shade on hardness of two composites. *Journal of Applied Oral Science*, *15*(6), 486–491.

Uhl, A., Sigusch, B. W., & Jandt, K. D. (2004). Second generation LEDs for the polymerization of oral biomaterials. *Dental Materials*, *20*(1), 80–87.

Vankerckhoven, H., Lambrechts, P., van Beylen, M., & Vanherle, G. (1981). Characterization of composite resins by NMR and TEM. *Journal of Dental Research*, *60*(12), 1957–1965.

Versluis, A., Douglas, W. H., Cross, M., & Sakaguchi, R. L. (1996). Does an incremental filling technique reduce polymerization shrinkage stresses? *Journal of Dental Research*, *75*(3), 871–878.

Xia, W. Z., & Cook, W. D. (2003). Exotherm control in the thermal polymerization of nona-ethylene glycol dimethacrylate (NEGDM) using a dual radical initiator system. *Polymer*, *44*(1), 79–88.

Yap, A. U., Tan, C. H., & Chung, S. M. (2004). Wear behavior of new composite restoratives. *Operative Dentistry*, *29*(3), 269–274.

5 Classifications and Applications of Biocomposite Materials in Various Biomedical Fields

*N. Bano,[1] S. S. Jikan,[2] H. Basri,[2] S. Adzila,[3]
N. A. Badarulzaman,[3] N. N. Ruslan,[3]
S. Abdullah,[2] and S. H. M. Suhaimy[2]*

[1]Department of Chemistry, Government Postgraduate College for Women, Raiwind, Lahore, Pakistan

[2]Faculty of Applied Sciences and Technology, Universiti Tun Hussein Onn Malaysia, Pagoh Educational Hub, Pagoh, Johor, Malaysia

[3]Faculty of Mechanical and Manufacturing Engineering, Universiti Tun Hussein Onn Malaysia, Parit Raja, Johor, Malaysia

CONTENTS

5.1 INTRODUCTION

In materials sciences, one of the most significant research areas is in medical applications. Examples of products used in medical fields include screws in orthopedics, orthodontic wires, sutures, pacemakers, breast implants, catheters, fixation plates for fractures, heart valves, dental, and nails filling materials in addition to replacement of total joint prostheses. Based on previous studies (Dorozhkin, 2011a; Khan et al., 2017), active lifestyle, sport activities, and long life duration have led to a surge of bone fractures and hard tissues-associated diseases. These bone fractures and hard tissues-associated diseases were successfully cured by grafting. It is essential to identify all implantable items from a distinct class of materials, also called biomaterials or biomedical materials, in order to be recognized by the living body (Dorozhkin, 2011b).

Biomaterials are broadly used in the form of therapeutic devices (artificial hearts, pacemakers, biosensors, and blood tubes) and implants (dental implants, vascular grafts, ligaments, intraocular lenses, heart valves, sutures, bone plates, and joint replacements). In order to recover the eminence of a patient's life, biomaterials are used to substitute, restore, and refurbish the function of degenerated and traumatized body tissues or any organs in the human body. Moreover, biomaterials are also helpful in recovering abnormalities, supporting healing, and improving the quality of a patient's life (Best et al., 2008; Tripathi et al., 2010).

This chapter discusses the definition and evolution of biomaterials, and then presents the introduction of classes of biomaterials such as metals and alloys, bioceramics, and biopolymers, and also discusses their basic properties and applications with examples. The development in biopolymer, metal and alloy, and bioceramic materials as matrices with the incorporation of either organic or inorganic fillers leads to formation of new biocomposites. These biocomposites are very beneficial in a wide range of biomedical applications, especially in the fields of orthopedic and dental surgeries. Biocomposite materials have outstanding biocompatibility, biodegradability, and greater mechanical properties. In addition, this review states the definition of biocomposites and summarizes the recent work on the development of biocomposite materials containing either biopolymer, bioceramic, or metal and alloy as the matrix with different bioactive fillers suitable for use in various biomedical fields. This chapter describes thoroughly the types of biocomposites based on reinforcement systems, numerous classifications, and applications in biomedical fields.

5.2 BIOMATERIALS

A large number of definitions have been established for the term "biomaterials". It has varying definitions encompassing substance to repair, replace, and regenerate organs (in terms of structure and functions). Biomaterial is "a nonviable material used in a medical device, intended to interact with biological systems" according to Black and Hastings (1998). In another definition, "Biomaterials are defined as synthetic or natural materials to be used to replace parts of a living system or to function in intimate contact with living tissues" (David, 1999). In addition, biomaterial is officially defined by The Clemson University Advisory Board for Biomaterials as "a systematically and pharmacologically inert substance designed for implantation within or incorporation with living systems" (Park & Lakes, 2007). An innovative description was presented in September 2009 that "A biomaterial is a substance that has been engineered to take a form which, alone or as part of a complex system, is used to direct, by control of interactions with components of living systems, the course of any therapeutic or diagnostic procedure, in human or veterinary medicine" (David, 2009). Material that has very close contact with a living system and is used to replace or repair the living system is generally considered as biomaterial (Mustafa, 2012). According to Migonney, the latest biomaterial definition is, "Material intended to supply/replace all or a part of a deficient organ". Conferring to the above-mentioned definitions, it is concluded that a biomaterial is "a material that is used for repair and reconstruction of loss" (Migonney, 2014).

The first recognized biomaterial was used during the Egyptian times, around 200 A.D., whereby the suture material was made from linen. It was not the result of heavy injuries in World War II that considerably drove the growth of biomaterials (Wong et al., 2012). In recent years, biomaterials for bone implants have gained much attention in the field of science due to a large number of rail and road accidents, new trauma and tumors cases, and decay of teeth due to poor diet and stress (Stefanut et al., 2015). The US Food and Drug Administration (FDA) sanctioned the biomaterials used for *in vivo* and *in vitro* studies due to their biocompatibility (Quezon, 2013). The biomaterials chosen can be shaped and functioned as extracellular matrix components, porous scaffolds, or carriers of cells, and bioactive agents that are common practice in repairing or replacing of hard tissues.

In general, biomaterials are limited to the field of health, which have an interface with tissues or tissue components. Nevertheless, any artificial materials that are in contact with skin such as wearable artificial limbs and hearing aids are not defined as biomaterials. The skin acts as a defensive barrier between the body and the outside world. Thus, the essential identification of biomaterials is based on the recognition by living organs or tissues of those used for replacement of organs or tissues (Jandt, 2007). A few investigations have been carried out in the last 30 years in order to understand the compatibility between the materials and the body tissues for the development of these biomaterials (de Moraes Porto, 2012).

5.2.1 BIOMATERIALS: EVOLUTION

Biomaterial applications have evolved through three generations. Upon development, these biomaterial generations vary based on complexity and their capability to interact with or modify the biological environment in the human body (Jandt, 2007). These biomaterials are available in the natural environment of the human being and used merely for restoration of organs (Ratner et al., 2013). The first generation begins in the 1950s until the 1960s to achieve a suitable combination of functional properties that could adequately match those of the replaced tissue without deleterious response by the host. The desirable combination of physical properties is specific to the intended clinical use because they were bioinert and considered as biocompatible. For example, artificial tissues being fabricated by placing cells within scaffold materials help in guiding the cell proliferation and differentiation (Ramalingam et al., 2013).

The second generation of biomaterials was intended in the 1980s to stimulate a controlled reaction with the tissues into which these biomaterials were implanted in order to induce a desired therapeutic effect. The bioactive materials such as glasses and ceramics in orthopedic and dental surgeries were used in controlled, local-ized drug-release applications (e.g., drug-eluting endovascular stents that have been shown to limit restenosis [blood vessel closure] following balloon angioplasty). The second generation of biomaterials also included the development of resorbable bio-materials, with rates of degradation that could be tailored to the requirements of a desired application. Thus, the discrete interface between the implant site and the host tissue could be eliminated in the long term. This is due to the fact that the foreign biomaterial is soluble and would ultimately be degraded as nontoxic products by the host (Narayan, 2010; Ramalingam et al., 2013).

The third generation of biomaterials focuses on stimulating the regeneration of functional tissue of damaged bladders, trachea, skin, corneal epithelium, and cartilage in humans. These types of biomaterials play an important role in the rapidly developing field of tissue engineering and regenerative therapeutics. These biomaterials act as a set of tools at the interface of the biomedical and engineering sciences that use living cells to aid tissue formation or regeneration, and thereby produce therapeutic or diagnostic benefit. For example, cells are seeded on a scaffold composed of synthetic polymer or natural materials such as bioresorb-able polymers in desired geometry of engineered tissue that matured *in vitro* and adhered to cells. The construct is implanted in a suitable anatomical site as pros-thesis (Bose et al., 2013).

5.2.2 CLASSES OF BIOMATERIALS

There are three major classes of biomaterials, namely (1) bioceramics, (2) bio-polymers, and (3) metals and alloy. The classifications are shown in Figure 5.1. The redevelopment of human tissues and replacement of human bones are very significant goals in the research of biomedical fields (Davis, 2003; Patel & Gohil, 2012).

5.2.3 METALS AND ALLOYS

Metals and alloys are used as biomaterials due to their excellent mechanical, elec-trical, and thermal properties (Binyamin et al., 2006). Metals are strong and mal-leable and have high wear resistance, which makes them suitable to be used for load-bearing implants such as knee or hip prostheses without causing large distor-tions and permanent dimensional variations. Metals and alloys are generally used in a variety of medical applications in order to substitute or fix dentin and bone mate-rials in maxillofacial implants and in cardiovascular surgeries. In addition, metals are used for an extensive range of therapeutic device applications such as for dental applications (dental root implants and braces), joint replacement, and fracture fixa-tion (screws, needles, wires, staples, stents, bone plates, and electrodes). However,

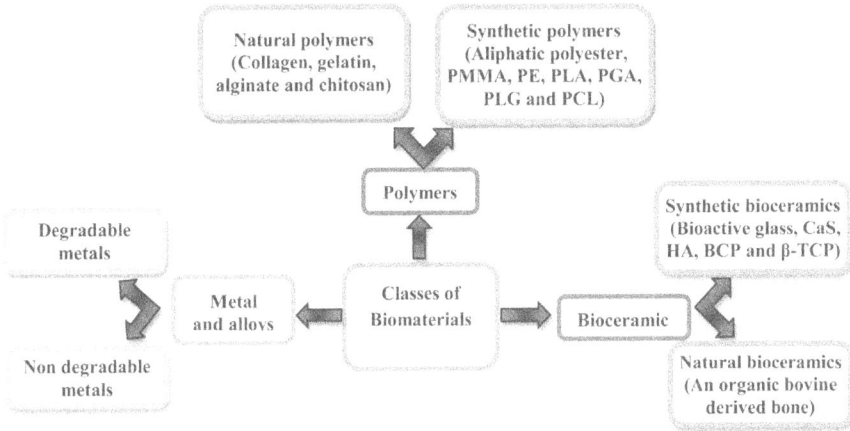

FIGURE 5.1 Classes of biomaterials.

metals are denser, corrosive, and difficult to be shaped as compared to other classes of materials (Kareem, 2018; Marti, 2000).

Various forms of metals have been used as implants. The first metal manufactured specifically for human uses, for making screws and bone fracture plates, was Sherman vanadium steel. Most of the metals such as nickel (Ni), iron (Fe), cobalt (Co), chromium (Cr), titanium (Ti), tantalum (Ta), niobium (Nb), tungsten (W), and molybdenum (Mo) are used to make alloys in order to develop implants. These metals can be accepted by the body in a small quantity. Occasionally, the naturally occurring metallic elements are significant in red blood cell functions (Fe), cross-linking of elastin in the aorta (Cu), and biosynthesis of a vitamin B_{12} (Co); however, B_{12} in enormous quantities cannot be tolerated in the body (Black & Hastings, 1998). The implant metals can corrode in the body environment, so the biocompatibility is of great concern. The corrosion leads to damage of the biomaterial that will deteriorate the graft. Eventually, this corroded product discharges into tissue, causing unwanted health effects (Park & Lakes, 2007). Generally, there are three major material clusters that lead to biomedical metals: titanium-based alloys, cobalt–chromium–molybdenum alloy, and stainless steel. These metal and alloys are used in various biomedical applications as shown in Table 5.1.

Other metals and alloys consist of nitinol (based on nickel and titanium, called shape-memory alloys [SMAs]), commercially pure titanium (Ti-Cp), and tantalum. Successively, outstanding biocompatibility (nontoxic), thermal and electrical conductivity, high wear resistance, high corrosion resistance, osseointegration (for bone prosthetics), appropriate mechanical properties, and reasonable cost are the foremost traits for choosing metals and alloys for biomedical utilizations for an extended period of time without any body rejection (Catledge et al., 2002). For any operation, it is essential to recognize the chemical and physical assets of the diverse metallic materials along with their relations with the human body host tissues (Chen & Thouas, 2015).

TABLE 5.1

Applications of Metals and Alloys as Biomaterials in Biomedical Fields (Chen & Thouas, 2015; Park & Lakes, 2007)

Types of Metallic Material	Applications
Titanium and its alloys	Heart pacemaker, artificial heart valves, dental implants, fragments for orthodontic operation and dental prosthetics, bone and joint replacement (knee, hip, shoulder, spine, elbow, and wrist), components in high-speed blood centrifuges, bone fixation materials (nails, screws, nuts, and plates), and surgical instruments for heart and eye surgery
Cobalt–chromium–molybdenum alloy	Orthopedic prosthesis, screws, dental root implant, mini plates, bone plate, pacer, suture, medical tools, and joint, bone, and joint replacements (knee and hip)
Stainless steel	Bone plate for fracture fixation, heart valve, stents, hip nails, wires, dental implant for tooth fixation, dental root implant, spinal instruments, joint replacements (hip, knee), surgical instruments, screws, pacer, fracture plates, and shoulder prosthesis
Nitinol	Stents, orthodontic wires, medical staples, shape memory plates, and guide wires
Tantalum	Suture wires, pins for the ligation of vessels, staples, pliable sheets, and plates

5.2.4 BIOCERAMICS

Ceramics are well defined as "the art and science of making articles using solid inorganic nonmetallic as their essential component materials" (Zakaria et al., 2013). Inorganic polycrystalline ancient ceramic materials are produced artificially, cast off in the form of earthenware for thousands of years. These ancient ceramics have superior compressive strength and biological inertness that make them suitable for biomedical applications. They include carbides, metallic oxides, silicates, selenides, sulfides, and various refractory hydrides. Oxides such as magnesium oxide (MgO), silicon dioxide (SiO_2), aluminum oxide (Al_2O_3), and zirconium dioxide (ZrO_2) contain nonmetallic and metallic elements, and ionic salts such as cesium chloride (CsCl), zinc sulfide (ZnS), and sodium chloride (NaCl). Relative electronegativity between positive and negative ions and the radius ratio of ceramic materials are two primary aspects that influence the structure and property of ceramic materials (Park & Lakes, 2007). It has been reported that the plaster of Paris was the first ceramic material used in clinical applications (Dorozhkin, 2011a).

The common application of ceramics as implantable biomaterials was limited because of their low impact strength, low ultimate strength, and inherent brittleness. However, the latest high-tech ceramics as biomaterials have drawn augmented

	Bioinert	Al_2O_3, ZrO_2 and Pyrolytic carbonAl_2O_3, ZrO_2 and Pyrolytic carbon
Bioceramis	Bioactive	Bioglass, glass ceramics, hydroxyapatite, and $Ca_3(PO_4)_2$
	Bioresorbable	Tricalcium phosphate implant (TCP), $Ca_3(PO_4)_2$

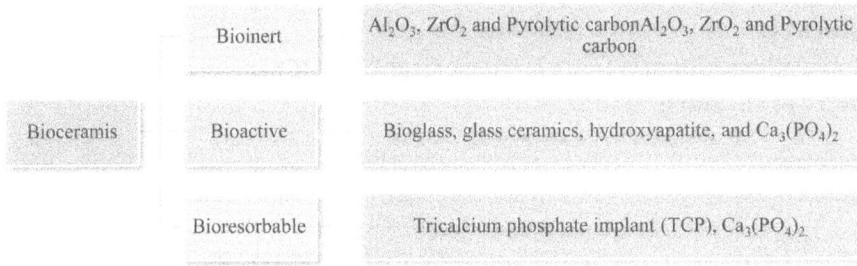

FIGURE 5.2 Types of bioceramics according to their bioactivity.

attraction owing to their high compressive strength and their relative bioinertness (Hench & Jones, 2005). These implantable ceramics are called bioceramics. Bioceramics are one of the classes of biomaterials to be designed for the replacement of fragments and defects in living tissues (Piconi et al., 2003). As a result, bioceramics have been utilized for restoration and implant, besides refurbishment of diseased or smashed body fragments and tissues. Based on the type of the atomic bindings, bioceramic materials are categorized into two groups: ionic and covalent bonds. Carbonaceous structural materials are covalently bonded, while ionic materials are generally oxide-based materials. Regardless of the presence of an extensive collection of ceramics, the selection of materials utilized in biomedical procedures is limited, and used in medical applications (Kapoor, 2017). One significant recent study (Hans & Lowman, 2002; Migonney, 2014) revealed that bioceramics can be categorized into three groups according to their bioactivity: (1) bioinert, for example, alumina (Al_2O_3), zirconia (ZrO_2), and pyrolytic carbon; (2) bioactive, for example, bioglass, glass ceramics, hydroxyapatite, and $Ca_3(PO_4)_2$; and (3) bioresorbable, for example, tricalcium phosphate implant (TCP) and $Ca_3(PO_4)_2$ (Figure 5.2).

Nowadays, bioceramic materials gain lots of attraction for some applications in science, medical, and other areas due to their certain desirable characteristics, that is, artistic appearance, biocompatibility, high compressive strength and moduli, inertness to body fluids, superior hardness, wear resistance, and high degradation resistance in corrosive atmospheres in contrast to metals. These properties make bioceramics suitable to restore or substitute rigid tissues such as bone and teeth (Migonney, 2014). Their applications are generally to repair hard tissues such as bones, joints, or teeth (Shastri, 2003). In addition, bioceramics used for different applications should be nonallergic, noninflammatory, nontoxic, noncarcinogenic, biocompatible, and biofunctional during their lifetime in the human host (Wong et al., 2012). Table 5.2 lists applications of bioceramics in biomedical fields.

5.2.5 BIOPOLYMERS

Polymers are made of small repeating units called monomers that are bonded together by covalent bonds to form long chain molecules. Polymers may be organized into linear, branched, and network arrangements, contingent on the repeating units' functionality (Sheikh et al., 2015). Polymers that are found in or synthesized

TABLE 5.2

Applications of Bioceramics in Biomedical Fields (Cheng et al., 2015; Derry et al., 2003)

Bioceramics	Applications in Biomedical Field
Zirconia	Repair for periodontal disease; replacement for teeth, hips, knees, tendons, and ligaments; bone grouts after tumor operation
Pyrolytic carbon	Heart valves, permanently implanted artificial limbs, end osseous tooth replacement implants
Bioglass–ceramics	Dental implants, artificial joint replacement, spinal fusion, wires, intramedullary nails, heart valves, bone plates, middle ear implants, screws, and tooth implants
Calcium phosphates	Skin treatments; facial surgery; jawbone reconstruction; orthopedics; dental implants; ear, nose, and throat repair

by living organisms are called biopolymers. Biopolymers gain broad spectrum consideration in the medical and research fields due to their multidirectional properties (Ratner et al., 2013). These types of polymers are synthesized from renewable resources and can be used to form biopolymers via a polymerization method (Wang, 2003). A biopolymer is a class of biomaterial that is made up of long chains of atoms connected by directional, covalent bonds. A monomer is the building block of a biopolymer, which is a repeated unit. Examples of biopolymers originating from renewable resources are corn-derived plastics such as polylactic acid (PLA), starch plastics (starch esters), and cellulosic plastics (cellulose acetate) (Mano et al., 2004).

Consequently, the other classes of biomaterials such as ceramics, metals, and alloys can be substituted by biopolymeric materials because of their distinctive usefulness as biomaterials. Moreover, in 2003, it has been reported that 88% of the total biomaterial market is the trades of polymeric biomaterials that surpassed $7 billion for that year. It has also been recorded that the market of biocompatible polymeric biomaterials gained $11.9 billion in 2008 (Quezon, 2013) and is projected to reach $24 billion by 2024 (Pulidindi & Chakraborty, 2017).

In 1936, polymethyl methacrylate (PMMA) was the first nondegradable polymer used in dental application, and since then, bioabsorbable polymers have had a great influence on health care. These biopolymers have been utilized in many studies and therapeutic fields, for instance drug delivery systems and fixation plates in addition to screws (Subramaniam & Sethuraman, 2014). Currently, biodegradable polymeric biomaterials such as PLA and their copolymers, poly(dioxane) (PDS), poly(trimethylene carbonate), poly(ε-caprolactone), and polyglycolic acid have been sanctioned by the FDA (Subramaniam & Sethuraman, 2014). However, in this biodegradable polymer series, PLA is a much more considerable and preferred polymer for biomedical procedures owing to its biocompatibility, biodegradability, and nontoxicity (Bohner, 2000).

Generally, there are two types of polymer: (1) natural polymers that are produced from renewable resources, such as plant, animal, and microorganisms, for example, DNA, proteins, collagen, chitosan, and starch; (2) synthetic polymers that are

| Polymer | Biodegradable | PLA, PCL, PGA, PHEMA, PLGA, PHB and PHBV |
| | Non-degradable | PVC, PET, PE, PSU, PP, PMMA, POM and PEEK |

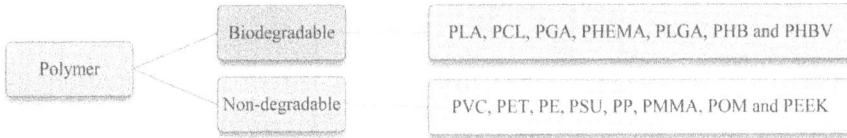

FIGURE 5.3 Types of polymer.

manufactured from petrochemicals such as PLA, polyvinyl chloride (PVC), polyethylene (PE), polypropylene (PP), and PMMA (Hans & Lowman, 2002; Mustafa, 2012; Park & Lakes, 2007). In the biomedical and research fields, synthetic polymeric materials have received significant attention because of their non-immunogenic responses, easy production processes, well measured, and reproducible molecular structures, as well as they have been broadly used in many biomedical applications. These applications comprise prosthetic materials, medical disposables, dental materials, dressings, implantable devices, injectable drug delivery, coatings on devices, vascular grafts, imaging systems, catheters and tubing, and extracorporeal devices (Mano et al., 2004; Subramaniam & Sethuraman, 2014; Wang, 2003). They are easily molded into many shapes such as films, rods, fibers, textiles, and gelatinous liquids (Subramaniam & Sethuraman, 2014). Synthetic and natural polymeric biomaterials, which are extensively used in biomaterial research, can be divided into two types. They are biodegradable and nondegradable biopolymers as shown in Figure 5.3.

Nonbiodegradable polymers are durable and biostable, so they do not break and change when used *in vivo*. These polymers are used in an extensive range of biomedical applications in soft tissue replacement, such as vessel wall cartilage, tendon, lens, and skin (Migonney, 2014). In fact, in bone tissue engineering the most significant synthetic nondegradable polymers are polyvinyl chloride (PVC), polyethylene (PE), polyethylene terephtalate (PET), polysulfone (PSU), polyamides (PAs), polypropylene (PP), polyurethanes (PU), polytetrafluoroethylene (PTFE), polyacetal resin, PMMA, polycarbonate (PC), polyoxymethylene (POM), and polyetheretherketone (PEEK) (Alonso & Sanchez, 2003). Table 5.3 presents the applications of nondegradable polymers in biomedical fields.

On the other hand, biodegradable polymers are types of polymers that contain functional groups that may be cleaved into fragments of polymeric chains and degrade within the body into nontoxic chemicals either hydrolytically or enzymatically (Migonney, 2014; Quezon, 2013). Biodegradable polymers either synthetically or biologically derived have been used in biomaterial research and medical fields. Polyglycolic acid was the first synthetic biodegradable polymer manufactured in the early 1970s, whereas polylactide was the next being introduced in 1985 (Nair & Laurencin, 2007). Synthetic biodegradable polymers are PLA, polyglycolide (PGA), poly(2-hydroxyethyl methacrylate) (PHEMA), poly ε-caprolactone (PCL), poly(lactide-co-glycolide) (PLGA), polyhydroxyl butyrate (PHB), and poly(3-hydroxybutyrate-co-3-hydroxyvalerate) (PHBV). On the other hand, chitosan, collagen, glycosaminoglycans, and hyaluronic acid are the natural biodegradable polymers. It has been reported that chitin and chitosan have been used for sustained-release drug delivery systems and wound bandages (Lv et al., 2009).

TABLE 5.3

Applications of Nondegradable Polymers in Biomedical Fields (Kato et al., 2003; Park & Lakes, 2007; Piconi et al., 2003)

Types of Polymer	Applications
Polytetrafluoroethylene (PTFE, Teflon®, Gore-Tex®)	Vascular grafts, introducers, catheters
Poly(ethylene terephthalate) (polyester, Ethibond, Dacron®)	Vascular graft, non-resorbable sutures, ligaments, drug delivery
Poly(etheretherketone) (PEEK)	Orthopedic applications
Poly(methyl methacrylate) (PMMA)	Bone cement, intraocular lens
Polycarbonate	Renal dialysis cartridge, trocars, heart–lung machine, tubing interconnectors
Silicone rubber, polydimethylsiloxane (PDMS)	Catheters, drainage tubes, introducer tips, flexible sheaths, feeding tubes, gas exchange membranes
Polyurethane	Catheters, tubing, wound dressing, heart valves, artificial hearts
Hydrogels (poly[ethylene glycol], poly[ethylene oxide], poly[vinyl alcohol], etc.)	Drug delivery, adhesion prevention, wound healing, hemostasis, contact lenses, reconstruction, extracellular matrices
Polypropylene (i.e., prolene)	Nonresorbable sutures, hernia mesh
Polyamides (nylon)	Nonresorbable sutures, catheter balloons in angioplasty procedure

However, only a select few of the synthetic biodegradable polymers, such as PLA, PLGA, PHBV, and PCL, have the high mechanical properties, biodegradability, controlled design, synthesis, and degradation rates that meet the requirements for clinical applications in bone repair and tissue regeneration. This is due to the fact that they improve bone remodeling by transferring mechanical load to the surrounding tissues. They can be totally degraded in the body and can be removed by normal metabolic routes.

Natural polymeric biomaterials that are under the category of biodegradable polymers are used for low-load-bearing applications such as the repair of small bone fractures. It is considered that in the future, biostable polymers may be substituted by these biodegradable polymers due to increased improvements in the area of degradable polymeric biomaterials (Mustafa, 2012). Moreover, the major comprehensive investigations and studies in the orthopedics fields were on synthetic biodegradable polymers relating to the class of poly(α-hydroxy acids). They are also known as polyesters, for example, PGA, PLA, and their copolymer PLGA (de Moraes Porto, 2012; Subramaniam & Sethuraman, 2014). The biological, mechanical, chemical, and physical properties of degradable biopolymers used in different medical applications change with time and must have tissue compatibility. However, biodegradable polymeric materials are normally used in a short period of time when the fracture or defect has healed such as (1) small implants

TABLE 5.4

Applications of Biodegradable Polymers in Biomedical Fields (Park & Lakes, 2007)

Types of Polymer	Applications
Polyglycolide (PGA)	Drug delivery systems and as sutures
Polylactide (PLA), their copolymers (PLGA)	Drug delivery vehicles, screws or pins in orthopedic uses, skin replacement materials, and suture reinforcements
Poly(ε-caprolactone) (PCL)	Scaffolds, staples, and drug delivery systems
Poly(propylene fumarate) (PPF)	Filling materials in bone defects and as injectable materials in tissue engineering
Poly(3-hydroxybutyrate) (PHB)	Devices in tissue engineering of nerves, cartilage, bone, tendon, and skin

like resorbable sutures, staples and nano- or micro-sized drug delivery devices and degradable scaffolds; (2) large fracture fixation implants like bone screws, bone plates, and contraceptive reservoirs; (3) porous structures or multifilament meshes for tissue engineering; and (4) plain membranes for tissue regeneration (Migonney, 2014; Nair & Laurencin, 2007). Table 5.4 shows the applications of biodegradable polymers in biomedical fields.

5.3 BIOCOMPOSITES

Biocomposites contain two or more different phases of materials in which one phase is a continuous phase called matrix (either biopolymer, metal and alloy, or bioceramic), whereas the other is the discontinuous phase. It is the incorporation of fillers in a biocompatible matrix material. Due to the outstanding properties of bioactive compounds, these biocomposites are broadly studied in order to accomplish osteointegration, the direct bonding of implant materials to the bone tissue (Jonge et al., 2008). For example, the biocomposite implant material can be designed by the reinforcement of the biofillers into the polymer matrix producing a polymer biocomposite. Other examples of natural biocomposites are bone and tooth. These biocomposites consist of an organic matrix and minerals, mainly hydroxyapatite, HAP: $Ca_{10}(PO_4)_6(OH)_2$.

5.3.1 TYPES OF BIOCOMPOSITES

The types of biocomposites can be categorized according to their reinforcement systems as shown in Figure 5.4. The categories are based on either short fibers, continuous fibers, or particulate (powder) forms of the fillers. All of these reinforcement systems have been used in the development of biocomposites for biomedical applications, such as screws and total hip replacement stems made from short fiber reinforcements, orthopedic bone plates fabricated using unidirectional (UD) laminate

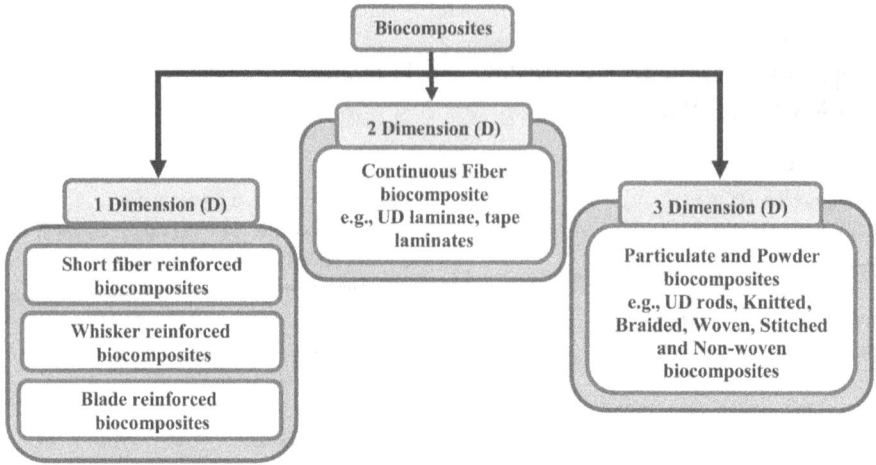

FIGURE 5.4   Types of biocomposites based on the reinforcement system.

or multidirectional tape laminates, and powder reinforced dental biocomposites (Ramakrishna et al., 2004).

5.3.2 PROPERTIES OF BIOCOMPOSITES

Biodegradation, bioactivity, porosity, mechanical, and biocompatibility are the main properties of biocomposite materials required for the replacement of bone and bone defects (Liu, 1997). The most important property of biocomposite that makes it superior to any other class of materials is its persistence in the biological environment without being damaged and while producing no harmful effects with the surroundings (Dorozhkin, 2011b). Figure 5.5 represents the ideal properties of biocomposite materials used in biomedical applications.

5.3.3 APPLICATIONS OF BIOCOMPOSITES

It is well documented that biocomposite materials have been commonly applied in numerous therapeutic applications, for examples, orthopedic (Cross et al., 2016), mandibular defects (Cho et al., 2017), cardiovascular (Goyal, 2016), occlusion devices for cardiac defects (Hendow et al., 2016), dental (Cho et al., 2017), urological (Goyal, 2016), gastrointestinal (Hendow et al., 2016), wound healing (Baino, 2011), ophthalmology (Kim & Evans, 2012), orbital implants (Boateng et al., 2008), orbital floor repair, plastic surgery, drug delivery, and tissue engineering (Singh, 2017). The biocomposites' performances in the human body can be explained through their types used in human body. The identification of problems is crucial in considering whether the problems are occurring at any organ or body system, as shown in Table 5.5. The role of biocomposite materials is governed by the interaction between the material and the body, specifically the effect of the

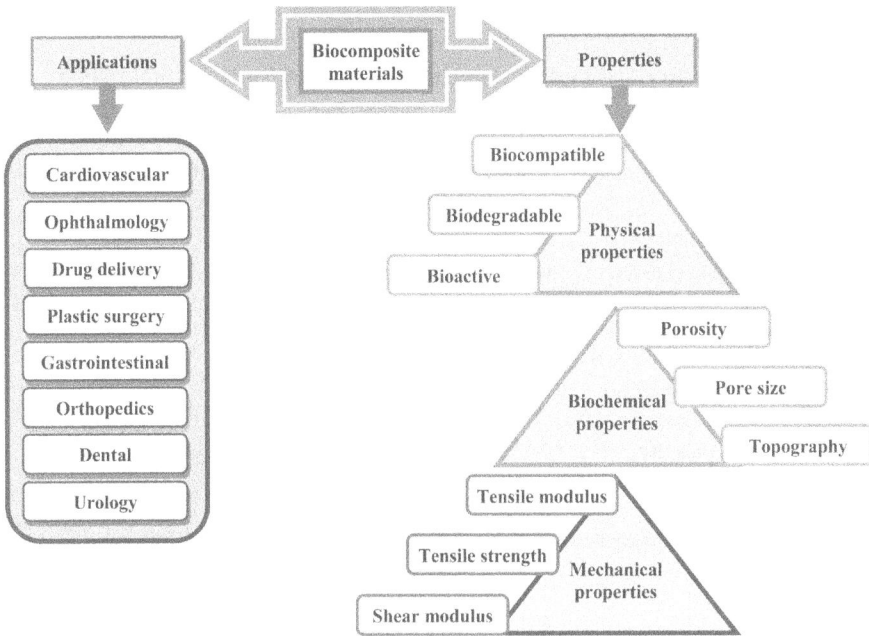

FIGURE 5.5   Properties and applications of biocomposite materials used as biomedical products.

biocomposites on the body and vice versa, the effect of the body environment on the biocomposites (Wong et al., 2012).

5.4   CONCLUSION

This chapter summarizes recent literature studies of different classes and potentials of biomaterials and biocomposite applications in numerous biomedical fields, which fascinate the attention of biomaterials' researchers and consultants. Extensive advances in biomaterials have been made since the early days of wooden teeth and gold dental implants, yet there is still significant development ahead. In the past, implementation of industrial biocomposites for biological applications was never proposed. Biocomposite materials have been established considerably over the last era. However, nowadays, biocomposites are being advanced to exactly interrelate through living tissue in body tissue restoration, regeneration, as well as replacement. Bioceramics and their composites are used as biological grafts due to their biocompatibility. Synthetic and natural biopolymers and their composites have been extensively utilized in biomedical fields due to their significant biocompatibility and biodegradability. In addition, metals and alloys are used as biomaterials and biocomposites because of their excellent mechanical, electrical, and thermal properties. In this chapter, an attempt has been made to increase the understanding on the utilization of biomaterials and biocomposites in therapeutic and biomedical fields.

TABLE 5.5

Biocomposite Materials' Performance in the Human Body (Ramakrishna et al., 2004)

Uses of Biocomposite Materials		Biocomposite Materials in Organs		Biocomposite Materials in Body Systems	
Problem Area	Examples	Organ	Examples	System	Examples
Replacement of diseased or damaged part	Kidney dialysis machine and artificial hip joint	Heart	Cardiac pacemaker, artificial heart valve, total artificial heart	Skeletal	Bone plate and total joint replacements
Assist in healing	Bone plates, sutures, and screws	Lungs	Oxygenator machine	Muscular	Sutures, muscle stimulator
Improve function	Cardiac pacemaker and intraocular lens	Eye	Contact lens, intraocular lens	Respiratory	Oxygenator machine
Correct functional abnormality	Cardiac pacemaker	Ear	Artificial stapes, cochlea implant	Endocrine	Microencapsulated pancreatic islet cells
Correct cosmetic problem	Chin augmentation and mammoplasty	Bone	Kidney, bone plate, intramedullary rod	Urinary	Catheters, stent, kidney dialysis machine
Aid to diagnosis	Probes and catheters	Kidney	Kidney dialysis machine	Nervous	Hydrocephalus drain, cardiac pacemaker, nerve stimulator
Aid to treatment	Catheters and drains	Bladder	Catheter and stent	Reproductive	Cosmetic replacements and augmentation mammoplasty

ACKNOWLEDGMENT

The authors would like to acknowledge Universiti Tun Hussein Onn Malaysia and Ministry of Higher Education Malaysia for the financial support through Fundamental Research Grant Scheme (FRGS Vote K199).

REFERENCES

Alonso, M. J., and Sanchez, A. (2003). The Potential of Chitosan in Ocular Drug Delivery. *Journal of Pharmacy and Pharmacology*, 55(11), 1451–1463.
Baino, F. (2011). Biomaterials and Implants for Orbital Floor Repair. *Acta Biomaterialia*, 7(9), 3248–3266.
Best, S. M., Porter, A. E., Thian, E. S., and Huang, J. (2008). Bioceramics: Past, Present and for the Future. *Journal of the European Ceramic Society*, 28(7), 1319–1327.
Binyamin, G., Shafi, B. M., and Mery, C. M. (2006). Biomaterials: A Primer for Surgeons. *Seminars in Pediatric Surgery*, 15(4), 276–283.

Black, J., and Hastings, G. (1998). *Handbook of Biomaterial Properties.* New York: Chapman & Hall.

Boateng, J. S., Matthews, K. H., Stevens, H. N. E., and Eccleston, G. M. (2008). Wound Healing Dressings and Drug Delivery Systems: A Review. *Journal of Pharmaceutical Sciences, 97*(8), 2892–2923.

Bohner, M. (2000). Calcium Orthophosphates in Medicine: From Ceramics to Calcium Phosphate Cements. *Injury, 31*(4), 37–47.

Bose, S., Bandyopadhyay, A., and Narayan, R. (2013). *Biomaterials Science: Processing, Properties and Applications III.* New Jersey: John Wiley and Sons, Inc.

Catledge, S. A., Fries, M. D., Vohra, Y. K., Lacefield, W. R., Lemons, J. E., Woodard, S., and Venugopalanc, R. (2002). Nanostructured Ceramics for Biomedical Implants. *Journal of Nanoscience and Nanotechnology, 2*(3), 293–312.

Chen, Q., and Thouas, G. A. (2015). Metallic Implant Biomaterials. *Materials Science and Engineering R: Reports, 87,* 1–57.

Cheng, K. C., Lin, Y. H., Guo, W., Chuang, T. H., Chang, S. C., Wang, S. F., and Don, T. M. (2015). Flammability and Tensile Properties of Polylactide Nanocomposites with Short Carbon Fibers. *Journal of Materials Science, 50*(4), 1605–1612.

Cho, Y., Seol, Y., Lee, Y., Rhyu, I., Ryoo, H., and Ku, Y. (2017). An Overview of Biomaterials in Periodontology and Implant Dentistry. *Journal of Advances in Materials Science and Engineering, 2017*(2).

Cross, L. M., Thakur, A., Jalili, N. A., Detamore, M., and Gaharwar, A. K. (2016). Nanoengineered Biomaterials for Repair and Regeneration of Orthopedic Tissue Interfaces. *Acta Biomaterialia, 42,* 2–17.

Davis, J. (2003). *Handbook of Materials for Medical Devices.* Cleveland, OH: ASM International.

De Jonge, L. T., Leeuwenburgh, S. C. G., Wolke, J. G. C., and Jansen, J. A. (2008). Organic-Inorganic Surface Modifications for Titanium Implant Surfaces. *Pharmaceutical Research, 25*(10), 2357–2369.

Derry, C., Schultz, P., and Vautier, D. (2003). Biomaterials in Laryngotracheal Surgery: A Solvable Problem in the Near Future? *The Journal of Laryngology and Otology, 117*(2), 113–117.

Dorozhkin, S. V. (2011a). Medical Application of Calcium Orthophosphate Bioceramics. *Bio, 1*(1), 1–51.

Dorozhkin, S. V. (2011b). Biocomposites and Hybrid Biomaterials Based on Calcium Orthophosphates. *Biomatter, 1*(1), 3–56.

Goyal, S. K. (2016). Current Status of Nanotechnology in Urology. *International Journal of Research in Medical Sciences, 4*(8), 3114–3120.

Hans, M., and Lowman, A. (2002). Biodegradable Nanoparticles for Drug Delivery and Targeting. *Current Opinion in Solid State and Materials Science, 6*(4), 319–327.

Hench, L. L., and Jones, J. R. (2005). *Biomaterials, Artificial Organs and Tissue Engineering.* Boca Raton: CRC Press Boca, 241–248.

Hendow, E. K., Guhmann, P., Wright, B., Sofokleous, P., Parmar, N., and Day, R. M. (2016). Biomaterials for Hollow Organ Tissue Engineering. *Fibrogenesis and Tissue Repair, 9*(1), 3.

Jandt, K. D. (2007). Evolutions, Revolutions and Trends in Biomaterials Science – A Perspective. *Advanced Engineering Materials, 9*(12), 1035–1050.

Kapoor, D. (2017). Nitinol for Medical Applications: A Brief Introduction to the Properties and Processing of Nickel Titanium Shape Memory Alloys and their Use in Stents. *Johnson Matthey Technology Review,* (1), 66.

Kareem, M. M. (2018). *Composite Bone Tissue Engineering Scaffolds Produced by Coaxial Electrospinning.* University of Glasgow: PhD Thesis. Retrieved from http://theses.gla.ac.uk/30822/1/2018KareemPhD.pdf

Kato, Y., Onishi, H., and Machida, Y. (2003). Application of Chitin and Chitosan Derivatives in the Pharmaceutical Field. *Current Pharmaceutical Biotechnology, 4*(5), 303–309.

Khan, M. N., Biswas, S., Islam, M. S., Rashid, T., Sharmeen, S., Shahruzzaman, M., Islam, M., Rahman, M., Mallik, A., Asaduzzaman, A. K., Haque, P., and Rahman, M. (2017). Green biocomposites from renewable biopolymers and their biomedical applications. In *Biocomposites: Properties, Performance and Applications*. New York: Nova Science Publishers.

Kim, J. J., and Evans, G. R. D. (2012). Applications of Biomaterials in Plastic Surgery. *Clinics in Plastic Surgery, 39*(4), 359–376.

Liu, Q. (1997). *Hydroxyapatite/Polymer Composites for Bone Replacement*. University of Twente: PhD Thesis.

Lv, Q., Nair, L., and Laurencin, C. T. (2009). Fabrication, Characterization, and in Vitro Evaluation of Poly(Lactic Acid Glycolic Acid)/Nano-Hydroxyapatite Composite Microsphere-Based Scaffolds for Bone Tissue Engineering in Rotating Bioreactors. *Journal of Biomedical Materials Research - Part A, 91*(3), 679–691.

Mano, J. F., Sousa, R. A., Boesel, L. F., Neves, N. M., and Reis, R. L. (2004). Bioinert, Biodegradable and Injectable Polymeric Matrix Composites for Hard Tissue Replacement: State of the Art and Recent Developments. *Composites Science and Technology, 64*(6), 789–817.

Marti, A. (2000). Inert Bioceramics (Al_2O_3, ZrO_2) for Medical Application. *Injury, 31*, 33–36.

Migonney, V. (2014). *Biomaterials*. Hoboken, NJ: John Wiley and Sons, 10–45.

Mustafa, Z. B. (2012). *Multiaxial Fatigue Characterization of Self-Reinforced Polylactic Acid-Calcium Phosphate Composite*. University of Glasgow: PhD Thesis.

Nair, L. S., and Laurencin, C. T. (2007). Biodegradable Polymers as Biomaterials. *Progress in Polymer Science, 32*(8–9), 762–798.

Narayan, R. J. (2010). The Next Generation of Biomaterial Development. *Philosophical Transactions of the Royal Society A, 368*, 1831–1837.

Park, J., and Lakes, R. S. (2007). *Biomaterials an Introduction*. Madison: Springer, 152–155.

Patel, N., and Gohil, P. (2012). A Review on Biomaterials: Scope, Applications and Human Anatomy Significance. *International Journal of Emerging Technology and Advanced Engineering, 2*(4), 91–101.

Piconi, C., Maccauro, G., Muratori, F., and Prever, E. B. Del. (2003). Alumina and Zirconia Ceramics in Joint Replacements Alumina and Zirconia Ceramics in Joint. *Journal of Applied Biomaterials and Biomechanics, 1*(1), 19–32.

de Moraes Porto, I. C. C. (2012). *Polymer Biocompatibility*. Alagoas: Intech, 44–56.

Pulidindi, K. and Chakraborty, S. (2017). Medical Polymers Market Size By Product (Medical Fibers & Resins [PVC, PP, PE, Polystyrene], Medical Elastomers [SBC, Rubber Latex], Biodegradable Medical Plastics), By Application (Medical Devices & Equipment, Medical Packaging), Industry Analysis Report, Regional Outlook, Growth Potential, Price Trends, Competitive Market Share & Forecast, 2016 – 2024, Industry Report, *Global Market Insight*, 2017.

Quezon, O. N. (2013). *Preparation of a Polylactic Acid with Hydroxyapatite Reinforcement Composite*. California Polytechnic State University, San Luis Obispo: Master's Thesis.

Ramakrishna, S., Huang, Z. M., Kumar, G. V., Batchelor, A. W., and Mayer, J. (2004). An Introduction to Biocomposites, Series on Biomaterials & Bioengineering: Volume 1. London: Imperial College Press.

Ramalingam, M., Wang, X., Chen, G., Ma, P., and Cui, F. (2013). *Biomimetics*. Austin, TX: Wiley-Scrivener.

Ratner, B. D., Hoffman, A. S., Schoen, F. J., and Lemons, J. E. (2013). *Biomaterials Science: An Introduction to Materials in Medicine* (Third Ed.). Amsterdam: Elsevier Inc.

Shastri, V. P. (2003). Non-Degradable Biocompatible Polymers in Medicine: Past, Present and Future. *Current Pharmaceutical Biotechnology, 4*(5), 331–337.

Sheikh, Z., Najeeb, S., Khurshid, Z., Verma, V., Rashid, H., and Glogauer, M. (2015). Biodegradable Materials for Bone Repair and Tissue Engineering Applications. *Materials*, *8*(9), 5744–5794.

Singh, A. (2017). Biomaterials Innovation for Next Generation Ex Vivo Immune Tissue Engineering. *Biomaterials*, *130*, 104–110.

Stefanut, M. N., Cata, A., Ienascu, I. M., Tanasie, C., Ursu, D., and Dobrescu, M. C. (2015). Different Methods for Obtaining of Some Bio-Composite Materials with Dental Use. In *The 5th IEEE International Conference on E-Health and Bioengineering – EHB 2015* (pp. 1–4).

Subramaniam, A., and Sethuraman, S. (2014). *Biomedical Applications of Nondegradable Polymers*. San Diego, CA: Elsevier Inc., 301–307.

Tripathi, G., Choudhury, P., and Basu, B. (2010). Development of Polymer Based Biocomposites: A Review. *Materials Technology*, *25*(3–4), 158–176.

Wang, M. (2003). Developing Bioactive Composite Materials for Tissue Replacement. *Biomaterials*, *24*(13), 2133–2151.

Wong, J. Y., Bronzino, J. D., and Peterson, D. R. (2012). *Biomaterials Principles and Practices* (Vol. 1). Boca Raton: CRC Press, 270–280.

Zakaria, S. M., Zein, S. H. S., Othman, M. R., Yang, F., and Jansen, J. A. (2013). Nanophase Hydroxyapatite as a Biomaterial in Advanced Hard Tissue Engineering : A Review. *Tissue Engineering: Part B*, *19*(5), 431–441.

6 Conceptual Design of Composite Crutches

S. M. Sapuan,[1,2] F. N. Shafiqa,[1] M. T. Mastura,[3] and R. A. Ilyas[1,2]

[1]Advanced Engineering Materials and Composites Research Centre (AEMC), Department of Mechanical and Manufacturing Engineering, Faculty of Engineering, Universiti Putra Malaysia

[2]Laboratory of Biocomposite Technology, Institute of Tropical Forestry and Forest Products (INTROP), Universiti Putra Malaysia, Malaysia

[3]Fakulti Teknologi Kejuruteraan Mekanikal dan Pembuatan (FTKMP), Universiti Teknikal Malaysia Melaka, Malaysia

CONTENTS

6.1 INTRODUCTION

Crutches have been used for 5,000 years (Epstein, 1937). From the fallen tree branches used as crutches to assist balance and ambulation, they have developed into their new configurations of underarm and forearm crutches. With the development and research on new technologies, crutches' geometry and materials have changed, but the general design of the crutch is still the same. They are basically sticks with hand and underarm or forearm supports. Crutches are used as mobility aids, allowing body weight to be transferred from the legs to the upper body to the crutch(es). They are often used by patients who cannot support their body weight using legs either for short-term injuries or lifelong disabilities. These users are in the process of healing or recovering from leg injuries or suffering from stroke. Crutches can assist in balancing and enable the users to perform their daily activities safely. Therefore, crutches must have a design that is high strength, robust, durable, easy to handle, and lightweight. Many types of crutch designs exist in the market, such as auxiliary crutch, forearm crutch, cane, platform, and others. However, most existing crutches are made by using aluminum, wood, and plastic.

In this chapter, new composite crutches are developed to replace the existing crutches in order to reduce crutch weight while improving its structural frame and performance. Hence, using a composite material makes a product lightweight and more durable than current crutches (Shortell, Kucer, Neeley, & LeBlanc, 2001). From an engineering point of view, composite materials are an evolution in materials technology nowadays because of their influence in engineering and technology fields as they reveal combined properties of reinforcement and matrix to form stronger and stiffer materials (Mahajan & Aher, 2012). Composite materials are often used in automotive, consumer goods, aerospace, textile, medical, building, sports, and electronic industries. Composite materials are an amalgamation of two or more constituent materials; the matrix and reinforcement results in better properties than those of the individual constituent materials acting alone. In addition, each material retains its separate physical, chemical, and mechanical properties. High strength and stiffness, combined with low density when compared with bulk materials, allow for weight reduction in the finished part; these are the main advantages of composite materials. The various types of composites available include a particle in matrix, fibers in matrix, or a combination of them (Salit, 2014).

A conceptual design is the first phase of the design process. The conceptual design can divide into two phases, which are concept generation and concept evaluation (Sapuan, 2017). Besides, conceptual design is one of the methods approached by Pugh (1991), which is a total design method. The total design method can help and guide designers and engineers to develop new products via six processes, such as market investigation, product design specification (PDS), conceptual design, detail design, product manufacture, and sales of final products. The final concept design generated from the conceptual design process is the output in the conceptual design part before it continues to the next stage, which is embodiment design and detail design (Pahl, Beitz, Feldhusen, & Grote, 2007). Creative idea generation is an important thing to apply in the conceptual design process, where there are many strategies that can be used in creating ideas. Some of the methods are brainstorming, concept map, functional decomposition and synthesis, morphological chart, biomimetic, and others (George & Schmidt, 2007).

6.2 METHODOLOGY

In Pugh's total design method, there are several components that are known as a design core. Several components in the design core are the market investigation, the development of the PDS, the conceptual design, the detailed design, the product's manufacture, and product sales (Villanueva, Lostado Lorza, & Corral Bobadilla, 2016). Conceptual design is undertaken in the first phase of product development and the objective is to observe the function of a new product. However, at the conceptual design stage, the information about the function of the product designed may be inexact and insufficient. Therefore, this problem can improve by Pugh's controlled convergence, where team members are encouraged to present and interpret their design related to criteria given. Besides, it helps to build a creative work on the development process of new design (Frey et al., 2009).

The development of composite crutches was performed on the basis of a total design approach. Figure 6.1 shows the total design of a composite crutch and focuses on the relationship between the concept design and design tools used. An approach for the development of design of composite crutches consists of market investigation, PDS, conceptual design, detailed design, production, and sales as presented in Pugh's model (1991). In this chapter, the methodology of design of composite crutches is covered only from market investigation until detail design stage, as shown in Figure 6.1. Initially, as shown in Figure 6.1, market investigation is conducted to gain information about design requirements of composite crutches, where the product details are collected through a Google survey to 180 respondents. Later, relevant elements were identified as key elements for PDS based on the result requirements that were obtained from market investigation. Next, a conceptual design was generated using a creative thinking method that is based on the concept map, functional decomposition models, biomimetic method, and morphological chart. Finally, selection of the final conceptual design was performed using Pugh's method and stress analysis, which is based on the design that was obtained from the conceptual design development with design evaluation that was performed by Catia v5 software.

6.2.1 MARKET INVESTIGATION

Market investigation is the first step in the total design method. In this chapter, market investigation has been conducted. The aspects of a crutch, including materials, design aesthetics, cost, and ergonomic issues that are related to user demands, are the main focus of this research. In the markets, there are many types of crutch. Manufacturers are always designing many different styles and types of

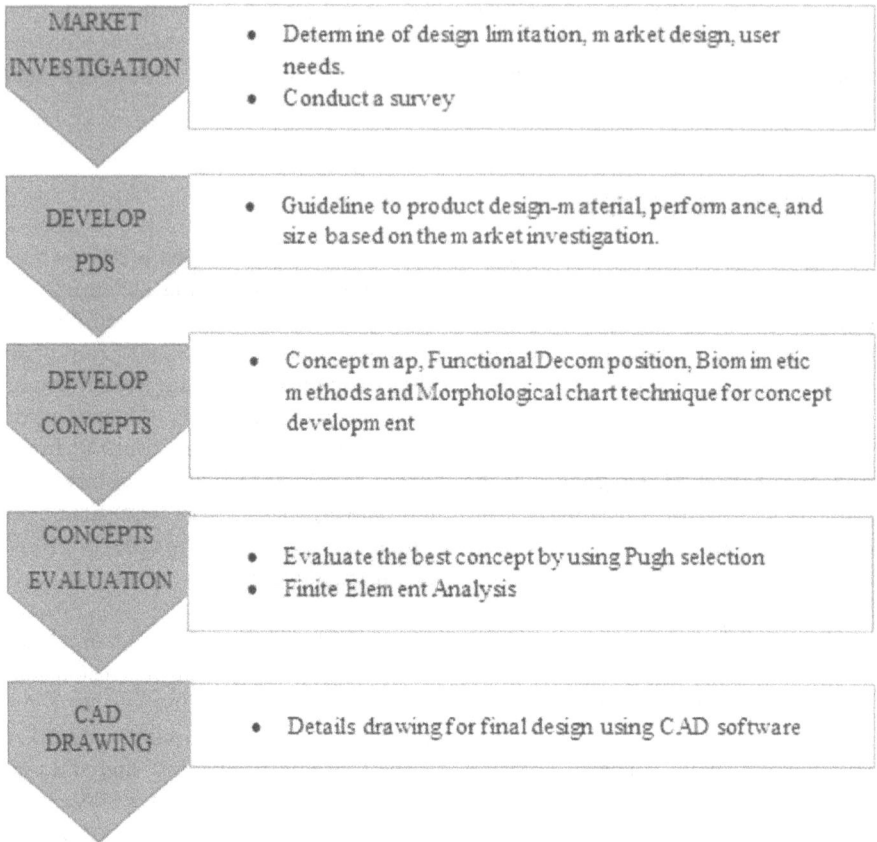

FIGURE 6.1   Process flow of the research.

crutch based on customer requirements and satisfaction. Thus, market investigation would be a great idea to get a clear view of the types of specification and demand for the product design.

The establishment of the market/user need situation in considerable depth should be the starting point for any design process. It is common practice to produce a device or a document, usually referred to as a "brief" at this stage before we go further for design to proceed. Market investigation is a successful way of observing the market and its fluctuation and a source of reporting news and updating customers/ users and interested parties with competitors' activities.

Several designs of crutches were investigated and analyzed to determine user need and performance of each design in this book. In addition, from the survey, designers also could get information and feedback from users of crutches to complete the market investigation. Any limitation on the project was considered and listed. The result and analysis of this market research is then used for the preparation of a PDS.

6.2.2 PRODUCT DESIGN SPECIFICATION (PDS)

A PDS is a document created during the problem definition activity very early in the design process after a market investigation has been done. In the PDS, all details and requirements must be considered in order for the product or the process to be successful. The elements that were used in this study are performance issues, market issues, capability issues, material used, size and dimensions, ergonomics, and cost.

The PDS thus acts as the monitor for the total design activity, because it places the boundaries on the subsequent designs. After the PDS, a conceptual design was developed within the boundary of PDS and so with the next process. Based on the market investigation, eight elements have been deduced for this book.

6.2.3 CONCEPTUAL DESIGN

The conceptual phase of the design core is primarily concerned with the generation of solutions to meet the stated needs from PDS and evaluation of these solutions to select the one that is most suitable and meets the PDS. Conceptual design is an initial design of the product design process, where drawings and other illustrations or models are used. Several methods can be used to develop conceptual design, for example, brainstorming method, morphology method, combination, and attribute listing. In this chapter, the conceptual design used two stages. The first stage was known as concept generation and the second stage was concept selection. In concept generation, the creative thinking method that is based on the concept map, functional decomposition models, biomimetic method, and morphological chart was used to generate several concepts. After the development of several designs, few designs were selected by using the Pugh selection method. The selected design was evaluated in detail based on criteria that were deduced from PDS.

6.2.4 DETAILED DESIGN

Further enhancement or modification on the design is required in this stage. Moreover, design modification can also be done to complete or fulfill other criteria in PDS. After the selected design had been analyzed and improved, a technical drawing was constructed.

6.3 RESULTS AND DISCUSSIONS

6.3.1 MARKET INVESTIGATION

Questionnaires were distributed to 178 respondents. The feedback was analyzed and the results are shown in Figures 6.2–6.6. The questions mainly asked about the end users and translated to design specifications. Figure 6.2 shows the gender of respondents that participated in the survey. Figure 6.3 shows the age range of respondents; most survey respondents were 18–24 years old. These users have experience using crutches. The questions shown in Figures 6.4 and 6.5 were asked to obtain data

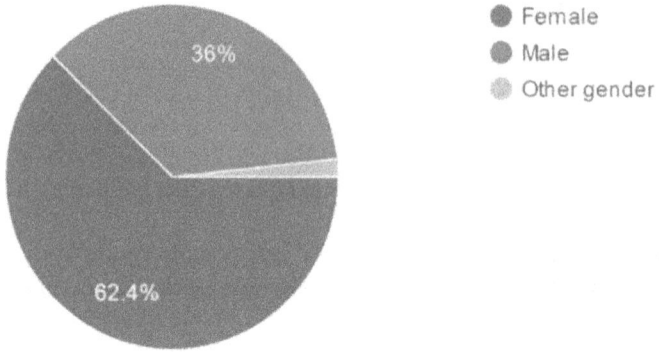

FIGURE 6.2   Results on gender distribution.

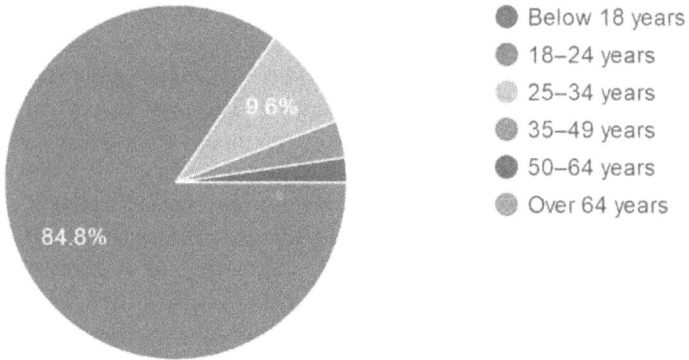

FIGURE 6.3   Results on age distribution.

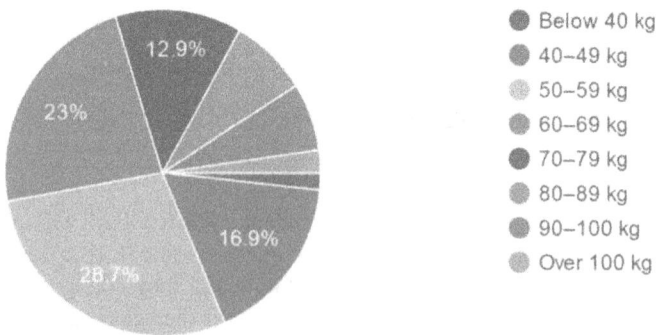

FIGURE 6.4   Results on weight distribution.

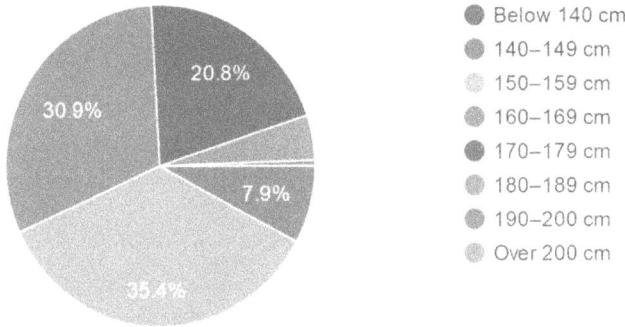

FIGURE 6.5   Results on height distribution.

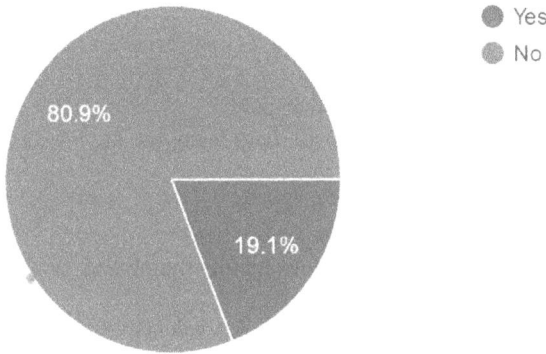

FIGURE 6.6   Results on crutches experience.

for ergonomic design purposes. From the results shown in Figure 6.4, it is known that the average weight of respondents who have ever experienced using crutches is around 50–59 kg. In order to design an ergonomic crutch, the weight and height of the user must be considered to make sure the crutch can withstand their load. Results of the users who have experience on crutches are shown in Figure 6.6.

Respondents also were asked about any issues that they have ever faced when using crutches. The results were interpreted and clarified as the following:

  i. Hard to adjust the height.
 ii. Hands started to hurt after some time and the crutches were slippery on ice.
iii. Hurts the arms.
 iv. Slipping.
  v. Having troubles when attending class and using public transport.
 vi. Blister in one's palm, the handle padding was not comfortable at all, and the handle will become useless within 3 months' usage time.
vii. Too short and unstable.
viii. Insufficient ability to support the body for long-term use.

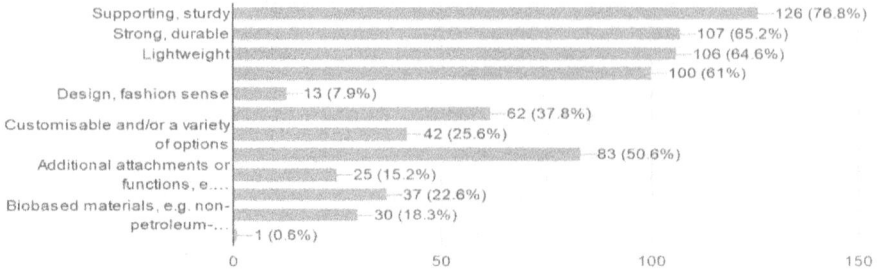

FIGURE 6.7 Results on the importance of bio-based materials used for the creation of crutches.

ix. Pain at armpit.
x. Crutches are heavy.
xi. Pain and discomfort in the armpit area. Some crutches might not be suitable for people of certain height(s).
xii. Balancing while trying to go downstairs.
xiii. They make one's arms hurt due to them being too big for certain people.
xiv. Broke one's leg.
xv. Base of crutches.
xvi. Lower arm felt sore after a while using them.
xvii. Getting abrasion to one's hand between the thumb and forefinger.

In the next section of the questionnaire, the respondents were asked about the design and materials of the crutches. Figures 6.7 and 6.8 show the results from this section. Figure 6.7 exhibits a number of respondents who most likely chose biomaterials as materials for crutches. In total, 45.1% of the respondents answered it is important to consider biomaterials in the design of a crutch. Figure 6.8 shows the results from respondents on the important criteria that should be considered in design of crutches.

How important is it for you that crutches are made from biobased materials, such as natural fibres, woods or bioplastics?
164 responses

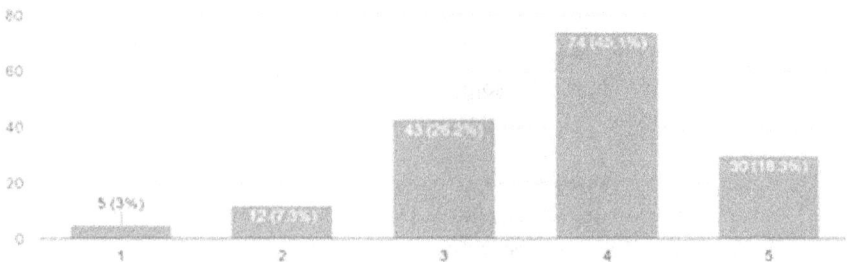

FIGURE 6.8 Results on the important features of crutches and/or the materials.

TABLE 6.1
PDS Elements of Composite Crutches

PDS Element	Description
Materials	• The materials used for the product are from bio-based composite
	• The material must not be brittle
Performance	• Able to withstand maximum 1,500 N of body weight
Weight	• The weight must have low specific weight to strength ratio compared with metallic materials
	• Lightweight
Size	• Easy to bring
	• Adjustable height
Aesthetics	• The surface will have smooth finishing
	• The color will follow the color of composites
Ergonomics	• Product must provide comfort
	• Product must be safe to the user
	• Foldable
Cost	• Total cost of the product must be affordable
Manufacturing	• Low-cost manufacturing process
	• Manual process for complex shape

From the results, respondents require strong and sturdy design of crutches, followed by durable and lightweight crutches.

6.3.2 PRODUCT DESIGN SPECIFICATION (PDS)

PDS is a set of dynamics used as a guideline in this total design method. Based on the market investigation, eight elements have been deduced for this book. Table 6.1 shows the detailed explanation of PDS elements of crutches.

6.3.3 CONCEPTUAL DESIGN

Conceptual design is one of the parts in the total design. This part is divided into two stages: concept generation and concept selection. For concept generation, ideas and solutions are generated by using a gallery technique, which come from a concept map, functional decomposition, biomimetic, morphological chart and are presented via sketching and CATIA v5 modeling. Then for concept selection, the best concept is then selecting by using the Pugh method. Figure 6.9 shows the procedure that was used to generate the conceptual design.

6.3.3.1 Concept Generation

The most critical stage in the engineering design process is concept generation to produce multiple design ideas. If this stage is not conducted, no design options will be generated. Concept generation is primarily concerned with the generation of solutions to meet the stated needs from PDS and evaluation of these solutions

Problem introduction
Understand the problems, product specification,
limitations and issues related

Idea generation stage
Generates several ideas and presented in drawings on papers

Association of ideas stage
Combined or collaborate concepts if necessaries. New
concepts generated or focus on continue develop existing
ideas.

Idea generation stage 2
Generate new ideas or develop existing ideas from the
previous. Stage 3 and Stage 4 can be repeated if necessary.

Idea selection
Final ideas are selected for further generation.

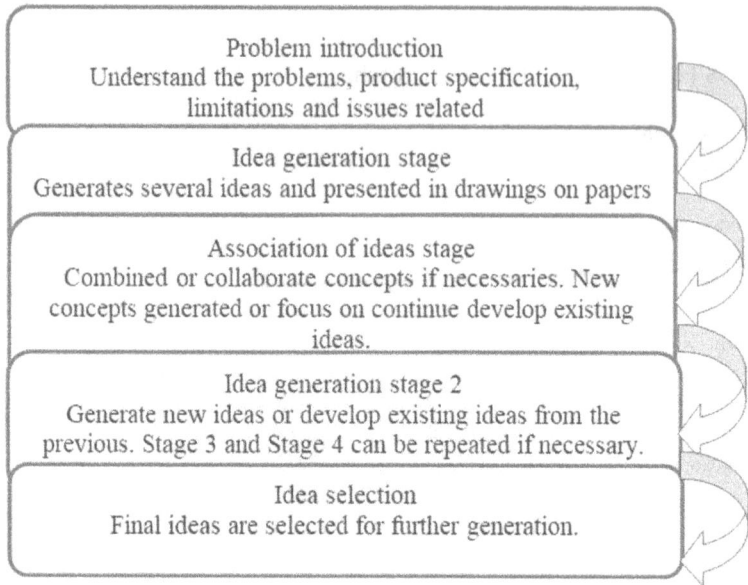

FIGURE 6.9   Procedure used to generate the conceptual design.

to select the one that is most suitable and meets the PDS and customer requirements. Conceptual design is the initial design of the product design process, where drawings and other illustrations or models are used. Several methods can be used to develop or generate conceptual design, for example, a concept map, functional decomposition, biomimetic, and morphological chart. A final design will be selected from the alternatives.

A mind map of suggestions for the conceptual design of the composite crutch is shown in Figure 6.10. The functional decomposition shows the effectiveness of the proposed models in improving the crutch for disabled individuals. There are two effective and efficient models: the Black Box model diagram and the functional model/functional decomposition. A product or a device can be modeled as a single-component entity that transforms inputs of materials, energy, and signal into desired outputs. It also aims to increase the performance of the device or the product, to enhance the crutch for individuals with powerlessness to walk. The functional model illustrates the Black Box model.

The Black Box model, shown in Figure 6.11, was created to imply the fundamental capability of people who are disabled who can walk normally by our proposed design. In this Black Box model, the bold line was defined as the material in this model; a thin line was used to state the energy for this model and a spotted line to identify the signal. The importance of this model was to show the functionalism of this device, while keeping in mind to reach the goal, which was to help people who are disabled to walk like able-bodied people.

The functional model structure shows the complete workings of any device/product, as it has the main function, material, energy, and sound. The functional

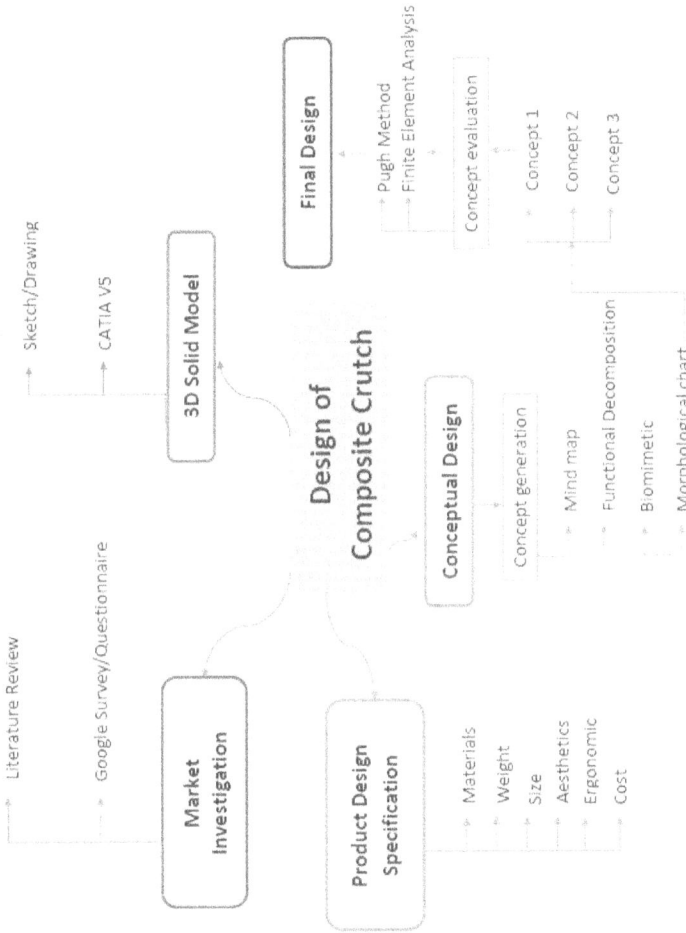

FIGURE 6.10 Concept map of the composite crutch.

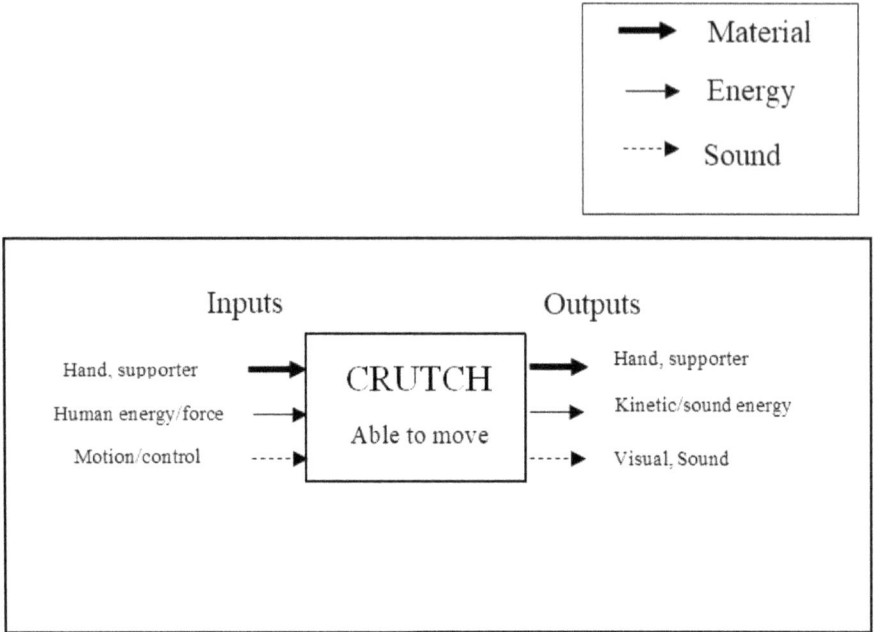

FIGURE 6.11   Function structure Black Box for composite crutch.

model demonstrates the flows and main functions for the potential device and specified subfunctions, functions, and flows that the device/product contains. The final function of the device/product can be patterned from the main function used in the Black Box model "Able to Move." This function is related to customer needs and PDSs. The functional model structure was an important learning device on how to divide the product into functions and flows that were more accessible to the customer needs. Figure 6.12 shows the function structures for a crutch.

The functions of crutches are to reduce human load on one leg and to widen a support base to enhance balance and stability by transferring the weight of one's lower body to the upper body. When using the crutch, the crutch supports the full weight of the body, so the crutch body/frame must have high strength to prevent bending or buckling. In addition to having a high-strength frame, the tip (base) of the crutch also must be considered as it protects crutches from slipping, absorbs shock, and makes them more energy efficient. By reducing shock, the tip (base) will increase stability and result in softer and smoother gait, stimulating the psychological and physical well-being of a user experiencing the normal walking step. In other words, the crutch functions as a support to the human; use the keyword biomimetic while researching to learn more. As explained in the framework, two steps can be used to find the nature strategies in the AskNature.org database. The keywords used for this research work was "support", "absorb", and "prevent", as shown in Table 6.2. From the table, it can be concluded that *prevent* gives more results for collections, inspired ideas, and resources compared to the *absorb* and *support* keywords, except for biological

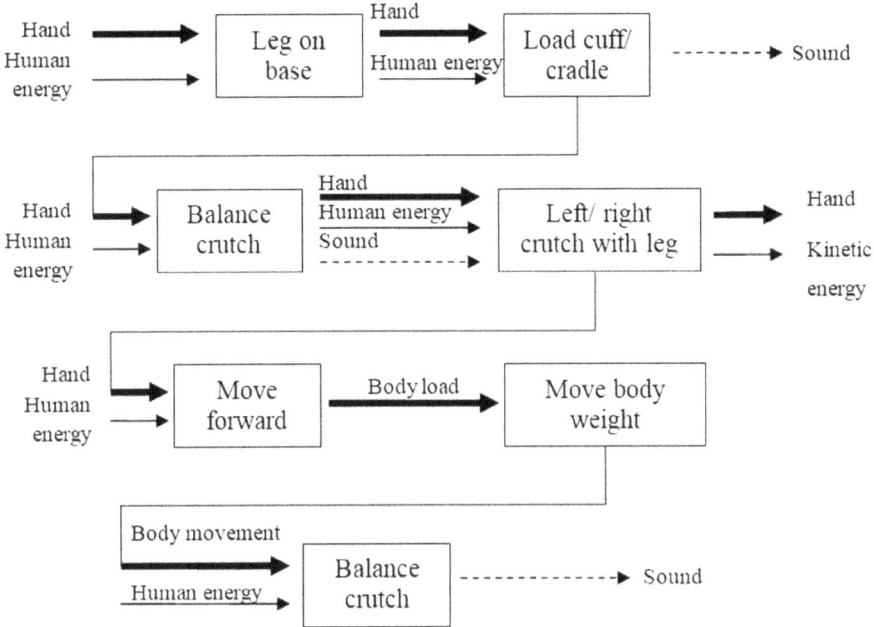

FIGURE 6.12 Function structure for a composite crutch.

strategies. From the keyword searching, there were a lot of lists in the database to get ideas for solving the problems of how nature supports, absorbs, and prevents.

Aside from searching a keyword using AskNature.org, another step using the keyword can be looking in the Biomimicry Taxonomy and finding the solutions in solving the problems of crutches. The Biomimicry Taxonomy has eight groups, 30 subgroups, and 160 functions, where the function can be looking directly and getting ideas of natural strategies from searching AskNature.org. As shown in Table 6.3, there were one group, two subgroups, and three functions of the nature related to the research work from Biomimicry Taxonomy.

It can be seen from Table 6.3 that both strategies, AskNature.org search and Biomimicry Taxonomy, can give ideas to suggest how nature can prevent, absorb, and support. From Tables 6.2 and 6.3, we can conclude that using the AskNature.org

TABLE 6.2

Results of the Keywords Using AskNature.org Search

Keyword	Prevent	Absorb	Support
Biological strategies	391	455	150
Inspired ideas	46	31	14
Resources	27	2	12
Collection	3	12	4

TABLE 6.3

Numbers of Biomimicry Cases Search by Function Using Biomimicry Taxonomy

Group	Maintain Physical Integrity		
Subgroup	Manage Structural Forces		Prevent Structural Failure
Function	Compression	Buckling	Fatigue (rupture)
Biological strategies	87	39	4
Inspired ideas	7	4	1

keyword search gives more results compared to using the Biomimicry Taxonomy function search. Next, some ideas on biomimicry cases that relate to the problems listed from the database can be chosen to solve the problems. Therefore, only one function of the Biomimicry Taxonomy, which is "buckling", was discussed as the sample of the strategies.

As shown in Table 6.3, there are 39 biological strategies and 4 inspired ideas from the results. The related biological strategies are listed in Table 6.4. Table 6.4 shows two biological strategies chosen to solve problems related to improving the strength of the crutch, discuss the potential solutions, and generate ideas in the conceptual design of the crutch.

From the previous studies by researchers, the buckling mode is corresponding to the bamboo's node structure, where the bamboo node gives benefit only when the buckling mode is local buckling. Besides, this finding concludes that the bamboo node plays a main role in resisting local buckling (Xing, Chen, Xing, & Yang, 2016). They also found that a structure that consists of both internode and nodes can perform better than a conventional thin-walled cylindrical shell.

A Venus basket sponge, also called *Euplectella aspergillum,* is a marine animal that lives anchored to the deep ocean floor with tube-shaped sea sponges typically standing 10–30 cm tall and filtering tiny food particles from the seawater as it flows through their bodies. From previous studies, researchers observed that the elevated rigid sponge cage attached to the ocean floor will undergo bending stresses that are concentrated at the anchor point. They found that the bundled spicules arranged horizontally and vertically into a square-grid cylindrical

TABLE 6.4

Biological Strategies from AskNature.org to Solve the Problem

Nature	Findings
Bamboo	Thin-walled tubular stems resist buckling
Venus's flower basket	Silica skeleton is tough and stable

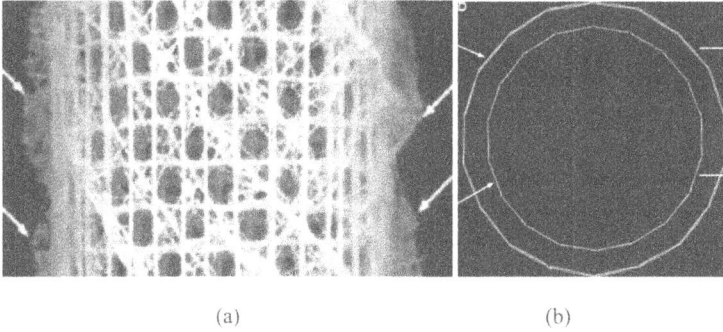

(a)   (b)

FIGURE 6.13   (a) Fragment of the cage structure showing the square-grid lattice of vertical and horizontal struts (Aizenberg et al., 2005), (b) Lateral ray overlap between neighboring spicules (Weaver et al., 2007).

cage reinforced by ancillary diagonal fibers running in both directions can make a structure stable and much stronger (Aizenberg et al., 2005). As shown in Figure 6.13, in this arrangement, the horizontal struts arranged on the interior surface, while the vertical struts are mostly arranged on the exterior lattice surface and become like the grid sandwiched between the two. Hence, this design strategy improves the toughness of the framework by presuming uniform support to the base structural framework (Weaver et al., 2007).

A morphological chart provides an organized method to generate ideas through sketches. By doing this chart, we are able to draw a few sketches of each crutch part. The morphology chart of the composite crutch can be seen in Tables 6.5–6.7. As shown in Tables 6.5–6.7, three concept designs are created by combining the design features of crutches. The concept design should not concentrate on the same design features or be called as "verbs" only because that would reduce the perspective and limit the options of the design. An ideal morphology chart uses diverse or different verbs as choices. Thus, designs with different patterns are compared while discussing the advantages and disadvantages of every design.

The morphology chart for Concept 1 of the composite crutch is shown in Table 6.5. There are five design features and three options to generate a concept for the composite crutch. Concept 1 was an underarm crutch. The design shape chosen from Concept 1 is a simple design with the use of a curved bamboo structure frame assembled with adjusted clip to adjust height for the main body frame. This design used a cylinder shape handle and cylinder shape cradle with underarm curve shape to comfort the user's armpit. This design has a hexagon shape rubber tip at the bottom part of the crutch for grip to the land surface. The drawing design for Concept 1 was drafted using a sketch and CATIA v5 software and is shown in Figure 6.14.

Table 6.6 shows the morphology chart for Concept 2. There are three options based on design features to generate the composite crutch. Concept 2 was a forearm crutch. As shown in Table 6.6, the shape of the Concept 2 composite crutch is

TABLE 6.5

Morphology Chart of Composite Crutch (Concept 1)

Design features	Options		
	1	2	3
Shape	Simple	Medium	Complex
Main body frame	Curve bamboo structure frame with adjusted clip	Straight bamboo structure frame with adjusted clip	Sea sponge structure frame with adjusted clip
Handle	Cylinder shape handle	Rectangular shape handle with finger grip	Cylinder shape handle with finger grip
Cuffs & cradles	-	Cylinder shape cradle with underarm curve shape	Round cuff
Tips	Round crutch tip	Hexagon crutch tip	Tripod crutch tip

considered medium complexity as it was a design inspired by bamboo that has a part by part section to assemble to make a main body frame. This design used a cylindrical handle with finger grip and open round cuff that allows it to grip the user's hand, for example when opening a door, without losing the crutch. This design has a round shape rubber tip at the bottom part of the crutch for grip to the land surface. The drawing design for Concept 2 was drafted using sketch and CATIA v5 software and is shown in Figure 6.15.

The morphology chart for Concept 3 for the composite crutch is shown in Table 6.7. Table 6.7 shows five design features and three options to generate a concept for the composite crutch. The design shape for Concept 3 is complex because it was design inspired by a sea sponge structure such as a skyscraper for a main body frame and assembled with adjusted clips. This design used a rectangular shape handle with finger grip, and because Concept 3 was a stick crutch, there was no cuff and cradle design in this concept. This design has a tripod tip at the bottom part of the crutch for extra stability and grip to the land surface. The drawing design for Concept 3 was drafted using computer design software and can be seen in Figure 6.16.

TABLE 6.6

Morphology Chart of Composite Crutch (Concept 2)

Design features	Options		
	1	2	3
Shape	Simple	Medium	Complex
Main body frame	Curve bamboo structure frame with adjusted clip	Straight bamboo structure frame with adjusted clip	Sea sponge structure frame with adjusted clip
Handle	Cylinder shape handle	Rectangular shape handle with finger grip	Cylinder shape handle with finger grip
Cuffs & cradles	-	Cylinder shape cradle with underarm curve shape	Round cuff
Tips	Round crutch tip	Hexagon crutch tip	Tripod crutch tip

6.3.3.2   Design Evaluation

Based on the PDS, the maximum load that the crutch can withstand is 1,500 N. Using Catia v5 software, a stress analysis has been done for each concept design main body frame to forecast or predict its structural performance based on the generated stress. The material of biocomposite, bamboo fiber reinforced polypropylene matrix, was chosen to perform this analysis. Based on the previous studies, researchers found that the yield strength and young modulus of the bamboo fiber reinforced with polypropylene matrix from the stress distribution was 3.03×10^7 and 3.66×10^9, respectively (Shito, Okubo, & Fujii, 2002). As shown in Table 6.8, the highest ranking position for the stress analysis is Concept 2. In terms of performance, Concept 2 obtains the lowest von Mises stress value compared to

TABLE 6.7
Morphology Chart of Composite Crutch (Concept 3)

Design features	Options		
	1	2	3
Shape	Simple	Medium	Complex
Main body frame	Curve bamboo structure frame with adjusted clip	Straight bamboo structure frame with adjusted clip	Sea sponge structure frame with adjusted clip
Handle	Cylinder shape handle	Rectangular shape handle with finger grip	Cylinder shape handle with finger grip
Cuffs & cradles	-	Cylinder shape cradle with underarm curve shape	Round cuff
Tips	Round crutch tip	Hexagon crutch tip	Tripod crutch tip

TABLE 6.8
The von Mises Stress Result for Each Concept

Concept	1	2	3
Yield strength (N/m^2)		3.03×10^7	
Young's modulus (N/m^2)		3.66×10^9	
Max von Mises stress (N/m^2)	6.67×10^7	5.96×10^7	5.97×10^9

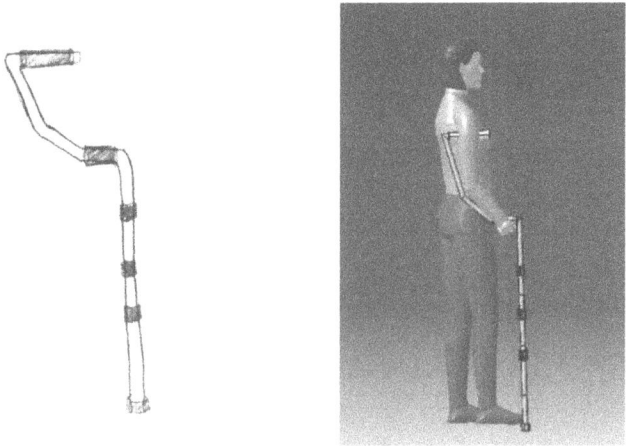

FIGURE 6.14 Concept 1.

Concepts 1 and 3. Stress distributions of Concept 2 that were simulated from the software are shown in Figure 6.17. A concept that generated the lowest von Mises stress value is able to withstand the applied load better (Mansor, Sapuan, Zainudin, Nuraini, & Hambali, 2014).

There are three concept designs. A design that is currently available in the market is set as a reference design, also known as datum, to do a comparison among other designs. This design satisfies all the basic requirements for a crutch. By using the Pugh selection method, three design concepts are evaluated with datum design.

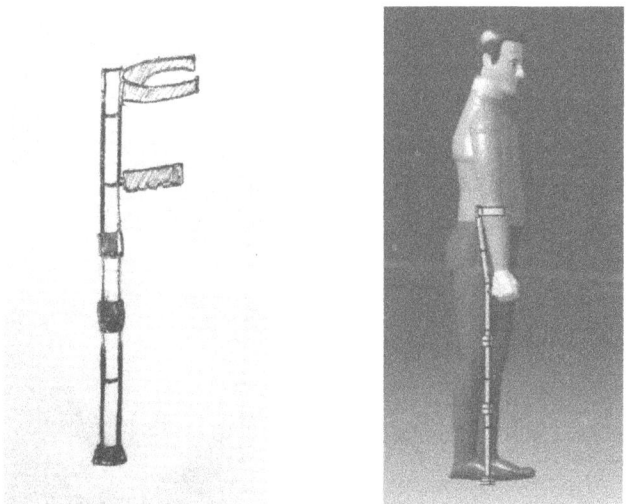

FIGURE 6.15 Concept 2.

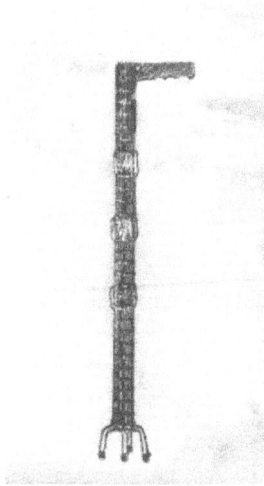

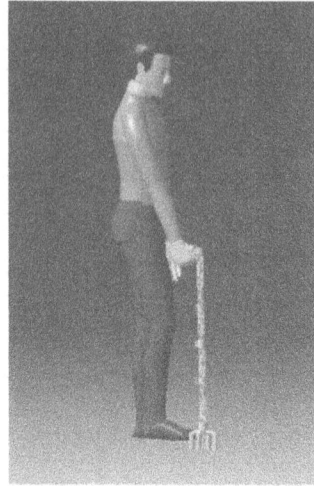

FIGURE 6.16   Concept 3.

Von Mises stress (nodal values).2
  N_m2
    5.96e+007
    5.36e+007
    4.77e+007
    4.17e+007
    3.58e+007
    2.98e+007
    2.39e+007
    1.79e+007
    1.2e+007
    6.03e+006
    8.29e+004
On Boundary

FIGURE 6.17   Stress distributions for Concept 2.

TABLE 6.9
Pugh Selection Method

Selection criteria	Reference	Concept 1	Concept 2	Concept 3
Design flexibility	D	+	+	+
Lightweight	A	+	+	+
Adjustable/Foldable	T	+	+	+
Ergonomic	U	0	+	0
Ease of fabrication	M	0	0	-
Total pluses		3	5	3
Total minuses		0	0	2
Total zeros		3	1	1
Net score		3	5	1
Rank		2	1	3

(+) = 1 POINT, (-) = 0 POINT

Basically, the parameters and criteria chosen are based on the PDS that can be developed from market investigation. Table 6.9 shows that Concept 2 has the highest net score compared to the other concepts and it was selected as the final design for the composite crutch.

6.4 CONCLUSION

From this work, several conclusions can be made.

1. The objective of this chapter, which is to design a composite crutch, was successfully achieved. The product has designed on product design specifications (PDS) and based on user requirements. A new composite crutch has been designed in this chapter. Generally, the Concept 2 composite crutch is an acceptable design where it includes the important design characteristics of being ergonomic, lightweight, and safe.
2. The total design process was the main part in this chapter. The total design method assists designers with a systematic design process.

Market investigation was performed initially to investigate and define the specifications of the product design. A guideline for this entire design process was created based on results from the market investigation. Some design tools have been used in order to generate ideas for crutch designs, such as concept maps, functional decomposition, biomimetic, and morphological charts before the best concept was selected. Selection of the concept design was conducted using the Pugh selection method. Finally, the best concept was chosen from the selection method based on the PDS.

REFERENCES

Aizenberg J., Weaver J.C., Thanawala M.S., Sundar V.C., Morse D.E., and Fratzl. P. (2005). Skeleton of *Euplectella Sp.*: Structural Hierarchy from the Nanoscale to the Macroscale. *Science*, 309(5732) 275–278.

Epstein, S. (1937). Art, History, and the Crutch. *Annals of Medical History* 9:304–313.

Dieter G.E., and Schmidt L.C. (2017). *Engineering Design*, 4th ed., New York: McGraw-Hill Higher Education.

Frey, D.D., Herder M.P., Wijnia Y., et al. (2009). The Pugh Controlled Convergence Method: Model-Based Evaluation and Implications for Design Theory. *Research in Engineering Design*, 20(1) 41–50.

Mahajan, G.V., and Aher V.S. (2012). Composite Material: A Review over Current Development and Automotive Application. *International Journal of Scientific and Research Publications*, 2(11) 2250–3153.

Mansor, M.R., Sapuan S.M., Zainudin E.S., Nuraini A.A., and Hambali A. (2014). Conceptual Design of Kenaf Fiber Polymer Composite Automotive Parking Brake Lever Using Integrated TRIZ–Morphological Chart–Analytic Hierarchy Process Method. *Materials and Design*, 54: 473–482.

Pahl, G., Beitz W., Feldhusen J., and Grote K. H. (2007). *Engineering Design: A Systematic Approach*. London: Springer-Verlag London Limited.

Pugh, S. (1991). *Total Design: Integrated Method for Successful Product Engineering*, 1st ed. Essex, England: Addison-Wesley Publishers Ltd.

Salit M.S. (2014). *Tropical Natural Fibre Composites: Properties, Manufacture and Applications*. Singapore: Springer.

Sapuan, S.M. (2017). Conceptual Design in Concurrent Engineering for Composites, in *Composite Materials: Concurrent Engineering Approach*, 1st ed. Oxford: Butterworth-Heinemann, 141–207.

Shito, T., Okubo K., and Fujii T. (2002). Development of Eco-Composites Using Natural Bamboo Fibers and Their Mechanical Properties. *WIT Transactions on The Built Environment*, Vol. 59.

Shortell, D., Kucer J., Neeley W.L., and LeBlanc M. (2001). The Design of a Compliant Composite Crutch. *Journal of Rehabilitation Research and Development*, 38(1) 23–32.

Villanueva P.M., Lostado Lorza R., and Corral Bobadilla M. (2016). Pugh's Total Design: The Design of an Electromagnetic Servo Brake with ABS Function – A Case Study. *Concurrent Engineering Research and Applications*, 24(3) 227–239.

Weaver J.C., Aizenberg J., Fantner G.E., Kisailus D., Woesz A., Allen P., Fields K., et al. (2007). Hierarchical Assembly of the Siliceous Skeletal Lattice of the Hexactinellid Sponge *Euplectella Aspergillum*. *Journal of Structural Biology*, 158(1) 93–106.

Xing D., Chen W., Xing D., and Yang T. (2016). Lightweight Design for Thin-Walled Cylindrical Shell Based on Action Mechanism of Bamboo Node. *Journal of Mechanical Design*, 135 (1) 1–6.

7 Conceptual Design of Kenaf Fiber Reinforced Polymer Composite Chair with Input from Anthropometric Data

A. Mahmood,[1,3] *S. M. Sapuan,*[1,4] *K. Karmegam,*[5]
A. S. Abu,[2] *S. Sivasankar,*[5] *and R. A. Ilyas*[1,4]

[1]Laboratory of Biocomposite Technology,
Institute of Tropical Forestry and Forest Products,
Universiti Putra Malaysia, UPM Serdang, Selangor, Malaysia

[2]Department of Mechanical and Manufacturing Engineering,
Universiti Malaysia Sarawak, Kota Samarahan, Sarawak

[3]Department of Mechanical Engineering, Politeknik Port
Dickson, Port Dickson, Negeri Sembilan, Malaysia

[4]Advanced Engineering Materials and Composites
Research Centre (AEMC), Department of Mechanical
and Manufacturing Engineering, Universiti Putra
Malaysia, UPM Serdang, Selangor, Malaysia

[5]Department of Environmental and Occupational Health,
Faculty of Medicine and Health Sciences, Universiti
Putra Malaysia, UPM Serdang, Selangor, Malaysia

CONTENTS

7.1   INTRODUCTION

Anthropometry is defined as "the science of measurement and the art of application that establishes the physical geometry, mass properties, and strength capabilities of the human body". In simple terms, anthropometry can be defined as the study that deals with body dimensions, that is, body size, shape, strength and working capacity for design purposes, and body composition. Anthropometry is also defined as a technique of expressing the quantitative form of the human body. It is recognized as the single most universally applicable, inexpensive, and noninvasive technique for assessing the size and proportions of the human body. This technique has been used by anthropologists worldwide to estimate body size and stature for many years. Thus, developing a correct anthropometric profile in stature is an important step during the design of user-friendly products (Nor et al., 2013; Taifa & Desai, 2017).

Anthropometry plays a main role in design development to create well-designed products. Measurement size reference is inseparable from the human body based on furniture design, in order for the user to meet the people's requirements for the size of furniture (Wang & Yang, 2012). There is a conflict between the natural tendency to unrestricted physical movement and the need to maintain a sitting posture for a longer period of time in polytechnic students (Troussier, 1999). This study is of particular interest as it concerns polytechnic students rather than younger children reported in most previous studies in Politeknik Kuching Sarawak, Malaysia, and elsewhere.

A composite material consists of two or more chemically distinct constituents, on a macroscopic level, having a distinct interface separating them (Sanjay & Yogesha, 2017). A composite is a structural material that is made up of two or more combined components that are combined at a macroscopic (visible to the naked eye) level and do not dissolve in each other. The component where the material is embedded is known as the matrix and the other component is known as the reinforcing phase. The matrix phase materials are usually continuous. The reinforcing phase materials are usually in the form of particles, fibers, or flakes. Fiber reinforced polymer materials are composites that are made up of high-strength fibers (reinforcement) embedded in polymeric matrices. The fibers for these types of materials are meant to be the load-carrying elements and to provide rigidity and strength; whereas polymer matrices are meant to maintain the fibers' alignment (orientation and position) and to protect them against possible damage and against the environment. The required mechanical strength for applications in diverse fields is not provided by a pure polymer. Hence, in order to make the fiber reinforced polymer composites (FRPCs) suitable for a large number of various implementations ranging from sports equipment to aerospace, the reinforcement by high-strength fibers provides the polymer substantially enhanced mechanical properties (Arpitha & Yogesha, 2017).

A combination of plant-derived fibers with a plastic binder makes up natural fiber composites. The natural fiber components may be cotton, wood, coconut, kenaf, sisal, flax, hemp, jute, banana leaf fibers, wheat straw, abaca, bamboo, or other fibrous material (Arpitha & Yogesha, 2017). Due to the fact that natural fiber composites have characteristics that are similar to conventional materials, the use of natural fiber composites has become more popular in engineering applications (Sapuan & Maleque, 2005). Natural fibers have gained considerable attention in recent times due to benefits such as biodegradability, low density, fewer health hazards, high strength, reduction in cost and weight, and renewability. The positive impact to environment is the most important feature of natural fibers. Based on energy efficiency, natural fibers will have an important role in the future emerging "green" economy (Sahari, Sapuan, Zainudin, & Maleque, 2013). Due to its good mechanical properties, kenaf was chosen as a material of choice. The kenaf plant can be harvested two to three times a year whereas wood is harvested once in 20 or 25 years. Hence kenaf, as well as other natural fibers, can play an important role in the substitution of wood going forward. Kenaf usually grows up to 3–4 meters within a short period of 4–5 months. The kenaf plant has three layers, which are the bast, core, and pith. The bast of the kenaf makes up one-third of the plant. The core and pith of the kenaf represent the rest of the plant. Compared to other parts of the kenaf plant, the bast fiber has been described to have better mechanical properties. In order to utilize kenaf reinforced composites, it is necessary to get a correct balance between toughness, stiffness, and strength. When designing composite materials, this is the first important step. Much work has been undertaken to study the mechanical properties of fiber loading and the fiber content on thermal properties. In order to better understand the behavior of the raw materials and the final product, it is important to understand the thermal properties of the material as well. Abrasion and hardness resistance are relevant as the hardness property gives material high resistance to various kinds of change of shape when force is applied accordingly (El-Shekeil, Sapuan, Abdan, & Zainudin, 2012).

Without exhaustive use of herbicides or pesticides, kenaf is a drought tolerant relative of hibiscus, which grows rapidly up to 4 meters in 7 months. Kenaf fiber (*Hibiscus cannabinus L.*) has a prospective as an alternative for partial replacement of synthetic fibers or conventional materials as composite reinforcement. It is reported in the literature that kenaf is already being used in hybrid form with synthetic materials such as glass, carbon, and polyethylene terephthalate (PET) (Yahaya, Sapuan, Jawaid, Leman, & Zainudin, 2016).

Usually, for the fabrication of products that use natural fiber polymer composites, traditional manufacturing techniques are utilized. These traditional manufacturing techniques have been utilized for conventional FRPCs with thermoplastics and thermosets. These techniques are injection and compounding molding, resin transfer molding (RTM), compression molding, vacuum infusion, and direct extrusion (Ho et al., 2012). Injection, compression, and extrusion molding techniques are often utilized to introduce fibers into the thermoplastic matrix (Salleh, Hassan, Yahya, & Azzahari, 2014).

The central idea for this research was inspired by the fact that chairs act as important equipment in the field of education in classrooms for every institution in the world. Previous studies carried out have highlighted that several mismatches were found with the current furniture, and this leads to the need that the design of furniture used in the Malaysian institutions of higher learning classrooms must be improved to better accommodate the studying process of the students (Aminian & Romli, 2012). Therefore, this research is valuable in providing scientific baseline information not available in Malaysia currently about the anthropometry database and seating posture among the users using that equipment as well at the same time to design and develop new chairs utilizing biocomposite materials with input from anthropometric data (AD).

The aim of this research is to design and develop a new polytechnic classroom chair using biocomposites with input from AD to replace the existing plastic-based backrest and seat of the chairs while maintaining the required structural strength for safety and functionality performance. Based on the project requirements, a new concurrent engineering approach with input from AD utilizing the total design process, morphological chart, and weighted objective methods approach was applied in the design and development of this new chair.

7.2   METHODS

This study was undertaken to design, develop, and evaluate the design of a new biocomposite chair to be used by students of polytechnics in Malaysia. The new virtual prototype biocomposite chair with inputs from AD was tested and verified with input from AD analysis. Pugh's total design process model was utilized for the design and development of the biocomposite chair with inputs from AD. This study was conducted with the intention to design, develop, evaluate, and virtually test with inputs from AD a prototype biocomposite chair used by polytechnic students. This study involves evaluation of AD of chair used by polytechnic students, conceptual design (generation), design, development, and evaluation of the chosen design of a new biocomposite chair. Therefore, this study was undertaken as per Figure 7.1.

7.3   RESULTS AND DISCUSSION

This study utilized a modified version of the Pugh design model where the process starts with the evaluation of AD of chairs used by polytechnic students. The Pugh total design method is utilized where the product design specifications (PDSs) set out for the design of the new biocomposite chair. Next the conceptual design (generation using brainstorming, mind mapping, and morphological chart) of the chair and the evaluation of the design with input from AD are carried out with the weighted objectives method.

The ability and capability of humans to design a product must be adapted to human features. When there is a mismatch between a product and human features, automatically the feeling of discomfort will arise in users of the product. Therefore, all products must be manufactured according to users' AD. Comfort assessment can be evaluated objectively and subjectively. Objective evaluation was carried out based on the existing parameters of the biocomposite polytechnic chair (Deros, Hassan, Daruis, & Tamrin, , 2015).

In order to design the furniture for students, the 5th and 95th percentile values of the anthropometry dimensions were selected. Due to that, anthropometry dimensions obtained in this research study are important in order to provide the best design for Malaysian students. One of the most important aspects of this design is that the chairs must provide a good learning environment to the students. The biocomposite

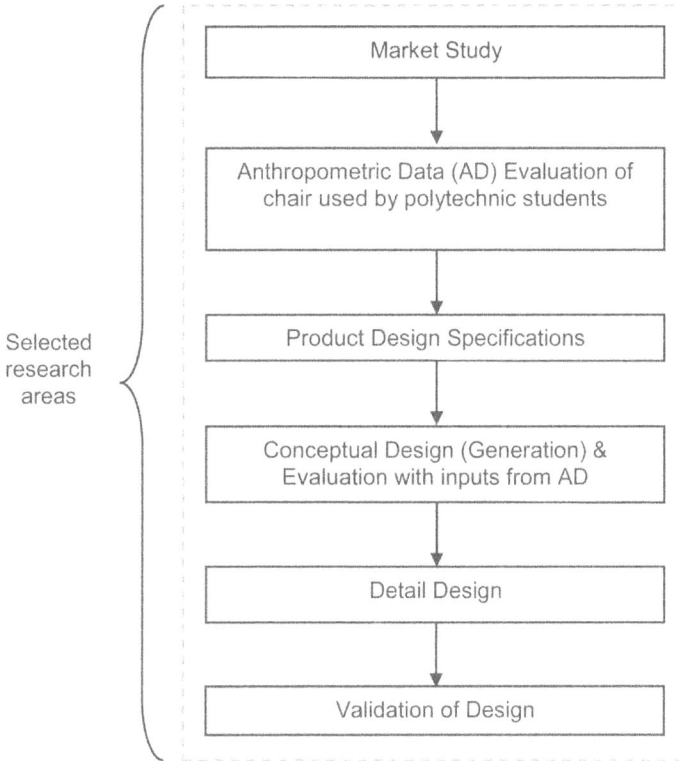

FIGURE 7.1   Selected research area in the product development process.

polytechnic chair can be designed to create comfort and satisfaction in the learning environment (Dawal et al., 2015).

7.3.1 ANTHROPOMETRIC DATA ANALYSIS

The collected AD was analyzed based on MS ISO 15535:2008 standard (SIRIM, 2008) and using SPSS program. The MS ISO standard method was used in order to check for irregular and outlier AD. The statistical data performed includes descriptive statistics (mean value, standard deviation, minimum, maximum, and 5th and 95th percentile) (Klamklay, Sungkhapong, Yodpijit, & Patterson, 2008) with the current findings of the study carried out using bivariate independent t-test analysis. Collected data was divided into several categories to facilitate analysis such as male student data, female student data, and overall data. Anthropometric measurements for each respondent were compared to the dimensions of furniture to identify the match or mismatch between the two.

7.3.1.1 Anthropometric Data Evaluation of Malaysian Adults

A total of 500 students, 213 males and 287 females, were involved in the study. Average age of respondents was between 18–24 years old. A total of 33 AD had been taken when the students were standing and sitting. Collected data was then processed for analysis in the form of mean, standard deviation, 5th percentile and 95th percentile. Tables 7.1 and 7.2 summarize the AD obtained to represent the current student population within the Politeknik Kuching Sarawak, Malaysia. Figure 7.2 shows the overall AD of students, aged 18–24 years old.

TABLE 7.1
Overall Anthropometric Data of Students, Aged 18–24 Years Old (N = 500)

No.	Dimension	Mean	Std Dev	5th Percentile	95th Percentile
1	Age (years)	20.13	1.02	19.00	22.00
2	Weight (kg)	64.69	16.83	46.00	99.00
3	Stature	168.23	6.04	159.61	178.10
4	Eye height	156.51	6.56	146.80	166.70
5	Shoulder height	139.23	6.16	131.11	148.99
6	Elbow height	106.29	4.55	99.81	114.08
7	Fist (grip axis) height	72.00	4.80	64.40	78.88
8	Shoulder (biacromial) breadth	43.54	3.20	39.33	49.88
9	Elbow-to-elbow breadth	45.77	4.42	39.90	54.38
10	Thigh clearance	15.03	2.33	11.71	19.69
11	Abdominal depth, sitting	18.52	3.70	14.11	27.67
12	Knee height	52.10	4.78	44.26	59.95
13	Hip breadth, sitting	31.30	3.80	27.11	39.15
14	Hand length	18.62	1.23	17.20	20.10
15	Hand breadth at metacarpals	7.02	0.62	6.30	8.39
16	Hand thickness	2.94	0.35	2.31	3.50
17	Thumb breadth	2.03	0.17	1.80	2.30
18	Index finger breadth, proximal	1.58	0.14	1.40	1.80
19	Sitting height (erect)	84.21	3.90	77.91	90.60
20	Eye height, sitting	72.53	4.05	66.71	78.69
21	Shoulder height, sitting	56.11	3.27	51.20	61.89
22	Elbow height, sitting	19.45	3.21	14.92	25.60
23	Elbow grip length	33.65	2.48	29.81	37.69
24	Grip reach; forward reach	73.87	4.22	67.50	80.50
25	Vertical grip reach, standing	200.79	8.39	186.52	215.37
26	Buttock-popliteal length (seat depth)	49.20	3.29	43.20	54.40
27	Buttock knee length	60.69	2.99	55.60	65.90
28	Buttock heel length	109.01	5.19	100.90	117.59
29	Lower leg length (popliteal height)	41.20	1.40	39.21	44.00
30	Foot length	25.53	1.22	23.60	27.60
31	Foot breadth	9.70	0.65	8.80	10.80
32	Head length	18.01	0.82	16.90	19.30
33	Head breadth	15.07	0.78	13.91	16.30
34	Head height	23.89	1.38	21.20	25.90
35	Head circumference	55.49	1.77	52.61	58.40

TABLE 7.2
Anthropometric Data for Male (N = 213) and Female (N = 287)

No.	Dimension	Male				Female			
		Mean	Std Dev	5th Percentile	95th Percentile	Mean	Std Dev	5th Percentile	95th Percentile
1	Age (years)	20.13	1.02	19.00	22.00	20.05	1.21	19.00	22.00
2	Weight (kg)	64.69	16.83	46.00	99.00	54.95	11.04	40.00	75.00
3	Stature	168.23	6.04	159.61	175.10	155.81	5.45	146.31	163.60
4	Eye height	156.51	6.56	146.30	166.70	144.51	5.44	134.12	152.88
5	Shoulder height	139.23	6.16	131.11	148.99	123.41	5.77	119.80	138.50
6	Elbow height	106.29	4.55	99.31	114.08	98.42	4.90	89.73	106.20
7	Fist (grip axis) height	72.00	4.80	64.40	78.88	66.26	4.66	59.10	73.00
8	Shoulder (biacromial) breadth	43.54	3.20	39.33	49.88	37.31	2.65	33.53	42.18
9	Elbow-to-elbow breadth	45.77	4.42	39.90	54.38	41.75	3.79	35.70	49.00
10	Thigh clearance	15.03	2.33	11.71	19.69	13.45	2.37	9.81	17.50
11	Abdominal depth, sitting	18.52	3.70	14.11	27.67	17.41	3.02	13.41	23.29
12	Knee height	52.10	4.78	44.26	59.95	50.10	2.31	46.31	53.89
13	Hip breadth, sitting	31.30	3.80	27.11	39.15	30.90	4.20	25.51	38.10
14	Hand length	18.62	1.23	17.20	20.10	16.73	1.20	14.50	18.50
15	Hand breadth at metacarpals	7.02	0.62	6.30	8.39	6.58	0.78	5.21	7.60
16	Hand thickness	2.94	0.35	2.31	3.50	2.39	0.29	2.00	2.80
17	Thumb breadth	2.03	0.17	1.80	2.30	1.74	0.18	1.41	2.00
18	Index finger breadth, proximal	1.58	0.14	1.40	1.30	1.52	0.16	1.30	1.80
19	Sitting height (erect)	84.21	3.90	77.91	90.60	78.37	4.18	72.20	34.79
20	Eye height, sitting	72.53	4.05	66.71	78.69	67.69	4.42	60.50	74.79

(Continued)

TABLE 7.2 (Continued)
Anthropometric Data for Male (N = 213) and Female (N = 287)

No.	Dimension	Male				Female			
		Mean	Std Dev	5th Percentile	95th Percentile	Mean	Std Dev	5th Percentile	95th Percentile
21	Shoulder height, sitting	56.11	3.27	51.20	61.89	52.07	3.90	44.41	59.05
22	Elbow height, sitting	19.45	3.21	14.92	25.60	19.00	3.27	13.22	24.10
23	Elbow grip length	33.65	2.48	29.81	37.69	34.31	4.26	28.91	41.89
24	Grip reach, forward reach	73.87	4.22	67.50	80.50	66.88	3.90	59.12	73.30
25	Vertical grip reach, standing	200.79	8.39	186.52	215.37	184.18	8.16	171.50	196.53
26	Buttock-popliteal length (seat depth)	49.20	3.29	43.20	54.40	45.73	3.70	40.30	52.68
27	Buttock knee length	60.69	2.99	55.60	65.90	56.32	3.60	49.73	62.99
28	Buttock heel length	109.01	5.19	100.90	117.59	100.17	4.88	93.12	103.69
29	Lower leg length (popliteal height)	41.20	1.40	39.21	44.00	39.39	2.48	33.53	42.70
30	Foot length	25.53	1.22	23.60	27.60	22.74	1.40	20.10	24.90
31	Foot breadth	9.70	0.65	8.80	10.80	8.58	0.84	7.20	9.70
32	Head length	18.01	0.82	16.90	19.30	18.26	1.88	17.00	22.50
33	Head breadth	15.07	0.78	13.91	16.30	14.45	1.18	12.80	17.07
34	Head height	23.89	1.38	21.20	25.90	22.05	1.30	20.00	24.30
35	Head circumference	55.49	1.77	52.61	58.40	54.59	2.73	50.15	58.60

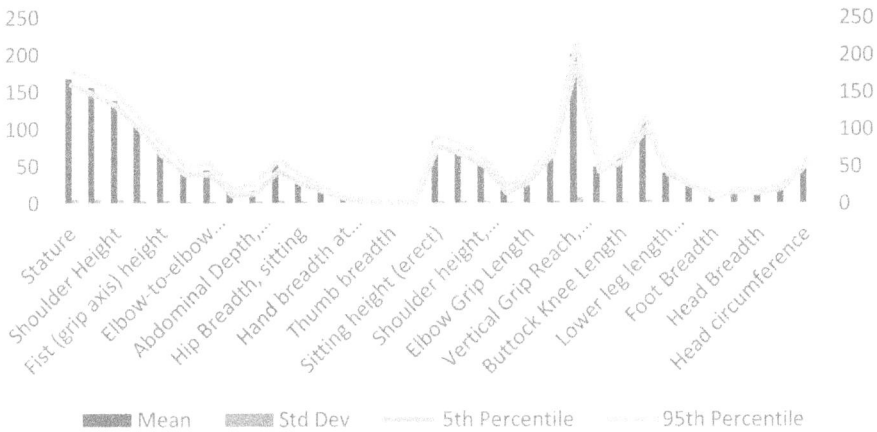

FIGURE 7.2 Overall anthropometric data of students, aged 18–24 years old (n = 500).

Table 7.1 presents the descriptive statistics of anthropometric dimensions for the overall population. Figure 7.2 is the graphical representation of Table 7.1. Hence from Table 7.1 and Figure 7.2, it can be seen that the average student height was 162.02 cm while the standard deviation for the student height was 8.46 cm. This shows the difference in standing height being relatively high for students, which is supported by a previous study carried out by Darliana (2008). For the 90th percentile of student height, the ranges are from 148.16 to 176.78 cm. From Table 7.1 and Figure 7.2, it is also known that the mean popliteal student height is 40.29 cm with standard deviation 2.21 cm, while the 5th percentile and 95th percentile are 35.50 cm and 43.20 cm, respectively. This means that a suitable chair design for polytechnic students should be within the range of the 90th percentile of the student population.

Table 7.2 presents the descriptive statistics of anthropometric dimensions for the males and females. Figure 7.3 is the graphical representation of Table 7.2. Hence Table 7.2 and Figure 7.3 show the comparison of AD values for male students against the female students. Overall, male AD exceeded female students, and this is normal as men are typically taller compared to women. The comparison involves data values for mean, standard deviation, 5th percentile, and 95th percentile. From the table we see a significant difference for mean weight of male students compared to the mean weight of females where the mean value is 64.69 kg and 54.95 kg, respectively. This can be used as a guide in selecting the type of material for seat design that can accommodate the weight among students in polytechnics. In addition, it can be seen that there is almost no difference in the mean value of the elbows while sitting between male and female students. The mean value is 19.45 cm for male students and 19.00 cm for female students. In addition, the standard deviation values between students are 3.21 cm and 3.27 cm, respectively.

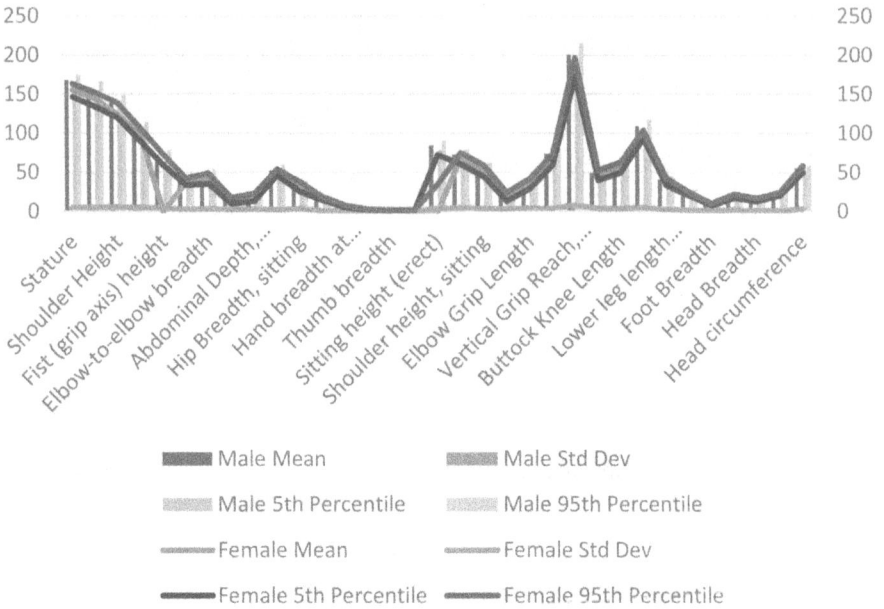

FIGURE 7.3 Anthropometric data for male (n = 213) and female (n = 287).

Figures 7.4 and 7.5 show the average width of the male students is 31.30 cm while the width of the female students is 30.90 cm. According to Dlugos (1999) and Deros et al. (2009), the width of the female hips is greater than that of male. The difference in the value of AD may be due to the age range used in this study between 18–24 years compared to 18–80 years by previous researchers.

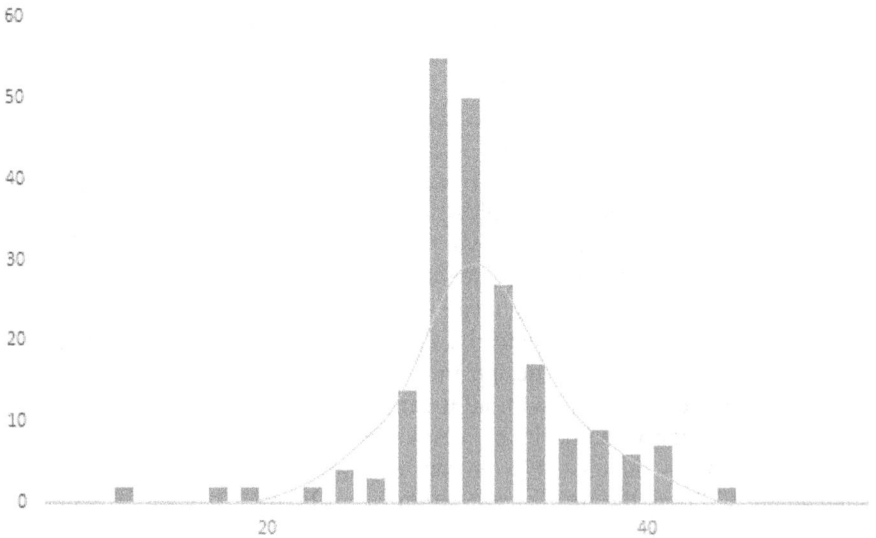

FIGURE 7.4 The normal distribution graph of the hip width for male students.

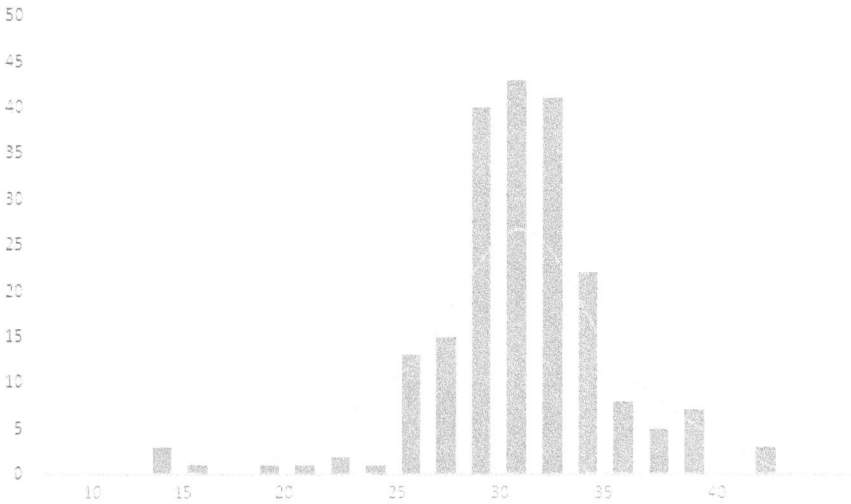

FIGURE 7.5   The normal distribution graph of the hip width for female students.

7.3.1.2   Anthropometric Data Comparison between Students with Chair Dimensions

7.3.1.2.1   Comparative Overview of Students

A comparison of AD among genders is made to look at the comfort level of the chairs used. The number of AD taken was 500 students, 213 males and 287 females.

To determine the comfort of the chairs that was used, the comparison is made between the AD of the students against the dimensions of the seat. This comparison includes AD of the popliteal height against the height of the chair (Figure 7.6)

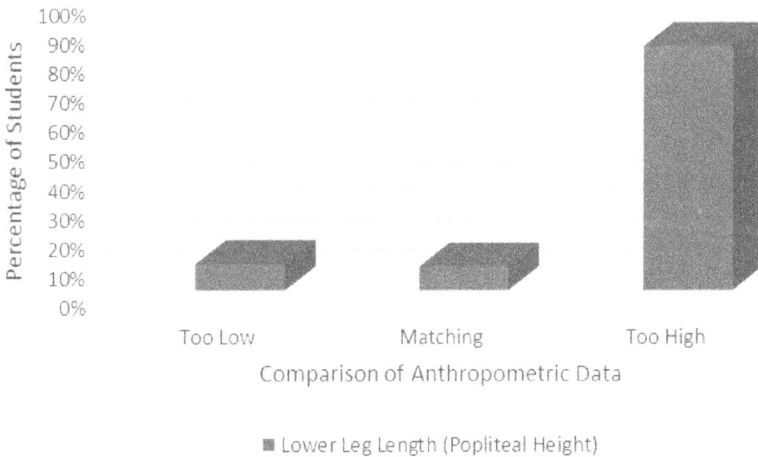

FIGURE 7.6   Percentage of total student matching height of the chair for popliteal height.

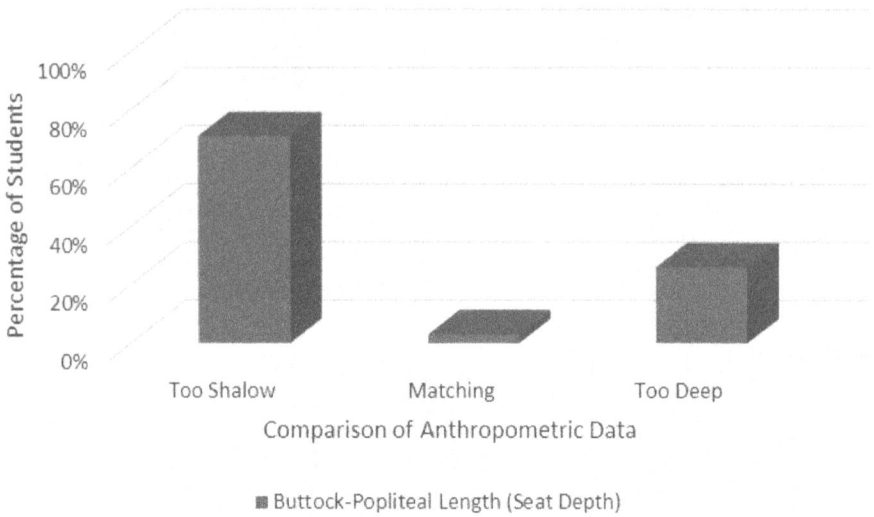

FIGURE 7.7   Percentage of total student matching chairs depth for buttock-popliteal length.

and buttock-popliteal length against the seat depth (Figure 7.7). Figure 7.6 shows the percentage of all students who match the seat for a popliteal height. From Figure 7.6, only 8% of the number of students are matching the height of the chair and are in the 90th percentile of the student population. Also, it has to be noted that, 83% of students are too tall at the popliteal height point. The low percentage of students indicates that the height of the chair in the classroom does not match the polytechnic students' AD.

Figure 7.7 shows the percentage of students who fit the seat for buttock-popliteal length. From the figure it is shown that only 3% of the 90th percentile student population fits with seat depth, whereas 71% of the population are matching for the 5th percentile of buttock-popliteal length. This small percentage value indicates that most of the students do not match the existing seat depth. Therefore, a new chair design should be proposed to provide comfort and safety to students using the chairs.

7.3.1.2.2   Comparison between Genders

Comparison of AD among genders is made to look at the comfort level of seat use. The number of respondents consisted of 213 males and 287 females.

Figure 7.8 shows the percentage of male students matching the height of chair for their popliteal height. The percentage of male students who match the chair height is 15% male students of the 90th percentile of popliteal height. Meanwhile, 57% of male students are taller than the 95th percentile popliteal height. For the male students, there is no mismatch for the 5th percentile population.

Figure 7.9 shows the percentage of female students matching the seat for popliteal height. From the figure it is shown that only 3% of students are fitting with the height of the chair and 97% are matching at popliteal height for over

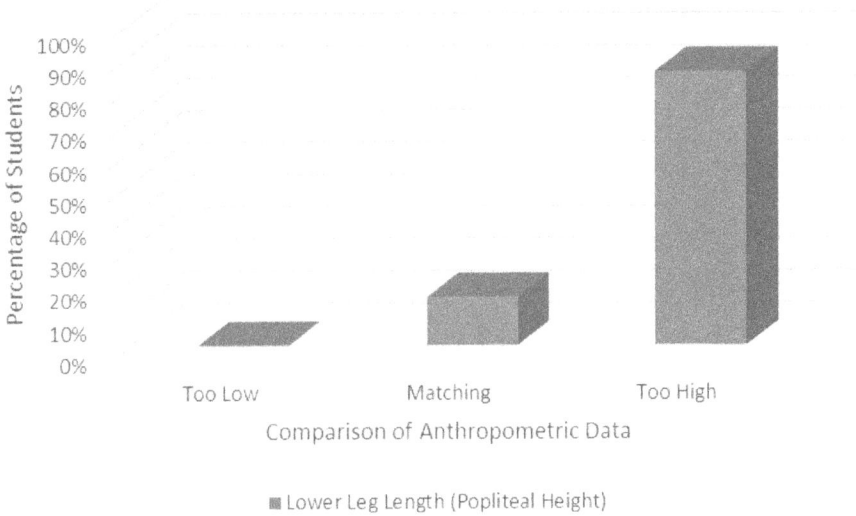

FIGURE 7.8 Percentage of male students matching the seat for a popliteal height.

95th percentile of the population. This shows that male and female students with popliteal height over 95th percentile should ideally be corresponding to the height of chairs. In addition, the difference in the percentage of mismatched male and female students is normal. This is because of the fact that men are usually taller than women.

Figures 7.10 and 7.11 show the percentage of male and female students who match the seats for the buttock popliteal length. The buttock popliteal length is compared

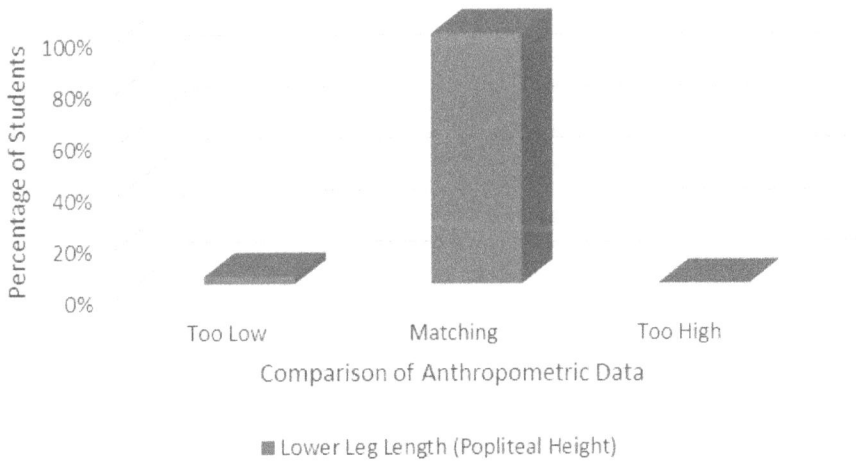

FIGURE 7.9 Percentage of female students who match the seats for popliteal height.

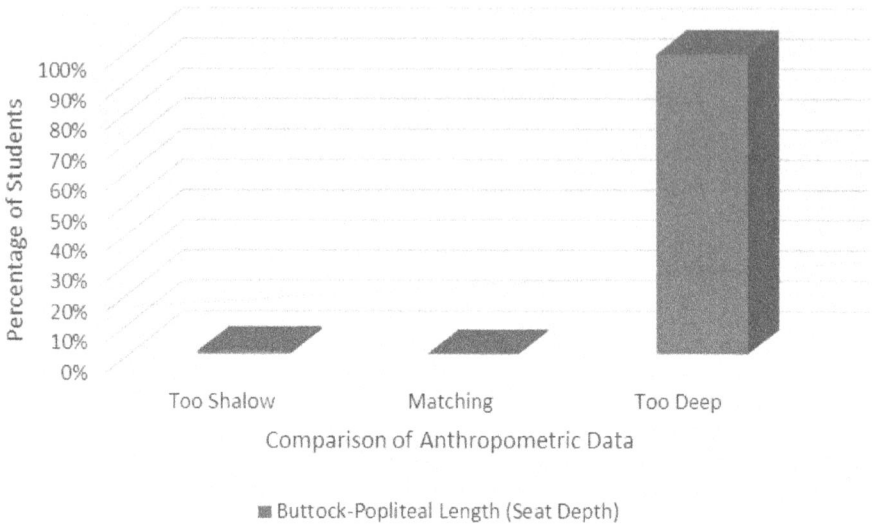

FIGURE 7.10 Percentage of male students matching the seat for the buttock popliteal length.

to the width of the seat, that is, the distance from the back to the front of the seat surface. From Figure 7.10, there is no percentage of male students matching the seat seating dimensions. In contrast, from Figure 7.11, there are 10% of female students who match the 95th percentile of the population of polytechnic students. This indicates that the female buttock popliteal length is shorter than men and this causes the male student to feel uncomfortable while sitting for a long time on the chair. For a student population of the 5th percentile, 1% of the male students and 5% for female students are within comfortable range.

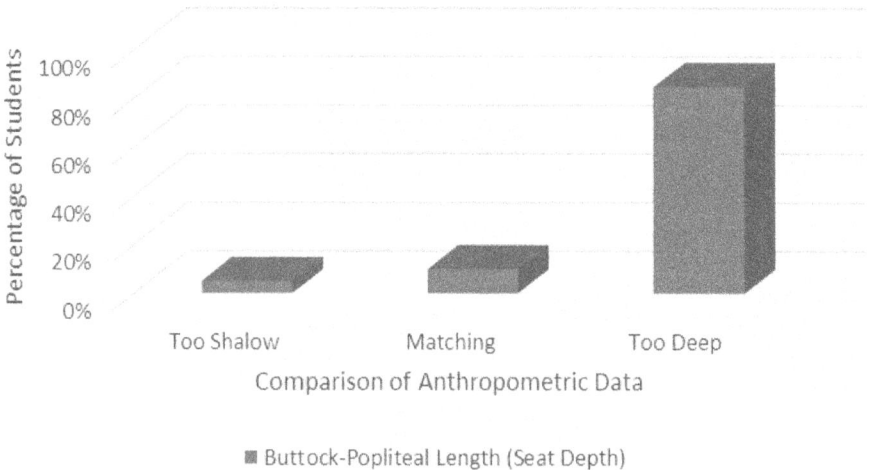

FIGURE 7.11 Percentage of female students matching the seat for the buttock popliteal length.

TABLE 7.3

Comparison of Anthropometric Data with Dimensions of Furniture in the Classroom

Furniture	Parameters	Dimension Anthropometry	Student Anthropometry Data (cm)	Current Furniture Dimensions (cm)
Chair	Chair width	Buttock popliteal length	40.30	40.00
	Chair height	Popliteal height	33.53–43.70	43.38
	Backrest height	Height of shoulder when sitting	44.41	34.00
	Saddle length	Hip width	39.15	48.50
	Backrest length	Shoulder width	49.88	48.00

7.3.2 MISMATCHES BETWEEN CHAIR

The combinations of chairs are not a proper fit and this causes discomfort and pain in some parts of the body, for example, the neck, shoulder, waist and arm of the body. The furniture dimension mismatch also makes the students constantly try to adjust the use of chairs with tables.

Table 7.3 shows the comparison between students' AD and existing seat dimensions.

All in all, comparison between students' AD and design dimensions of the classroom furniture is summarized in Table 7.4. It can be observed that there exist several mismatches based on the design standards for furniture.

Data obtained from anthropometric dimensions indicates that there are a lot of differences even between people from the same country who make up the population of students in polytechnics across Malaysia. There is a mismatch between the anthropometric dimensions and the current furniture used in classrooms of Politeknik Kuching, Sarawak, Malaysia. From the comparison result, several mismatches were found and this leads to the conclusion that the design of furniture used in the classroom needs to be improved to better accommodate the studying

TABLE 7.4

New Dimensions of the Furniture to Match the Anthropometric Data

Furniture	Parameters	Student Anthropometry Data (cm)	Dimension Anthropometry (Baba et al., 2008)	Recommended New Dimension of Furniture (cm)
Chair	Chair width	40.00	39.00	40.00
	Chair height	33.00–44.00	38.00–53.00	36.00–49.00
	Backrest height	44.00	46.44	45.00
	Saddle length	39.00	48.00	44.00
	Backrest length	50.00	—	50.00

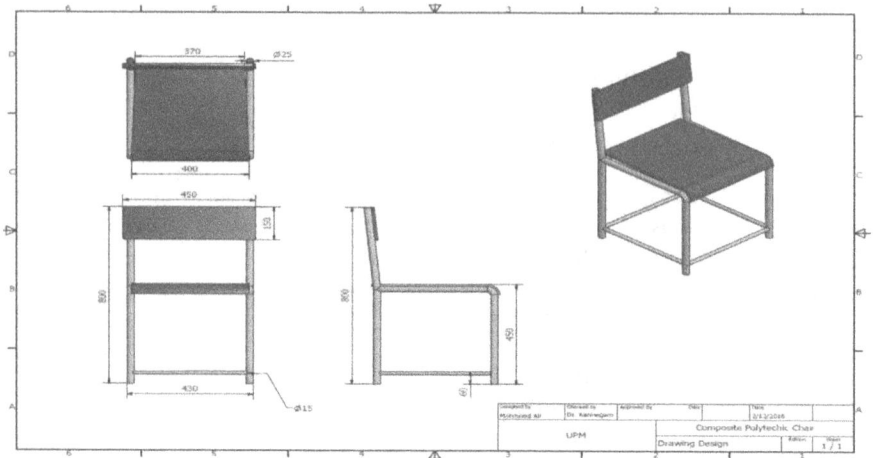

FIGURE 7.12   Finalized engineering drawing for composite polytechnic chair (units in mm).

process of the students. The scenario can also be extended to other polytechnics and universities in Malaysia. The study results provide the necessary support to the idea that there are ongoing problems with the design of chairs in the current polytechnic classroom, and this is the solid background for the design of new biocomposite chairs based on AD.

7.3.3   DETAIL DESIGN OF BIOCOMPOSITE CHAIR WITH INPUTS FROM AD

A detail drawing of the selected concept of biocomposite polytechnic chair with inputs from AD is shown in Figure 7.12. An important activity in this study is the development of biocomposite for polytechnic chairs. The density and mechanical properties of biocomposite material are shown in Table 7.5. The final selected

TABLE 7.5
Density and Mechanical Properties of Biocomposite Material (Sathishkumar, 2016)

Density (g/cm³)	Tensile Strength (MPa)	Tensile Modulus (GPa)	Elongation (%)
—	692	10.94	4.3
—	930	53	1.6
1.45	930	53	1.6
1.4	284–800	21–60	1.6
1.5	350–600	40	2.5–3.5
0.75	400–550	—	—
0.6	—	—	—
0.749	223–624	11–14.5	2.7–5.7
1.2	295	—	3–10

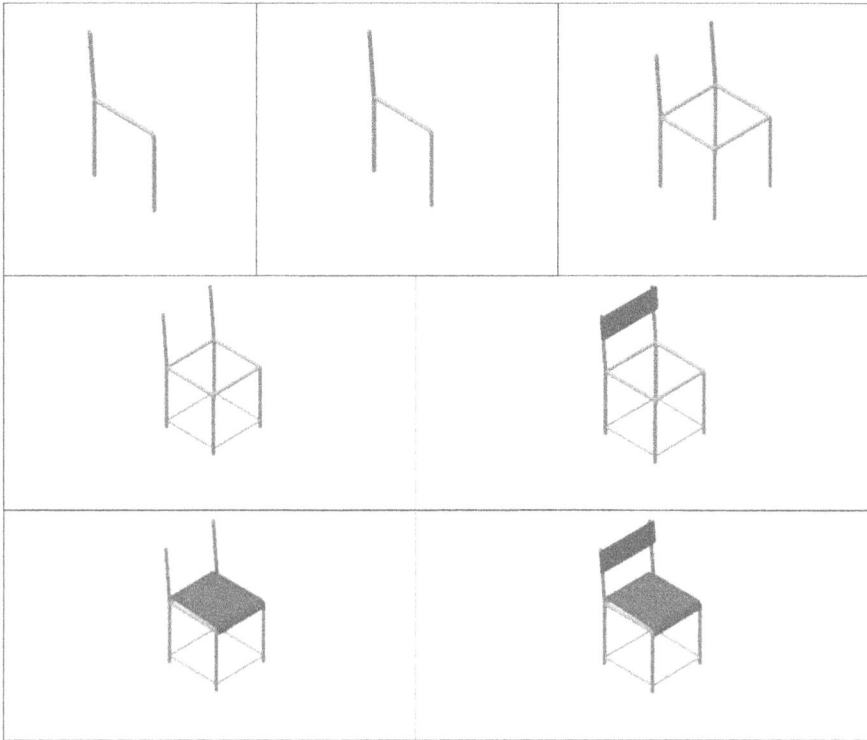

FIGURE 7.13   Subcomponents for the composite polytechnic chair.

concept was chosen because it scored the highest value of the weighted objectives and hence is the best solution. Figure 7.12 shows the engineering drawing of the composite polytechnic chair with dimensions using Autodesk Inventor 2015 software.

The subcomponents for the composite polytechnic chair have also been extensively designed. This is shown in Figure 7.13.

7.4   CONCLUSION

This study was undertaken to design an anthropometrically enhanced biocomposite chair to be used by students of polytechnics. The biocomposite chair was successfully designed according to anthropometric data (AD) inputs. The product design specification (PDS) of the biocomposite chair was based on inputs from AD. This type of study is a new type of way toward designing chairs that are anthropometric compliant for Malaysian students. Therefore, this study was undertaken in order to design chairs that are anthropometric compliant for Malaysian students in polytechnics. The design concept involved brainstorming, mind mapping, and a morphological chart. The evaluation of the best design concept was completed by using weighted objectives method. Prior to releasing the final design of the biocomposite chair, the design of the new chair was validated anthropometrically and via feedback from users under normal working conditions.

ACKNOWLEDGMENTS

This project was done as part of academic research work in the polytechnic and no funding was received from any government or private organization. The authors would like to thank the Department of Polytechnic Education, Ministry of Higher Education, Malaysia, for providing the scholarship award to the principal author in this project.

REFERENCES

Aminian, N. O., & Romli, F. (2012). Mismatch between anthropometric body dimensions and classroom furniture in Malaysian Universities. *Proceedings of the Canadian Engineering Education Association, 1*(1), 1–5.

Arpitha, G., & Yogesha, B. (2017). An overview on mechanical property evaluation of natural fiber reinforced polymers. *Materials Today: Proceedings, 4*(2), 2755–2760.

Darliana, M., Baba, M. D., & Ahmad, R. I. (2008). Development of an Anthroprometric Database for the Malaysian Population. *Universiti Kebangsaan Malaysia's Advanced Manufacturing Group Seminar Proceedings 2008 (AMReG 08).*

Dawal, S. Z. M., Ismail, Z., Yusuf, K., Abdul-Rashid, S. H., Shalahim, N. S. M., Abdullah, N. S., ... Kamil, N. S. M. (2015). Determination of the significant anthropometry dimensions for user-friendly designs of domestic furniture and appliances–experience from a study in Malaysia. *Measurement, 59*, 205–215.

Deros, B. M., Hassan, N. H. H., Daruis, D. D. I., & Tamrin, S. B. M. (2015). Incorporating Malaysian's population anthropometry data in the design of an ergonomic driver's seat. *Procedia-Social and Behavioral Sciences, 195*, 2753–2760.

Deros, B. M., Mohamad, D., Ismail, A. R., Soon, O. W., Lee, K. C. & Nordin, M. S. (2009). Recommended chair and work surfaces dimensions of VDT tasks for Malaysian citizens. *European Journal of Scientific Research, 34*, 156–167.

Dlugos, C. (1999). Lecture 19-Pelvic Cavity and Organs. Available from http://www.smbs. buffalo.edu/ana/newpage45.htm. State University of New York Buffalo, pp. 332–388.

El-Shekeil, Y., Sapuan, S., Abdan, K., & Zainudin, E. (2012). Influence of fiber content on the mechanical and thermal properties of Kenaf fiber reinforced thermoplastic polyurethane composites. *Materials & Design, 40*, 299–303.

Ho, M.-p., Wang, H., Lee, J.-H., Ho, C.-k., Lau, K.-t., Leng, J., ... Hui, D. (2012). Critical factors on manufacturing processes of natural fibre composites. *Composites Part B: Engineering, 43*(8), 3549–3562.

Klamklay, J., Sungkhapong, A., Yodpijit, N., & Patterson, P. E. (2008). Anthropometry of the southern Thai population. *International Journal of Industrial Ergonomics, 38*(1), 111–118.

Nor, F. M., Abdullah, N., Mustapa, A.-M., Wen, L. Q., Faisal, N. A., & Nazari, D. A. A. A. (2013). Estimation of stature by using lower limb dimensions in the Malaysian population. *Journal of Forensic and Legal Medicine, 20*(8), 947–952.

Sahari, J., Sapuan, S., Zainudin, E., & Maleque, M. A. (2013). Mechanical and thermal properties of environmentally friendly composites derived from sugar palm tree. *Materials & Design, 49*, 285–289.

Salleh, F. M., Hassan, A., Yahya, R., & Azzahari, A. D. (2014). Effects of extrusion temperature on the rheological, dynamic mechanical and tensile properties of kenaf fiber/ HDPE composites. *Composites Part B: Engineering, 58*, 259–266.

Sanjay, M., & Yogesha, B. (2017). Studies on natural/glass fiber reinforced polymer hybrid composites: An evolution. *Materials Today: Proceedings, 4*(2), 2739–2747.

Sapuan, S., & Maleque, M. (2005). Design and fabrication of natural woven fabric reinforced epoxy composite for household telephone stand. *Materials & Design*, *26*(1), 65–71.

Sathishkumar, T. (2016). Development of snake grass fiber-reinforced polymer composite chair. *Proceedings of the Institution of Mechanical Engineers, Part L: Journal of Materials: Design and Applications*, *230*(1), 273–281.

SIRIM, Standard and Industrial Research Institute of Malaysia. (2008). MS ISO 15535:2008 General requirements for establishing anthropometric databases (ISO 15535:2006, IDT) *Malaysian Standard*. Kuala Lumpur, Malaysia: Department of Standards Malaysia.

Taifa, I. W., & Desai, D. A. (2017). Anthropometric measurements for ergonomic design of students' furniture in India. *Engineering Science and Technology, an International Journal*, *20*(1), 232–239.

Troussier, B. (1999). Comparative study of two different kinds of school furniture among children. *Ergonomics*, *42*(3), 516–526.

Wang, Q., & Yang, A. (2012). The Furniture Design Strategies Based on Elderly Body Size. In *Soft Computing in Information Communication Technology* (pp. 501–509). Springer, Berlin, Heidelberg.

Yahaya, R., Sapuan, S., Jawaid, M., Leman, Z., & Zainudin, E. S. (2016). Effect of fibre orientations on the mechanical properties of kenaf–aramid hybrid composites for spall-liner application. *Defence Technology*, *12*(1), 52–58.

8 A Review on Nanocellulose Composites in Biomedical Application

N. S. Sharip,[1] T. A. T. Yasim-Anuar,[2]
M. N. F. Norrrahim,[3] S. S. Shazleen,[1]
N. Mohd. Nurazzi,[4] S. M. Sapuan,[5,6] and R. A. Ilyas[5,6]

[1]Laboratory of Biopolymer and Derivatives,
Institute of Tropical Forestry and Forest Products
(INTROP), Universiti Putra Malaysia, Malaysia

[2]Department of Bioprocess Technology, Faculty
of Biotechnology and Biomolecular Sciences,
Universiti Putra Malaysia, Malaysia

[3]Research Centre for Chemical Defence, Universiti
Pertahanan Nasional Malaysia, Malaysia

[4]Center for Defence Foundation Studies, Universiti
Pertahanan Nasional Malaysia, Malaysia

[5]Advanced Engineering Materials and Composites (AEMC),
Department of Mechanical and Manufacturing Engineering,
Faculty of Engineering, Universiti Putra Malaysia, Malaysia

[6]Laboratory of Biocomposite Technology,
Institute of Tropical Forestry and Forest Products
(INTROP), Universiti Putra Malaysia, Malaysia

CONTENTS

8.1   INTRODUCTION

Cellulose is the most abundant component in biomass and finds applications in many spheres of modern industry (Ilyas et al., 2017; Ilyas & Sapuan, 2020; Ilyas, Sapuan, & Ishak, 2018; Kalia, Avérous, et al., 2011; Sanyang et al., 2018). Cellulose is a carbohydrate polymer comprising a repeating unit of β-D-glucopyranose units linked by β-1, 4 glycosidic bonds (Qin et al., 2008). Cellulose has grabbed the attention of researchers due to its potential for several applications such as nanocellulose, biosugars, biocomposites, pulp and paper, and bioethanol (Aisyah et al., 2019; Asyraf et al., 2020; Atiqah et al., 2019; Azammi et al., 2020; Jumaidin, Ilyas, et al., 2019; Jumaidin, Khiruddin, et al., 2019; Jumaidin, Saidi, et al., 2019; Norizan et al., 2020; Nurazzi, Khalina, Sapuan, Ilyas, et al., 2019; Nurazzi, Khalina, Sapuan, & Ilyas, 2019). There is a growing interest in the production of nanocellulose because of its interesting properties such as high specific surface area, high crystallinity, low density, non-abrasive and combustible nature, non-toxicity, low cost, and biodegradability (Ilyas, Sapuan, Sanyang, et al., 2018; Kalia, Dufresne, et al., 2011; Norrrahim, Ariffin, Yasim-Anuar, Ghaemi et al., 2018). Nanofiber including nanocellulose can be described as a fiber that has a diameter of 100 nm or less with extremely high specific area and high porosity which contribute to its excellent pore interconnectivity (Abral et al., 2019, 2020; Ilyas, Sapuan, Ibrahim, Abral, et al., 2019; Syafri et al., 2019). The tailorable surface chemistry and functionality may enable achievement of desired characteristics of nanocellulose materials befitting their targeted applications.

Nanocellulose can be classified into cellulose nanofiber (CNF), cellulose nanocrystals (CNC), and bacterial nanocellulose (BNC) based on its dimensions, functions, and method of preparation (Ilyas, Sapuan, Ibrahim, Atikah, et al., 2019). Both CNF and CNC are considered plant-derived nanocellulose produced through the disintegration of plant cellulose using mechanical or chemical methods, accordingly. Subjected to chemical treatment or acid hydrolysis, CNC possesses a rod-like shape with near-perfect crystallinity (Abitbol et al., 2016; Hazrol et al., 2020; Hubbe et al., 2008; Ilyas et al., 2018b, 2018c). The size of CNC is approximately 5–70 nm wide and less than 100 nm in length (Kaboorani & Riedl, 2015). Meanwhile, CNF of micrometer-long fiber with a 20–40 nm diameter size consists of both amorphous and crystalline structure (Abitbol et al., 2016; Ilyas et al., 2018a; Ilyas, Sapuan, Ishak, et al., 2019; Ilyas et al., 2019a; Ilyas, Sapuan, Ishak, Zainudin, et al., 2018; Siró & Plackett, 2010). BNC, on the other hand, is secreted extracellularly by bacteria such as *Acetobacter* sp., *Agrobacterium* sp., *Alcaligenes* sp., *Pseudomonas* sp., *Rhibozium* sp., or *Sarcina* sp. (El-Saied et al., 2004; Jonas & Farah, 1998), with a general size of 20–80 nm in diameter (Dima et al., 2017; Jozala et al., 2016; Mohammadkazemi et al., 2015).

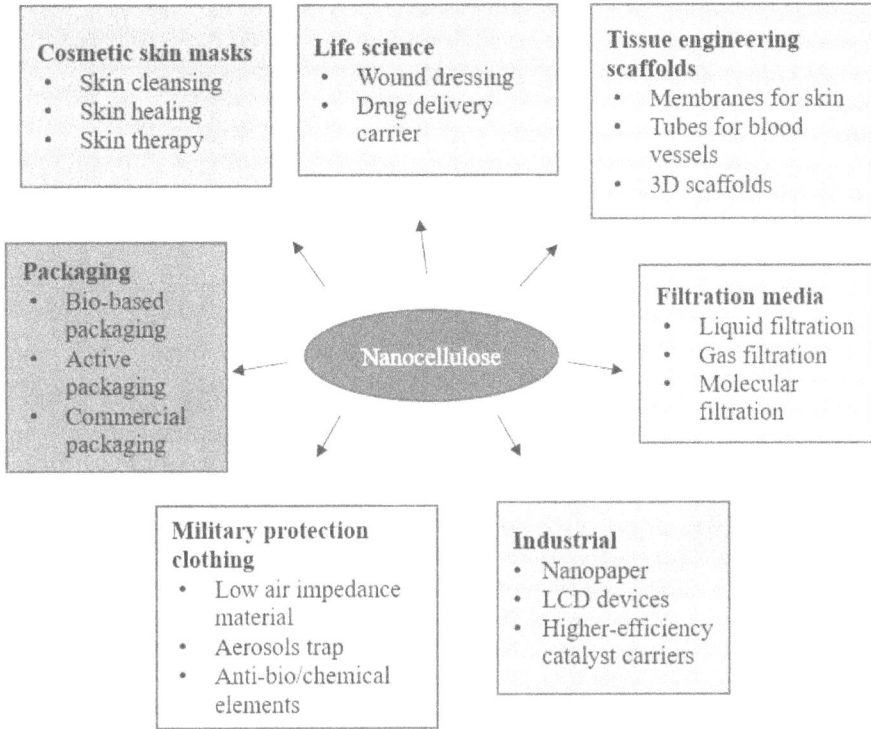

FIGURE 8.1   Applications of nanocellulose.

Interest in the use of nanocellulose is increasing for several applications relevant to the fields of materials science, biomedical engineering, cosmetics, pharmaceuticals, foods, and packaging (Ariffin et al., 2017; Haafiz et al., 2013; Yasim-Anuar et al., 2019). Figure 8.1 shows recent developments in the applications of nanocellulose. In terms of biomedical application, nanocellulose could provide mechanical support to the tissues in which it resides, despite being physically insoluble and inelastic. It also possesses excellent biocompatibility and biodegradability, captivating high interest among researchers and industries as a cost-effective advanced material for biomedical applications.

The high stiffness of nanocellulose enables increments of mechanical strength of general-purpose thermoplastics polymers such as polypropylene (PP), polyethylene (PE), and polylactic acid (PLA). Polymers are versatile, and they have been used in a wide range of applications for various purposes and in various fields. Nevertheless, some polymers require a reinforcement agent, also known as a filler, to improve their properties and fulfill the requirement of targeted applications. Owing to its renewable nature, anisotropic shape, outstanding mechanical properties, and good biocompatibility, nanocellulose has been gaining interest to be used as a filler for polymer composites (Norrrahim, Ariffin, Yasim-Anuar, Hassan et al., 2018; Yasim-Anuar et al., 2019). Table 8.1 shows several applications of composites from nanocellulose.

TABLE 8.1

Applications of Nanocellulose Composites

Composite	Findings	Reference
Packaging	– Nanocellulose composites from sugarcane bagasse and oil palm biomass cellulose were successfully produced with improved properties and suitable to be applied in packaging.	Ghaderi et al. (2014); Norrrahim, Ariffin, Yasim-Anuar, Hassan et al. (2018); Atikah et al. (2019); Ilyas et al. (2020); Ilyas, Sapuan, Atiqah, et al. (2019); Ilyas et al. (2019b)
Electronic	– Production of malleable displays, solar cells, smart cards, radio frequency tags, medical implants, and wearable computers. – Nanocellulose paper has high optical transparency and a low coefficient of thermal expansion.	Pandey et al. (2013); Hazrol et al. (2019, 2020)
Building material	– Production of load-bearing walls, stairs, roof systems, and subflooring. – High-performance material.	Uddin & Kalyankar (2011)
Automobile	– Replacement material for glass or carbon fiber polymer composites that are mainly used as door panels, package trays, and trunk liners in cars and trucks.	Masoodi et al. (2012)
Digital display	– Production of optically transparent plastic substrate for bendable displays. – Multiple advantages such as high paper-like reflectivity, flexibility, contrast, and biodegradability.	Nogi et al. (2009)
Pharmaceutical and medical	– Excellent compaction properties when blended with another pharmaceutical excipient. – Application as skin replacements for burns and wounds, drug-releasing system, blood vessel growth, nerves, gum, and dura mater reconstruction, scaffolds for tissue engineering, stent covering, and bone reconstruction.	Kalia, Dufresne, et al. (2011)

TABLE 8.2

Examples of Nanocellulose Polymer Composites for Biomedical Applications

Type of Polymer	Nanocellulose	Targeted Applications	References
Polylactic acid (PLA)	Cellulose nanocrystals (CNC)	Scaffold	Shi et al. (2012)
Polyvinyl alcohol (PVA) hydrogel	Cellulose nanofiber (CNF), CNC	To mimic collagenous soft tissues	Tummala et al. (2017)
Polypropylene	Bacterial nanocellulose (BNC)	For reconstructive surgery of soft and hard tissues	Ludwicka et al. (2019)
PVA	BNC	To mimic artificial blood vessels	Tang et al. (2015)
Poly(vinyl pyrrolidone)	CNC	*In vitro* wound dressing	Poonguzhali et al. (2017)

From the aspect of biomedical applications, biostability and biodegradability of polymers need to be considered in selecting a polymer (Francis, 2018). Polyamides, polyesters, polyanhydrides, poly(ortho esters), poly(amido amines), polyhydroxy-alkanoates, poly(β-amino esters), poly(lactic-co-glycolic acid), poly(glycolic acid), and PLA are among the important and safe biomedical polymers that have been mainly used in biotechnology and medicine. Table 8.2 shows some examples of nanocellulose polymer composites for biomedical applications.

8.2 BIOCOMPATIBILITY AND TOXICOLOGY OF NANOCELLULOSE COMPOSITES

The exploration of nanocellulose for biomedical applications is gaining attention mainly due to its favorable properties particularly its biocompatibility and low toxicology toward living cells. A search was done on lens.org using the keyword "nanocellulose biocompatibility" and it was found that the number of manuscripts focusing on the biocompatibility of nanocellulose has been increasing for years, as shown in Figure 8.2. Nevertheless, a similar search using another keyword "nanocellulose toxicology" showed an inconsistent trend of toxicology studies on nanocellulose, especially after 2015, as shown in Figure 8.3. This is worrisome as the toxicology properties of a material are one of the important factors that need to be considered for the development of safe biomedical applications.

Biocompatibility, in general, refers to the ability of an implanted material to perform its desired function without causing undesirable or systemic effects to the host (Afrin & Karim, 2015). In general, biocompatibility properties of materials for medical applications are widely tested either using *in vitro* or *in vivo* methods (Ferraz et al., 2012). According to Afrin and Karim (2015), the *in vitro* test is usually conducted at the initial stage to identify the host's responses toward the implanted

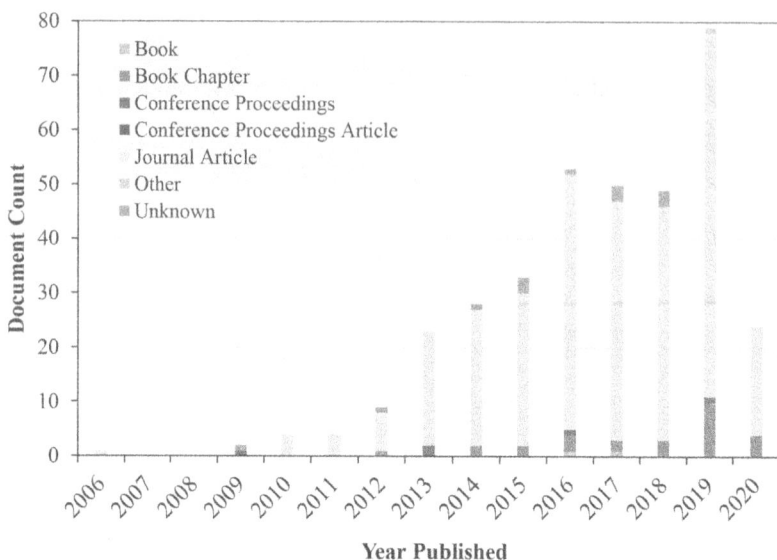

FIGURE 8.2   A chart of published manuscripts focused on nanocellulose compatibility.

material under controlled conditions. Nevertheless, the true responses may not be reflected, and hence, *in vivo* testing is required at later stages.

For biomedical polymeric-based composites, many factors can influence their biocompatibility such as copolymer ratio, chemical structure and functional groups of the polymer, as well as morphologies and processing of fillers (Kulkarni & Rao, 2013). All these factors may influence the cellular activity of the targeted host.

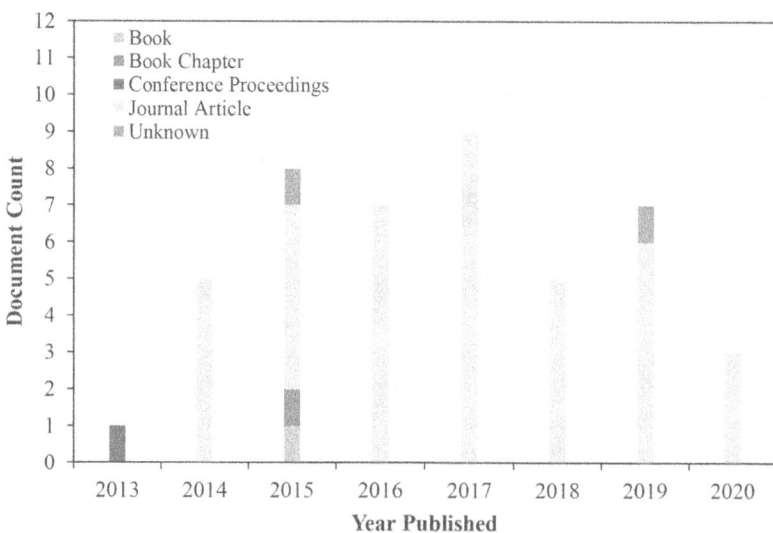

FIGURE 8.3   A chart of published manuscripts focused on nanocellulose toxicology.

In terms of cellulose, it cannot be degraded by the human body because it lacks cellulolytic enzymes; this may lead to some incompatibility (Lin & Dufresne, 2014). A similar effect is expected for nanocellulose.

Due to this issue, Zhang et al. (2019) tried to surface modify nanocellulose, particularly CNF, to improve its biocompatibility. The findings showed that crosslinking of CNF with adipic acid dihydrazide (ADH) and oxidized konjac glucomannan (OKGM) was able to improve the biocompatibility of CNF. It was reported that the cell viability kept was around 90% at all tested concentrations (1.25–5 mg/mL) for OKGM-ADH CNF, suggesting good biocompatibility of these materials. In fact, it was found that the OKGM-ADH CNF was able to support the growth of cells on the membrane surface. The cells survived, attached, and proliferated on the membrane.

In addition to biocompatibility, the toxicology properties of an implanted material need to be determined, to ensure it does not chemically interfere with the body system. Fundamentally, the main concept of toxicology is the "dose-response" relationship (Bourgeois et al., 2016). The severity of the toxic effect is related to the dose/amount of substance that enters the body. In terms of nanocellulose, its toxicology properties have not been widely studied and, in fact, it has yet to be accepted as a safe implanted material in living organisms, especially humans. Thus, investigating CNF toxicology for biomedical applications is of great importance.

Recently, Deloid et al. (2019) studied the toxicology of CNF when it is consumed by rats. *In vivo* toxicity was evaluated in rats gavaged twice weekly for 5 weeks with 1% w/w suspensions of CNF. The findings revealed that CNF is nontoxic and nonhazardous when ingested in small quantities. Other than the study by Deloid et al. (2019), Table 8.3 summarized the toxicology reports involving nanocellulose.

TABLE 8.3

Examples of the Toxicological Evaluation of Nanocellulose-Based Materials for Biomedical Applications

Nanocellulose Type	Toxicological Experiment	Findings
Cellulose nanofiber (CNF)	*In vitro* cytotoxicity test of CNFs with fibroblast 3T3 cells.	The CNFs did not exert toxic behavior on fibroblast cells and showed no effect on the cell membrane, mitochondrial activity, or DNA proliferation.
Cellulose nanocrystals (CNC)	*In vitro* cytotoxicity evaluation of CNCs with nine different cell lines.	No cytotoxic effects in the concentration range (0–50 µg/mL) and with an exposure time of 48 hours.
Bacterial nanocellulose (BNC)	*In vitro* and *in vivo* cytotoxicity of BNC in human umbilical vein endothelial cells (with viability and flow cytometric assays) and mouse model.	No toxicity in endothelial cells and no biochemical differences were observed after 7 days in animal experiments.

Source: Adapted from Jorfi and Foster (2015).

All in all, most of the reports suggested that nanocellulose is safe and does not cause damage at the cellular and genetic levels.

8.3 BIOMEDICAL APPLICATION OF NANOCELLULOSE COMPOSITES

In consideration of biocompatibility and the remarkable reinforcement effect of nanocellulose in composite applications, numerous studies related to various biomedical applications have been conducted. Generally, biomedical or biomaterial application can be classified into several categories including for uses in pharmaceutical field (i.e., drug deliveries), general surgeries (i.e., sutures, burn dressings, and skin substitutes), cardiovascular medical devices (i.e., stents, grafts), dental and orthopedic applications (i.e., implants, scaffolds), ophthalmologic applications (i.e., contact lenses, prosthetic retina), and bioelectrodes or biosensors. Several studies on nanocellulose for biomedical application have been reviewed elsewhere (Abdul Khalil et al., 2015; Abitbol et al., 2016; Jorfi & Foster, 2015; Sharip & Ariffin, 2019).

Nontoxic, biodegradable and biocompatible, cellulose-based hydrogels are highly hydrated porous cellulosic soft materials with good thermal and mechanical properties. These nanocellulose-based gels can be processed from plant or bacterial cellulose, which are nontoxic, highly hydrated porous soft, biocompatible, biodegradable, hydrophilic, and renewable. Nanocellulose, whether CNC, CNF, or BNC, has abundant hydroxyl groups, high surface area, large specific surface area, high crystallinity, high strength and stiffness, low weight, biodegradability, high aspect ratio, great mechanical properties and thermal resistance, and it can be chemically modified with functional groups or by grafting biomolecules. According to Curvello et al. (2019), there are five main categories of nanocellulose hydrogel applications: 3D cell culture, tissue engineering, drug delivery, diagnostics, and separation of biomolecules (Figure 8.4). Herein, recent findings on nanocellulose composites for biomedical application are highlighted.

8.3.1 PHARMACEUTICAL

An important aspect related to drug delivery systems to the target treatment site is a controlled release of the drug or biological agent (Rezaie et al., 2015). A regulated controlled system is essential to ensure the release of the right dosage at the right time and to prevent the early release of the drug prior to reaching the target site. The use of coating on the drug-embedded matrix or carrier and a hydrophilic matrix has been addressed as one of the popular approaches in developing extended-release dosage forms attributed to its flexible and well-designed structure allowing a reproducible release profile (Abdul Khalil et al., 2015).

For instance, the release of 5-fluoracil (5-FU), a drug for cancer treatment, was found controllable in polymer-modified CNF cryogel microspheres. The CNF acts as a microreactor possessing a robust and highly porous framework, enabling good stability of the temperature-sensitive monomer, N-isopropylacrylamide (NIPAm), thus resulting in a controllable drug release system that is dependent on temperature

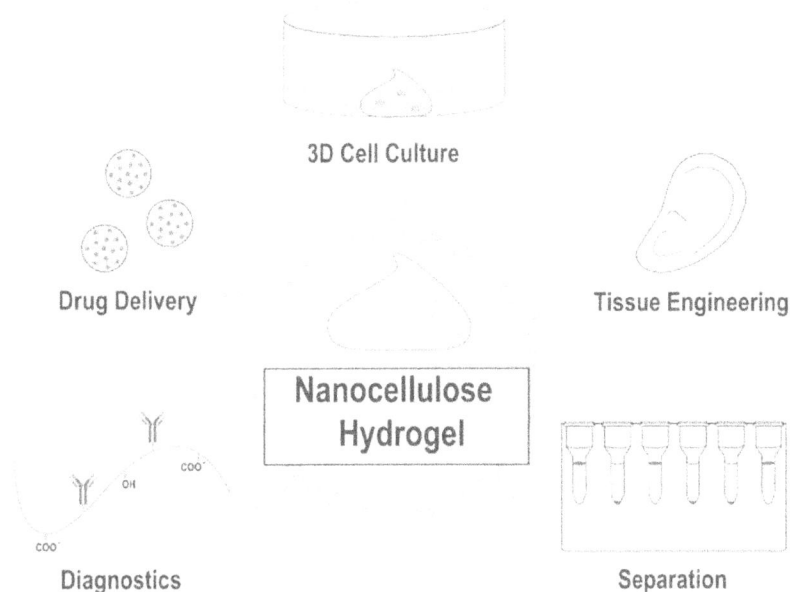

3D Cell Culture

Drug Delivery

Tissue Engineering

Nanocellulose Hydrogel

Diagnostics

Separation

FIGURE 8.4 Nanocellulose hydrogel is biocompatible, biodegradable, and shows potential for multiple biomedical applications. (Reproduced with permission from Curvello et al., 2019.)

changes (Zhang et al., 2016). Another study reported the uses of nanocellulose composite hydrogels for regulating drug releases through temperature and pH changes. Addition of poly-NIPAm with alterations of carboxyl charge level of the CNF during 2,2,6,6-tetramethylpiperidine-N-oxyl (TEMPO)-mediated oxidation resulted in dual-responsive behavior of drug releases, suggesting promising use of CNF–poly(NIPAm) composite hydrogels for drug delivery systems (Masruchin et al., 2018). A more recent study was conducted using CNF and TEMPO-mediated oxidized CNF with polyvinyl alcohol (PVA) for investigating the releases of acetaminophen. The alternated amorphous and crystalline structure of CNF in PVA aids in the entanglement of the drug in the matrix, hence enabling a controlled released of acetaminophen by over 144 hours (O'Donnell et al., 2020).

A study conducted by Müller et al. (2013) on the applicability of BNC as a drug delivery system for proteins, using serum albumin as a model drug, showed that the BNC is proven to be non-cytotoxic as well as environment friendly with an excellent biocompatibility (Figure 8.5). This was focused on proteins as they are introduced into the market as an increasing part of pharmaceutical compounds. In addition, there is a growing interest for BNC as additives in medical devices; for example, in growth factors, cell attractants for tissue engineering, or antibodies for wound healing.

Moreover, in order to identify the usage of novel materials in drug delivery applications, Mohd Amin et al. (2012) investigated the use of bacterial cellulose (BC) incorporated with different proportions of acrylic acid (AA) to produce hydrogels

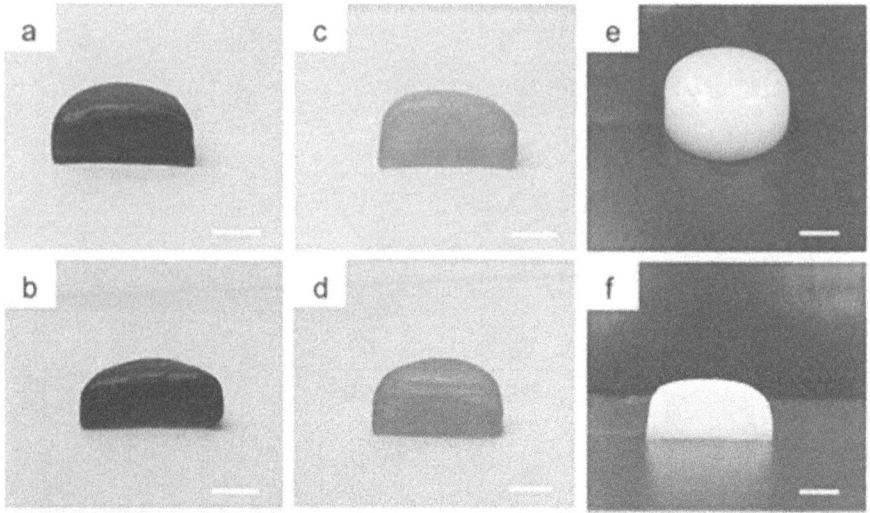

FIGURE 8.5 Visualization of bovine serum albumin (BSA) distribution in bacterial nanocellulose (BNC) (cross sections) by staining with BCA assay reagent. BNC loaded by adsorption (a) and high-speed technique (b), corresponding unloaded controls incubated only in buffer without BSA (adsorption, c; vortex treatment, d), and untreated, unstained BNC (e and f) (scale bars, 5 mm). (Reproduced with permission from Müller et al., 2013.)

by exposure to accelerated electron-beam irradiation at different doses. The result revealed that the morphology of the hydrogels is much dependent on the irradiation dose and composition of hydrogel. These morphological observations propose that the highly porous sponge-like structure of BC/AA hydrogels helps water diffusion in all directions, thus making these hydrogels suitable for drug release systems for protein-based drugs (Figure 8.6). The pore sizes of the 208035 (60–190 µm), 208050 (20–110 µm), and 307035 (2–50 µm) hydrogels recommend that they are more suitable compared to the other hydrogels for bovine serum albumin (BSA) loading.

8.3.2 Wound Dressings and Skin Substitutes

The utilization of nanocellulose as wound dressings and skin substitutes is highly applicable as compared to other biomedical purposes. Cellulose-based materials are capable of controlling wound exudates and providing a moist environment to the wound, resulting in better wound healing (Czaja et al., 2006). The process involves tissue regeneration, repair, and reconstruction in which clots form trigger recruitment of macrophage and neutrophils. This event is followed by fibroblast recruitment around the wound area, thus filling it with collagen. As a result, rebuilt tissue is attained, along with induction of endothelial cell proliferation and vessel generation prior to mature skin and finally wound closings (Halib et al., 2017; Sharpe & Martin, 2013). This whole process could be delayed by infections, particularly if the wound is formed through thermal sources or burns, which are responsible for up to 75% morbidity (Heo et al., 2013).

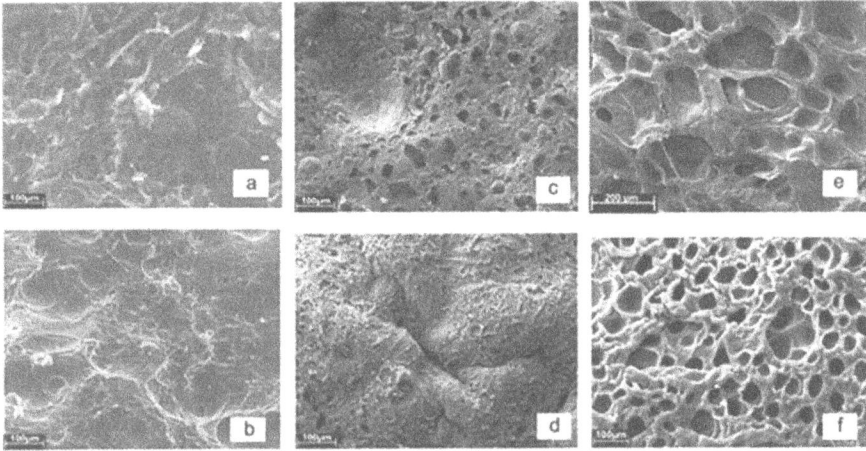

FIGURE 8.6   Scanning electron microscopy (SEM) images of thermo- and pH-responsive poly(acrylic) acid-bacterial nanocellulose hydrogel applied to drug delivery. (Reproduced with permission from Mohd Amin et al., 2012.)

In wound dressing applications, nanocellulose has often served as a matrix, where antimicrobial substances such as silver nanoparticles (AgNP) are incorporated into/or onto nanocellulose via several means including coating and chemical reduction (Maneerung et al., 2008). As such, AgNP/BNC composites were found to successfully exhibit remarkable inhibitory activities for wound infection prevention (Berndt et al., 2013; Hu et al., 2009; Jung et al., 2009; Maneerung et al., 2008; Sambhy et al., 2006). A CNF-*grafted*-titania nanocomposite fabricated through surface binding mechanisms was also found successful in inhibiting the growth of common pathogens, Gram-positive *Staphylococcus aureus* and Gram-negative *Escherichia coli* irradiation (Galkina et al., 2015). Other studies showed that development of transparent porous nanodiamond/cellulose nanocomposite membranes enabled doping of antimicrobial substance or drug onto the resulted products to serve as multifunctional biomaterials for wound dressing applications (Luo et al., 2016).

CNF could also be used as a matrix to guide the growth of collagen to form composite aerogels comprising interwoven CNF/collagen fibers. Modification of CNF into dialdehyde through oxidation enabled the formation of active sites on its surface for crosslinking reaction with collagen. Exhibited low densities between 0.02 and 0.03 g cm^{-3}, high porosities up to 95%, and excellent water absorption up to 4,000%, this material showed good biocompatibility and high levels of cell activity, indicating its proprieties suitable to be used for wound dressings and as tissue-engineered scaffolds (Lu et al., 2014). Similarly, owing to its excellent biocompatibility, BNC/acrylic acid hydrogel was found successful in increasing the healing rate of burn wounds. The *in vivo* experiment against mice treated with the composite hydrogel was 19% higher than control-treated mice, showing promising applications for nanocellulose composites as wound dressings and cell carriers (Mohamad et al., 2019).

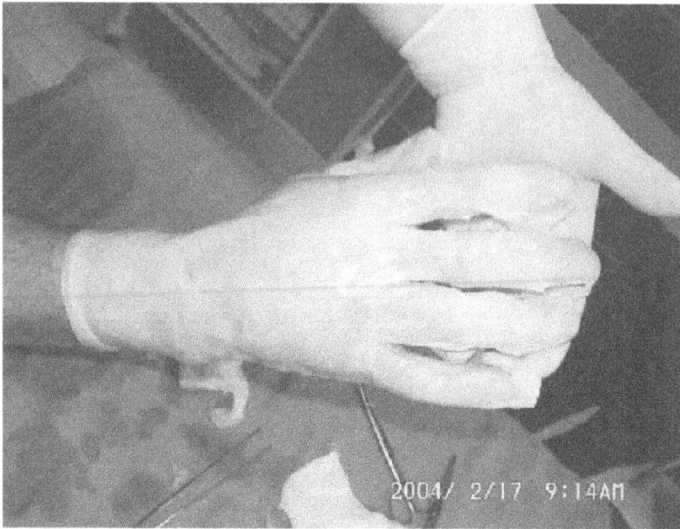

FIGURE 8.7   Bacterial cellulose dressing applied on a wounded hand. (Reproduced with permission from Czaja et al., 2006.)

BNC dressing had been proved to have great conformability and a high degree of adherence to wound sites, and its elastic properties allowed excellent molding to all moving parts (such as hands (Figure 8.7) (Czaja et al., 2006), facial contour (Figure 8.8) (Czaja et al., 2007), mouth, nose, and eyelids) during clinical trials on a large number of patients. A BNC membrane has been successfully utilized as a wound-healing device for severely damaged skin and as a small-diameter blood vessel replacement. According to Czaja et al. (2007), a complete closure of the wounded face was made with a single sheet of BNC in which holes for mouth, nose, and eyes were made after BNC was applied to a patient with severe deep second-degree burns of the facial surface. It was reported after 44 days, the wounded face was entirely healed with no need for skin grafting and no significant signs of extensive scarring (Czaja et al., 2007).

8.3.3   CARDIOVASCULAR MEDICAL DEVICES

Heart valves, endovascular stents, vascular grafts, stent-grafts, and other cardiovascular grafts are among the medical devices in cardiovascular applications. Complications involving aortic and/or the mitral valve such as severe age-related calcific aortic stenosis (narrowed opening of valve due to deposition of calcium) could be treated by aortic valve replacement (Lindman et al., 2016). The stents and vascular graft used are from various types of coated or noncoated polymers and metals with or without coating such as titanium, polytetrafluoroethylene, and cellulose (Ratner et al., 2004; Zahedmanesh et al., 2011).

The potential of BNC as a scaffold for overcoming reconstructive problems related to extended vascular disease was investigated by Wippermann et al. (2009). Hollow BNC grafts, 10 mm in length with a 3.0–3.7 mm inner diameter and 0.6–1.0 mm wall

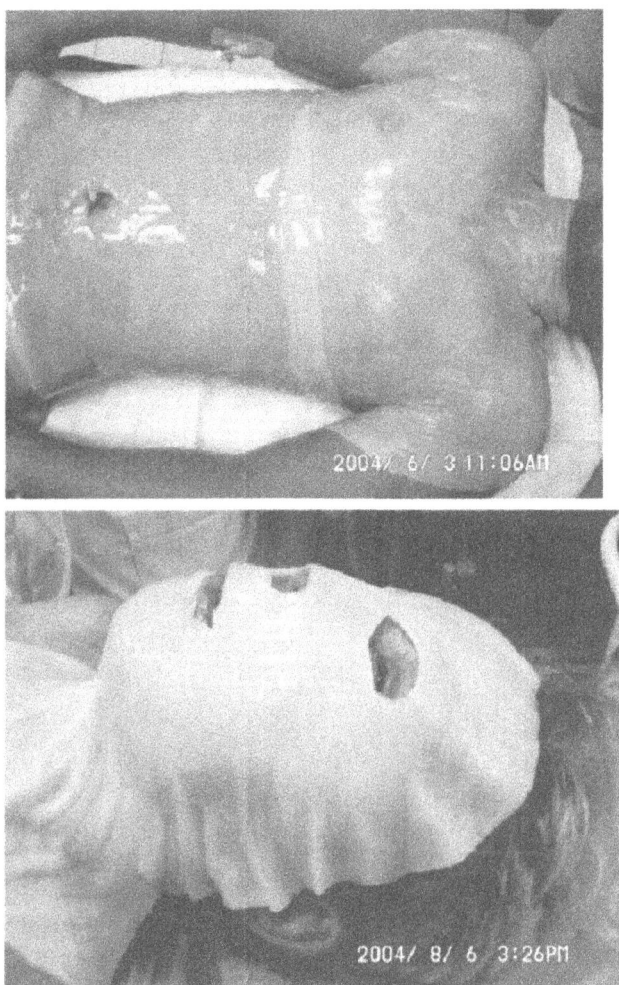

FIGURE 8.8   Bacterial cellulose dressing applied on wounded torso and face. (Reprinted (adapted) with permission from Czaja et al., 2007. Copyright (2007) American Chemical Society.)

thickness, were used to replace the carotid arteries of eight pigs. Analysis of the grafts after 3 months showed rapid re-cellularization by recipient endothelial cells, indicating stable vascular conduits, which exhibit attractive properties for their use in future tissue-engineered blood vessels (TEBVs) for vascular surgery (Wippermann et al., 2009).

Nevertheless, the compatible mechanical match between the implanted device and surrounding cardiovascular tissues is essential to prevent intimal hyperplasia (an increase in cell proliferation leading to the gross enlargement of an organ) and graft replacement failure (Xue & Greisler, 2003). Catering to this issue, anisotropic PVA-BNC composite, exhibiting a wide range of mechanical properties that closely match with that of the aorta in the physiological range, was reported (Millon et al., 2008).

Moreover, development of cytocompatible cellulose-chitosan hollow tube (small diameter of 4 mm) having properties closely similar to those of the human coronary artery was reported, by which the mechanical properties can be simply changed by alteration of the cellulose-chitosan compositional ratio (Azevedo et al., 2013). In another study, the effect of cellulose on heparin immobilization on cellulose-chitosan composites was investigated. A maximum of 4 mg/mL heparin concentration was successfully immobilized resulting in mechanically strong, reasonably compliant, and antithrombogenic properties implying promising application as a vascular bypass graft hollow tube.

Besides that, nanocellulose from pineapple leaf fibers (PALF) reinforced polyurethane (PU) has been used for the production of vascular prostheses, as shown in Figure 8.9 (Cherian et al., 2011). The elastic properties possessed by polyurethane

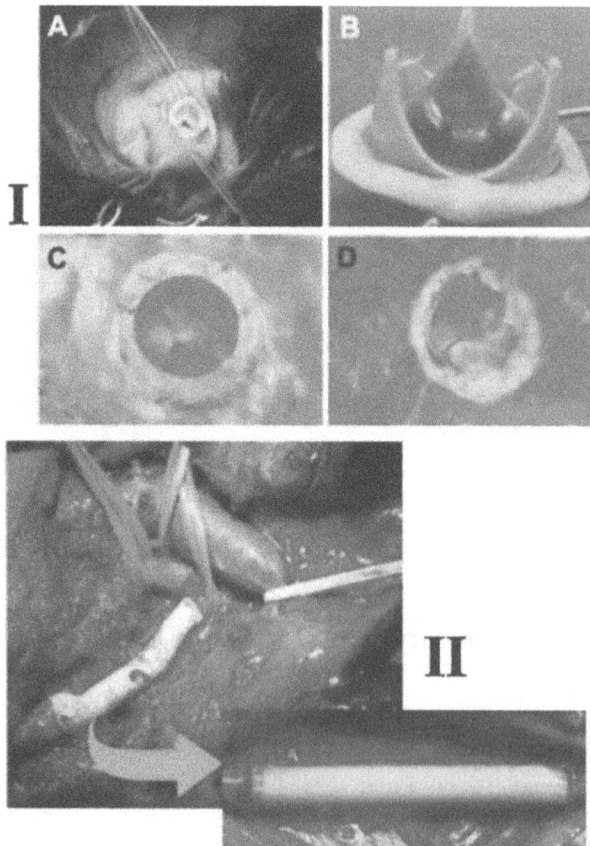

FIGURE 8.9 (I) Nanocellulose–polyurethane prosthetic heart valve: (a) valve implant, (b) heart valve, (c) viewed *in situ* immediately prior to explant (inflow surface), (d) viewed *in situ* immediately prior to explant (outflow surface). (II) Vascular prostheses made of nanocellulose-polyurethane placed between the brachiocephalic trunk and the right common carotid artery in a 26-year-old male patient with multiple endocrine neoplasia 2B (MEN 2B). (Reproduced with permission from Cherian et al., 2011.)

polymer, coupled with low thrombogenicity and exceptional physical as well as mechanical properties, have led to a considerable research effort aimed at the development of nanocellulose polyurethane vascular grafts. PALF reinforced PU vascular grafts with a wall thickness of 0.7–1.0 mm displayed elongation at a break of 800–1,200% and withstood hydraulic pressures up to 300 kPa.

8.3.4 ORTHOPEDICS

Bones, as part of the body's skeletal system, are capable of self-regeneration by actions of bone formation (osteogenesis) and bone lysis (osteolysis). However, severe defects could disrupt the efficiency of this function (Esmonde-White et al., 2013; Filipowska et al., 2017). Therefore, grafting as replacement is needed to help in providing mechanical or structural support, fill defective gaps, and enhance bone tissue formation (Bohner, 2010; Damien & Parsons, 1991). In addition to replacing missing bones, grafting materials also aid to regenerate lost bone. Conventionally, bone grafting is conducted by transferring natural bones from several sources such as autografting, allografting, and xenografting. In corresponding order, the sources of bone grafts are from the bones of another part of the recipients' body, bones from a different individual of the same species, and bones from a different species such as from animal to human. However, these three methods expose patients to the risk of rejection and transmission of infectious diseases, on top of donor shortage limitation for autografts and allografts (Damien & Parsons, 1991; Sheikh et al., 2015).

Another option is to use alloplastic of synthetic single- or multiphase materials. It does not provide the required essential features needed for bone growth, hence the use of multiphase materials that could provide similar composition and structure like natural bones comes to a need (Murugan & Ramakrishna, 2005; Nandi et al., 2010). As such, nanocomposites such as hydroxyapatite-collagen-based composites have gained acceptance as promising bone graft material that mimics the natural bone matrix (Shikinami et al., 2005; Zhang & Ma, 1999). Materials for synthetic bone grafts can be grouped into four different categories of metals and alloys, ceramics and polymers, composites and nanocomposites, and tissue-engineered composites. The utilization of these materials is with respect to the target therapies of hard tissue or soft tissue application. While mostly metals-ceramics and polymers are used in hard tissue and soft tissue application, respectively, composites can be used for both applications (Baino, 2017; Bohner, 2010; Murugan & Ramakrishna, 2005; Sabir et al., 2009; Sheikh et al., 2015).

Materials are considered to be good bone grafting materials if they can promote osteoconduction, osteoinduction, and osteogenesis activity of bone cells. Osteoconductivity of materials enables spreading and generation of new bone, utilizing the graft material as a framework or as scaffolds. In turn, the stimulation of osteoprogenitor cells to differentiate into osteoblasts followed by new bone formation is associated with osteoinductivity of the materials. Last, the materials could be considered ideal graftings when osteoblasts originating from them could contribute to new bone growth (Nandi et al., 2010). Equally important, the use of grafting material that does not remain in the biological environment for a very long time and does consequently degrade after having completed its purpose or functions is considerably favored. In this case, the biocompatibility of CNFs makes them a very relevant candidate as grafting material in bone regeneration. To date, several studies have reported on the use of cellulose-based materials for bone grafting scaffold (Ao et al., 2017; Gaihre & Jayasuriya, 2016; Saska et al., 2011, 2012; Singh et al., 2016).

Similar to bone, the regeneration and healing process of cartilage, an avascular tissue, is limited once it experiences degeneration, wounded or affected by genetic defects (Lu et al., 2014; Murugan & Ramakrishna, 2005) whereby reconstructive option can be opted through repairing or replacement. A current novel approach for cartilage replacement is through the bioprinting technology where a three-dimensional (3D) tissue structure is generated and incorporated with cells, resulting in new cartilage formation. Nevertheless, it is essential and crucial to have bio-ink with appropriate viscoelastic properties and structural stability (Halib et al., 2017). Taking CNF as an example, its combination with calcium alginate (CA) enables cross-linking between the two, thus resulting in a bio-ink with shear-thinning properties (Markstedt et al., 2015). Besides that, CNF/alginate composite has been reported to be a potential material for bilayer scaffold generation, portraying strength and resemblance with collagen fibrils in the extracellular matrix of different tissues (Ávila et al., 2015). Additionally, a more recent study showed that CNF/CA composites embedded in human bone marrow-derived mesenchymal stem cells (hBMSC) and human nasal chondrocytes (hNC) allowed chondrogenesis to occur after 60 days, indicating successful proliferation of cells within the prepared CNF/CA composites (Möller et al., 2017).

The 3D bioprinting is expected to transform the field of regenerative medicine and tissue engineering. Markstedt et al. (2015) conducted a study on a 3D bio-ink that combines the outstanding shear-thinning property of CNF with the fast cross-linking ability of alginate for cartilage tissue engineering applications. They anatomically shaped cartilage structures, such as a human ear and sheep meniscus, using a 3D printer with magnetic resonance imaging (MRI) and computed tomography (CT) images as blueprints (Figure 8.10). They summarized that the CNF-based bio-ink is a suitable hydrogel to be utilized in 3D bioprinting with living tissues and organs.

8.3.5 DENTAL

Metals, ceramics, polymers, and composites have been used as tooth crowns and root replacements in dental application. Similar to other biomedical intended applications, dental restorative materials' compatibility and permanent bond with tooth structure are essential aside from possessing closely matched properties like tooth enamel, dentin, or other some tissues. Unlike bone and cartilage grafting, it is crucial for the nanocomposites or materials in dental application to persist as long as possible.

For instance, addition of CNC significantly improved mechanical properties of a commercial dental glass ionomer cement (GIC); for example, compressive strength improved up to 110% compared to the control group, elastic modulus improved by 161%, diameter tensile strength increased by 53%, and mass loss decreased from

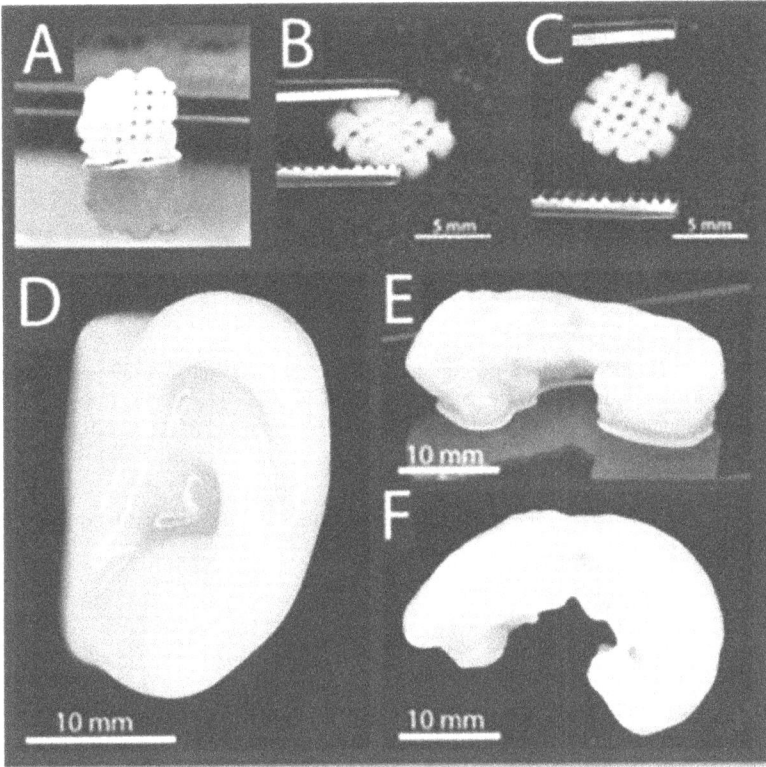

FIGURE 8.10 (a) Model 3D printed nanocellulose–alginate scaffolds. (b) The scaffold deforms when under mechanical force, but (c) then fully recovers its original shape. (d) Bioprinted human ear and (e and f) sheep meniscus scaffolds. (Reprinted (adapted) with permission from Markstedt et al., 2015. Copyright (2015) American Chemical Society.)

10.95% to 3.87%. This improvement signifies the promising potential of nanocellulose composites as dental restorative material (Silva et al., 2016). CNF could also be used in tooth-supporting tissue regeneration such as grafting for bone, cementum, and periodontal ligament (Halib et al., 2017). The reinforcement of CNC/ZnO nanohybrids on dental resins composites (DRCs) was investigated by Wang et al. (2019). Results showing higher compressive strength and flexural modulus as compared to DRCs along with outstanding antibacterial properties and a reduction in the bacterial amount of 78% indicated a positive impact on the mechanical and antibacterial properties of DRCs to overcome secondary caries and bulk fracture in the dental application (Wang et al., 2019).

Besides that, nanocellulose was also tested in dental tissue regeneration. The BNC was produced by the *Gluconacetobacter xylinus* strain to regenerate dental tissues in humans (Figure 8.11). Nanocellulose products such as Gore-Tex® and Gengiflex® have been observed as promising materials within the dental industry. These products were produced to help periodontal tissue recovery (Abdul Khalil et al., 2015). The advantages of using this BNC tissue included the reestablishment

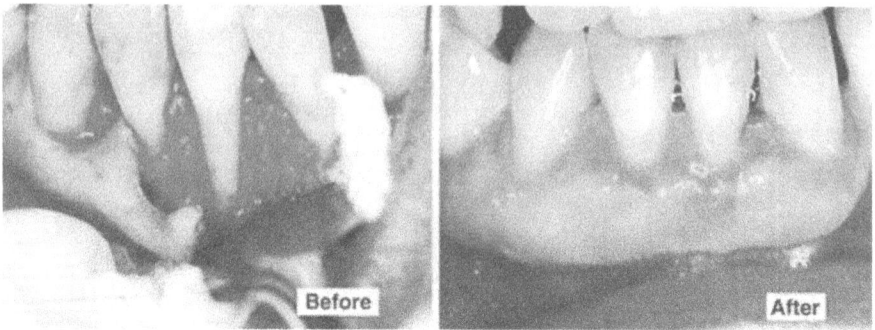

FIGURE 8.11 Nanocellulose used in dental tissue regeneration in a 39-year-old female patient. (Reprinted (adapted) with permission from Kalia et al., 2011a. Copyright (2016) Creative Commons Attribution License. Attribution 3.0 Unported (CC BY 3.0).)

of aesthetics and function of the mouth. In addition, this tissue reduced the number of surgical steps.

8.3.6 OPHTHALMOLOGIC APPLICATION

In ophthalmologic applications, implants or materials are used to treat diseases related to eyes such as vision impairment/low vision, blindness, refractive error (myopia and hyperopia), astigmatism, presbyopia, cataracts, primary open-angle glaucoma, age-related macular degeneration (AMD), and diabetic retinopathy.

Goncalves et al. (2015) conducted research to evaluate the viability of BNC as a novel substratum for the culture of the retinal pigment epithelium (RPE). Thin and heat-dried BNC substrates were surface-modified using chitosan and carboxymethyl cellulose by acetylation and polysaccharide adsorption. The results obtained showed that all surface-modified BNC substrates had equal coefficients of permeation, with solutes of up to 300 kDa. BNC acetylation reduced the swelling and endotoxin volume and its surface alteration significantly increased the adhesion of RPE cells and their proliferation. Besides, all samples displayed similar stress-strain behavior; BNC and acetylated BNC had the highest elastic modulus, but it showed a slightly lower tensile strength and break elongation relative to pure BC. Even though similar rates of proliferation among modified substrates were observed, higher initial cell adhesion was shown by the acetylated ones. The main reason for this variation may be the slightly hydrophilic coating produced after acetylation.

Another study conducted by Tummala et al. (2019) reported that the hydrogel material in the form of a contact lens made from PVA and CNC exhibited resistance toward protein adsorption, most likely due to the interactions between the positively charged lysozyme and the negatively charged CNC contained in the PVA matrix. The direct contact experiment showed that the physical presence of the lenses did not affect either the cohesion or metabolic function of the corneal epithelial cell

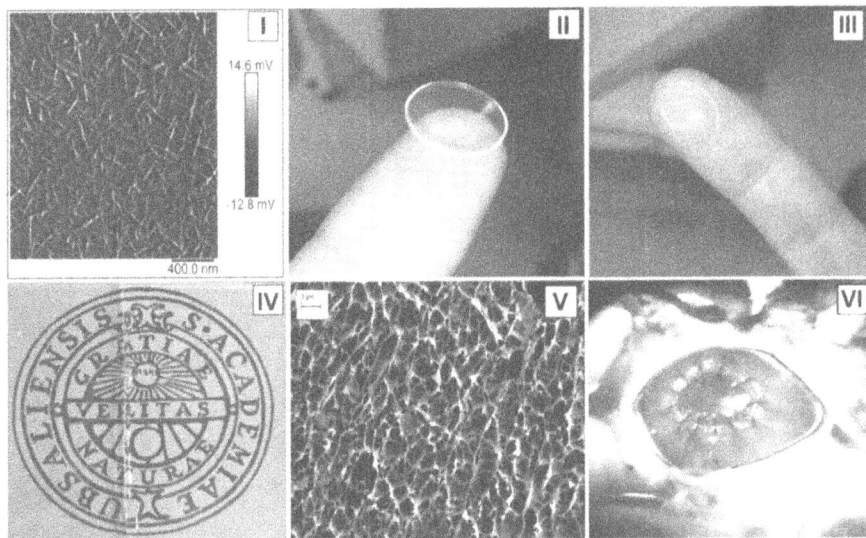

FIGURE 8.12 Schematic illustrations of NC-based hydrogels for liver TE and ophthalmic TE. CNC-PVA hydrogel lens. (I) AFM image of carboxylated CNC; (II) Self-standing lens; (III) Lens showing conformability; (IV) Transparent sheet of hydrogel; (V) SEM image of freeze-dried lens; (VI) CNC-PVA hydrogel implant sutured to an *ex vivo* porcine cornea. (Reprinted (adapted) with permission from Tummala et al., 2016. Copyright (2016) American Chemical Society.)

monolayers. Their findings proved the versatility of PVA-CNC hydrogel composite as an ophthalmic biomaterial that motivates future biocompatibility studies *in vitro* and *in vivo*.

The hyperelastic hydrogel was developed by reinforcing CNC within the PVA matrix. This is done to enhance the mechanical properties of the PVA matrix (Tummala et al., 2016). The result indicated that the CNC-PVA hydrogel displayed high water content and showed similar mechanical properties as compared to native tissues. Moreover, this hydrogel encouraged the growth of human corneal epithelial cells. This phenomenon was seen to be highly favorable for corneal implants, disposable contact lenses, and ophthalmic prostheses (Figure 8.12).

8.4 CONCLUSION

The application of nanocellulose composites in biomedical applications is highlighted in this chapter. Nanocellulose composites hold great promise in a wide variety of biomedical applications; these include pharmaceuticals, wound dressings and skin substitutes, cardiovascular applications, orthopedics, dental, as well as ophthalmologic applications. The versatility of nanocellulose and its excellent properties—biocompatibility, high mechanical strength, good flexibility and elasticity, low toxicity, shear-thinning properties, and ability to swell and hold water—enable it to be used in various biomedical applications. Despite its advantages, further studies should aim to investigate different effects of the three types of nanocellulose (BNC, CNC, and CNF) on the respective biomedical applications. This will indirectly pave the way for greater acceptance of nanocellulose as a commercially available nanomaterial in biomedical applications.

REFERENCES

Abdul Khalil, H. P. S., Bhat, A. H., Abu Bakar, A., Tahir, P. M., Zaidul, I. S. M., & Jawaid, M. (2015). Cellulosic Nanocomposites from Natural Fibers for Medical Applications: A Review. In Handbook of Polymer Nanocomposites. Processing, Performance and Application (pp. 475–511). Springer, Berlin Heidelberg. https://doi. org/10.1007/978-3-642-45232-1_72

Abitbol, T., Rivkin, A., Cao, Y., Nevo, Y., Abraham, E., Ben-Shalom, T., Lapidot, S., & Shoseyov, O. (2016). Nanocellulose, a tiny fiber with huge applications. Current Opinion in Biotechnology, 39(I), 76–88. https://doi.org/10.1016/j. copbio.2016.01.002

Abral, H., Ariksa, J., Mahardika, M., Handayani, D., Aminah, I., Sandrawati, N., Pratama, A. B., Fajri, N., Sapuan, S. M., & Ilyas, R. A. (2020). Transparent and antimicrobial cellulose film from ginger nanofiber. Food Hydrocolloids, 98, 105266. https://doi. org/10.1016/j.foodhyd.2019.105266

Abral, H., Ariksa, J., Mahardika, M., Handayani, D., Aminah, I., Sandrawati, N., Sapuan, S. M., & Ilyas, R. A. (2019). Highly transparent and antimicrobial PVA based bionano-composites reinforced by ginger nanofiber. Polymer Testing, October, 106186. https://doi.org/10.1016/j.polymertesting.2019.106186

Afrin, S., & Karim, Z. (2015). Nanocellulose as novel supportive functional material for growth and development of cells. Cell & Developmental Biology, 04(02). https://doi. org/10.4172/2168-9296.1000154

Aisyah, H. A., Paridah, M. T., Sapuan, S. M., Khalina, A., Berkalp, O. B., Lee, S. H., Lee, C. H., Nurazzi, N. M., Ramli, N., Wahab, M. S., & Ilyas, R. A. (2019). Thermal properties of woven kenaf/carbon fibre-reinforced epoxy hybrid composite panels. International Journal of Polymer Science, 2019(December), 1–8. https://doi. org/10.1155/2019/5258621

Ao, C., Niu, Y., Zhang, X., He, X., Zhang, W., & Lu, C. (2017). Fabrication and characterization of electrospun cellulose/nano-hydroxyapatite nanofibers for bone tissue engineering. International Journal of Biological Macromolecules, 97, 568–573. https://doi. org/10.1016/J.IJBIOMAC.2016.12.091

Ariffin, H., Norrrahim, M. N. F., Yasim-Anuar, T. A. T., Nishida, H., Hassan, M. A., Ibrahim, N. A., & Yunus, W. M. Z. W. (2017). Oil palm biomass cellulose-fabricated polylactic acid composites for packaging applications. Bionanocomposites for Packaging Applications, 95–105. https://doi.org/10.1007/978-3-319-67319-6_5

Asyraf, M. R. M., Ishak, M. R., Sapuan, S. M., Yidris, N., & Ilyas, R. A. (2020). Woods and composites cantilever beam: A comprehensive review of experimental and numerical creep methodologies. Journal of Materials Research and Technology. https://doi. org/10.1016/j.jmrt.2020.01.013

Atikah, M. S. N., Ilyas, R. A., Sapuan, S. M., Ishak, M. R., Zainudin, E. S., Ibrahim, R., Atiqah, A., Ansari, M. N. M., & Jumaidin, R. (2019). Degradation and physical properties of sugar palm starch/sugar palm nanofibrillated cellulose bionanocomposite. Polimery, 64(10), 27–36. https://doi.org/10.14314/polimery.2019.10.5

Atiqah, A., Jawaid, M., Sapuan, S. M., Ishak, M. R., Ansari, M. N. M., & Ilyas, R. A. (2019). Physical and thermal properties of treated sugar palm/glass fibre reinforced thermoplastic polyurethane hybrid composites. Journal of Materials Research and Technology, July. https://doi.org/10.1016/j.jmrt.2019.06.032

Ávila, H. M., Feldmann, E. M., Pleumeekers, M. M., Nimeskern, L., Kuo, W., de Jong, W. C., Schwarz, S., Müller, R., Hendriks, J., Rotter, N., & van Osch, G. J. (2015). Novel bilayer bacterial nanocellulose scaffold supports neocartilage formation in vivo and in vitro. Biomaterials, 44, 122–133. https://doi.org/10.1016/j.biomaterials.2014.12.025

Azammi, A. M. N., Ilyas, R. A., Sapuan, S. M., Ibrahim, R., Atikah, M. S. N., Asrofi, M., & Atiqah, A. (2020). Characterization Studies of Biopolymeric Matrix and Cellulose Fibres Based Composites Related to Functionalized Fibre-Matrix Interface. In Interfaces in Particle and Fibre Reinforced Composites (1st ed., Issue November, pp. 29–93). Elsevier. https://doi.org/10.1016/B978-0-08-102665-6.00003-0

Azevedo, E. P., Retarekar, R., Raghavan, M. L., & Kumar, V. (2013). Mechanical properties of cellulose: Chitosan blends for potential use as a coronary artery bypass graft. Journal of Biomaterials Science, Polymer Edition, 24(3), 239–252. https://doi.org/10.1080/09205063.2012.690273

Baino, F. (2017). Ceramics for Bone Replacement: Commercial Products and Clinical use. In Advances in Ceramic Biomaterials: Materials, Devices and Challenges: First Edition. Elsevier Ltd. https://doi.org/10.1016/B978-0-08-100881-2.00007-5

Berndt, S., Wesarg, F., Wiegand, C., Kralisch, D., & Müller, F. A. (2013). Antimicrobial porous hybrids consisting of bacterial nanocellulose and silver nanoparticles. Cellulose, 20(2), 771–783. https://doi.org/10.1007/s10570-013-9870-1

Bohner, M. (2010). Resorbable biomaterials as bone graft substitutes. Materials Today, 13(1–2), 24–30. https://doi.org/10.1016/S1369-7021(10)70014-6

Bourgeois, M., Johnson, G., & Harbison, R. (2016). Human Health Risk Assessment. In International Encyclopedia of Public Health (Second Edition, Vol. 5, Issue 2003). Elsevier. https://doi.org/10.1016/B978-0-12-803678-5.00388-X

Cherian, B. M., Leão, A. L., de Souza, S. F., Costa, L. M. M., de Olyveira, G. M., Kottaisamy, M., Nagarajan, E. R., & Thomas, S. (2011). Cellulose nanocomposites with nanofibres isolated from pineapple leaf fibers for medical applications. Carbohydrate Polymers, 86(4), 1790–1798. https://doi.org/10.1016/j.carbpol.2011.07.009

Curvello, R., Raghuwanshi, V. S., & Garnier, G. (2019). Engineering nanocellulose hydrogels for biomedical applications. Advances in Colloid and Interface Science, 267, 47–61. https://doi.org/10.1016/j.cis.2019.03.002

Czaja, W., Krystynowicz, A., Bielecki, S., & Brown, J. R. M. (2006). Microbial cellulose—the natural power to heal wounds. Biomaterials, 27(2), 145–151. https://doi.org/10.1016/j.biomaterials.2005.07.035

Czaja, W. K., Young, D. J., Kawecki, M., & Brown, R. M. (2007). The Future prospects of microbial cellulose in biomedical applications. Biomacromolecules, 8(1), 1–12. https://doi.org/10.1021/bm060620d

Damien, C. J., & Parsons, J. R. (1991). Bone graft and bone graft substitutes: a review of current technology and applications. Journal of Applied Biomaterials : An Official Journal of the Society for Biomaterials, 2(3), 187–208. https://doi.org/10.1002/jab.770020307

Deloid, G. M., Cao, X., Molina, R. M., Silva, D. I., Bhattacharya, K., Ng, K. W., Loo, S. C. J., Brain, J. D., & Demokritou, P. (2019). Toxicological effects of ingested nanocellulose in: In vitro intestinal epithelium and in vivo rat models. Environmental Science: Nano, 6(7), 2105–2115. https://doi.org/10.1039/c9en00184k

Dima, S.-O., Panaitescu, D.-M., Orban, C., Ghiurea, M., Doncea, S.-M., Fierascu, R., Nistor, C., Alexandrescu, E., Nicolae, C.-A., Tricǎ, B., Moraru, A., & Oancea, F. (2017). Bacterial nanocellulose from side-streams of kombucha beverages production: Preparation and physical-chemical properties. Polymers, 9(12), 374. https://doi.org/10.3390/polym9080374

El-Saied, H., Basta, A. H., & Gobran, R. H. (2004). Research progress in friendly environmental technology for the production of cellulose products (Bacterial cellulose and its application). Polymer-Plastics Technology and Engineering, 43(3), 797–820. https://doi.org/10.1081/PPT-120038065

Esmonde-White, K. A., Esmonde-White, F. W. L., Holmes, C. M., Morris, M. D., & Roessler, B. J. (2013). Alterations to bone mineral composition as an early indication of osteomyelitis in the diabetic foot. Diabetes Care, 36(11), 3652–3654. https://doi.org/10.2337/dc13-0510

Ferraz, N., Straømme, M., Fellström, B., Pradhan, S., Nyholm, L., & Mihranyan, A. (2012). In vitro and in vivo toxicity of rinsed and aged nanocellulose-polypyrrole composites. Journal of Biomedical Materials Research - Part A, 100A(8), 2128–2138. https://doi.org/10.1002/jbm.a.34070

Filipowska, J., Tomaszewski, K. A., Niedźwiedzki, Ł., Walocha, J. A., & Niedźwiedzki, T. (2017). The role of vasculature in bone development, regeneration and proper systemic functioning. Angiogenesis, 20(3), 291–302. https://doi.org/10.1007/s10456-017-9541-1

Francis PJ, J. (2018). Biomedical applications of polymers – An overview. Current Trends in Biomedical Engineering & Biosciences, 15(2), 44–45. https://doi.org/10.19080/ctbeb.2018.15.555909

Gaihre, B., & Jayasuriya, A. C. (2016). Fabrication and characterization of carboxymethyl cellulose novel microparticles for bone tissue engineering. Materials Science and Engineering: C, 69, 733–743. https://doi.org/10.1016/J.MSEC.2016.07.060

Galkina, O. L., Önneby, K., Huang, P., Ivanov, V. K., Agafonov, A. V., Seisenbaeva, G. A., & Kessler, V. G. (2015). Antibacterial and photochemical properties of cellulose nanofiber-titania nanocomposites loaded with two different types of antibiotic medicines. Journal of Materials Chemistry B, 3(35), 7125–7134. https://doi.org/10.1039/c5tb01382h

Ghaderi, M., Mousavi, M., Yousefi, H., & Labbafi, M. (2014). All-cellulose nanocomposite film made from bagasse cellulose nanofibers for food packaging application all-cellulose nanocomposite film made from bagasse cellulose nanofibers for food packaging application. Carbohydrate Polymers, 104(April), 59–65. https://doi.org/10.1016/j.carbpol.2014.01.013

Halib, N., Perrone, F., Cemazar, M., Dapas, B., Farra, R., Abrami, M., Chiarappa, G., Forte, G., Zanconati, F., Pozzato, G., Murena, L., Fiotti, N., Lapasin, R., Cansolino, L., Grassi, G., & Grassi, M. (2017). Potential applications of nanocellulose-containing materials in the biomedical field. Materials, 10(8), 1–31. https://doi.org/10.3390/ma10080977

Hazrol, M. D., Sapuan, S. M., Ilyas, R. A., Othman, M. L., & Sherwani, S. F. K. (2020). Electrical properties of sugar palm nanocrystalline cellulose, reinforced sugar palm starch nanocomposites. Polimery, 55(5), 33–40. https://doi.org/10.14314/polimery.2020.5.5

Hazrol, M. D., Sapuan, S. M., Zuhri, M. Y. M., & Ilyas, R. A. (2019). Electrical properties of sugar palm nanocellulose fibre reinforced sugar palm starch biopolymer composite. Prosiding Seminar Enau Kebangsaan, 57–62.

Heo, D. N., Yang, D. H., Lee, J. B., Bae, M. S., Kim, J. H., Moon, S. H., Chun, H. J., Kim, C. H., Lim, H.-N., & Kwon, I. K. (2013). Burn-wound healing effect of gelatin/polyurethane nanofiber scaffold containing silver-sulfadiazine. Journal of Biomedical Nanotechnology, 9(3), 511–515. https://doi.org/10.1166/jbn.2013.1509

Hu, W., Chen, S., Li, X., Shi, S., Shen, W., Zhang, X., & Wang, H. (2009). In situ synthesis of silver chloride nanoparticles into bacterial cellulose membranes. Materials Science and Engineering: C, 29(4), 1216–1219. https://doi.org/10.1016/J.MSEC.2008.09.017

Hubbe, M. a., Rojas, O. J., Lucia, L. a., & Sain, M. (2008). Cellulosic nanocomposites: A review. BioResources, 3(3), 929–980. https://doi.org/10.15376/biores.3.3.929-980

Ilyas, R. A., & Sapuan, S. M. (2020). The preparation methods and processing of natural fibre bio-polymer composites. Current Organic Synthesis, 16(8), 1068–1070. https://doi.org/10.2174/157017941608200120105616

Ilyas, R. A., Sapuan, S. M., Atiqah, A., Ibrahim, R., Abral, H., Ishak, M. R., Zainudin, E. S., Nurazzi, N. M., Atikah, M. S. N., Ansari, M. N. M., Asyraf, M. R. M., Supian, A. B. M., & Ya, H. (2019). Sugar palm (Arenga pinnata [Wurmb .] Merr) starch films containing sugar palm nanofibrillated cellulose as reinforcement: Water barrier properties. Polymer Composites, 1–9. https://doi.org/10.1002/pc.25379

Ilyas, R. A., Sapuan, S. M., Ibrahim, R., Abral, H., Ishak, M. R., Zainudin, E. S., Asrofi, M., Atikah, M. S. N., Huzaifah, M. R. M., Radzi, A. M., Azammi, A. M. N., Shaharuzaman, M. A., Nurazzi, N. M., Syafri, E., Sari, N. H., Norrrahim, M. N. F., & Jumaidin, R. (2019). Sugar palm (Arenga pinnata (Wurmb.) Merr) cellulosic fibre hierarchy: a comprehensive approach from macro to nano scale. Journal of Materials Research and Technology, 8(3), 2753–2766. https://doi.org/10.1016/j.jmrt.2019.04.011

Ilyas, R. A., Sapuan, S. M., Ibrahim, R., Abral, H., Ishak, M. R., Zainudin, E. S., Atikah, M. S. N., Mohd Nurazzi, N., Atiqah, A., Ansari, M. N. M., Syafri, E., Asrofi, M., Sari, N. H., & Jumaidin, R. (2019a). Effect of sugar palm nanofibrillated cellulose concentrations on morphological, mechanical and physical properties of biodegradable films based on agro-waste sugar palm (Arenga pinnata (Wurmb.) Merr) starch. Journal of Materials Research and Technology, 8(5), 4819–4830. https://doi.org/10.1016/j.jmrt.2019.08.028

Ilyas, R. A., Sapuan, S. M., Ibrahim, R., Abral, H., Ishak, M. R., Zainudin, E. S., Atikah, M. S. N., Mohd Nurazzi, N., Atiqah, A., Ansari, M. N. M., Syafri, E., Asrofi, M., Sari, N. H., & Jumaidin, R. (2019b). Effect of sugar palm nanofibrillated cellulose concentrations on morphological, mechanical and physical properties of biodegradable films based on agro-waste sugar palm (Arenga pinnata (Wurmb.) Merr) starch. Journal of Materials Research and Technology, 8(5), 4819–4830. https://doi.org/10.1016/j.jmrt.2019.08.028

Ilyas, R A, Sapuan, S. M., Ibrahim, R., Abral, H., Ishak, M. R., Zainudin, E. S., Atiqah, A., Atikah, M. S. N., Syafri, E., Asrofi, M., & Jumaidin, R. (2020). Thermal, biodegradability and water barrier properties of bio-nanocomposites based on plasticised sugar palm starch and nanofibrillated celluloses from sugar palm fibres. Journal of Biobased Materials and Bioenergy, 14(2), 234–248. https://doi.org/10.1166/jbmb.2020.1951

Ilyas, R. A., Sapuan, S. M., Ibrahim, R., Atikah, M. S. N., Atiqah, A., Ansari, M. N. M., & Norrrahim, M. N. F. (2019). Production, Processes and Modification of Nanocrystalline Cellulose from Agro-Waste: A Review. In Nanocrystalline Materials (pp. 3–32). IntechOpen. https://doi.org/10.5772/intechopen.87001

Ilyas, R. A., Sapuan, S. M., & Ishak, M. R. (2018). Isolation and characterization of nanocrystalline cellulose from sugar palm fibres (Arenga Pinnata). Carbohydrate Polymers, 181, 1038–1051. https://doi.org/10.1016/j.carbpol.2017.11.045

Ilyas, R. A., Sapuan, S. M., Ishak, M. R., & Zainudin, E. S. (2017). Effect of delignification on the physical, thermal, chemical, and structural properties of sugar palm fibre. BioResources, 12(4), 8734–8754. https://doi.org/10.15376/biores.12.4.8734-8754

Ilyas, R. A., Sapuan, S. M., Ishak, M. R., & Zainudin, E. S. (2018a). Water transport properties of bio-nanocomposites reinforced by sugar palm (Arenga Pinnata) nanofibrillated
cellulose. Journal of Advanced Research in Fluid Mechanics and Thermal Sciences
Journal, 51(2), 234–246.

Ilyas, R. A., Sapuan, S. M., Ishak, M. R., & Zainudin, E. S. (2018b). Development and characterization of sugar palm nanocrystalline cellulose reinforced sugar palm starch
bionanocomposites. Carbohydrate Polymers, 202, 186–202. https://doi.org/10.1016/j.
carbpol.2018.09.002

Ilyas, R. A., Sapuan, S. M., Ishak, M. R., & Zainudin, E. S. (2018c). Sugar palm nanocrystalline cellulose reinforced sugar palm starch composite: Degradation and water-barrier
properties. IOP Conference Series: Materials Science and Engineering, 368(1). https://
doi.org/10.1088/1757-899X/368/1/012006

Ilyas, R. A., Sapuan, S. M., Ishak, M. R., & Zainudin, E. S. (2019). Sugar palm nanofibrillated cellulose (Arenga pinnata (Wurmb.) Merr): Effect of cycles on their
yield, physic-chemical, morphological and thermal behavior. International
Journal of Biological Macromolecules, 123, 379–388. https://doi.org/10.1016/j.
ijbiomac.2018.11.124

Ilyas, R.A., Sapuan, S. M., Ishak, M. R., Zainudin, E. S., & Atikah, M. S. N. (2018).
Nanocellulose reinforced starch polymer composites : A review of preparation, proper-
ties and application. Proceeding: 5th International Conference on Applied Sciences and
Engineering (ICASEA, 2018), 325–341.

Ilyas, R.A., Sapuan, S. M., Sanyang, M. L., Ishak, M. R., & Zainudin, E. S. (2018).
Nanocrystalline cellulose as reinforcement for polymeric matrix nanocomposites and
its potential applications: A Review. Current Analytical Chemistry, 14(3), 203–225.
https://doi.org/10.2174/1573411013666171003155624

Jonas, R., & Farah, L. F. (1998). Production and application of microbial cellulose.
Polymer Degradation and Stability, 59(1–3), 101–106. https://doi.org/10.1016/
S0141-3910(97)00197-3

Jorfi, M., & Foster, E. J. (2015). Recent advances in nanocellulose for biomedical applications. Journal of Applied Polymer Science, 132(14), 1–19. https://doi.org/10.1002/
app.41719

Jozala, A. F., de Lencastre-Novaes, L. C., Lopes, A. M., de Carvalho Santos-Ebinuma,
V., Mazzola, P. G., Pessoa-Jr, A., Grotto, D., Gerenutti, M., & Chaud, M. V. (2016).
Bacterial nanocellulose production and application: a 10-year overview. Applied
Microbiology and Biotechnology, 100(5), 2063–2072. https://doi.org/10.1007/
s00253-015-7243-4

Jumaidin, R., Ilyas, R. A., Saiful, M., Hussin, F., & Mastura, M. T. (2019). Water transport
and physical properties of sugarcane bagasse fibre reinforced thermoplastic potato
starch biocomposite. Journal of Advanced Research in Fluid Mechanics and Thermal
Sciences, 61(2), 273–281.

Jumaidin, R., Khiruddin, M. A. A., Asyul Sutan Saidi, Z., Salit, M. S., & Ilyas, R. A. (2019).
Effect of cogon grass fibre on the thermal, mechanical and biodegradation proper-
ties of thermoplastic cassava starch biocomposite. International Journal of Biological
Macromolecules. https://doi.org/10.1016/j.ijbiomac.2019.11.011

Jumaidin, R., Saidi, Z. A. S., Ilyas, R. A., Ahmad, M. N., Wahid, M. K., Yaakob, M. Y.,
Maidin, N. A., Rahman, M. H. A., & Osman, M. H. (2019). Characteristics of cogon
grass fibre reinforced thermoplastic cassava starch biocomposite: Water absorption and
physical properties. Journal of Advanced Research in Fluid Mechanics and Thermal
Sciences 62, 62(1), 43–52.

Jung, R., Kim, Y., Kim, H.-S., & Jin, H.-J. (2009). Antimicrobial properties of hydrated cellulose membranes with silver nanoparticles. Journal of Biomaterials Science, Polymer
Edition, 20(3), 311–324. https://doi.org/10.1163/156856209X412182

Kaboorani, A., & Riedl, B. (2015). Surface modification of cellulose nanocrystals (CNC) by a cationic surfactant. Industrial Crops and Products, 65, 45–55. https://doi.org/10.1016/j. indcrop.2014.11.027

Kalia, S., Avérous, L., Njuguna, J., Dufresne, A., & Cherian, B. M. (2011a). Natural fibers, bio- and nanocomposites. International Journal of Polymer Science, 2011. https://doi. org/10.1155/2011/735932

Kalia, Susheel, Dufresne, A., Cherian, B. M., Kaith, B. S., Avérous, L., Njuguna, J., & Nassiopoulos, E. (2011b). Cellulose-based bio- and nanocomposites: A review. International Journal of Polymer Science, 2011. https://doi.org/10.1155/2011/837875

Kulkarni, A. A., & Rao, P. S. (2013). Synthesis of Polymeric Nanomaterials for Biomedical Applications. In Nanomaterials in Tissue Engineering: Fabrication and Applications. Woodhead Publishing Limited. https://doi.org/10.1533/9780857097231.1.27

Lin, N., & Dufresne, A. (2014). Nanocellulose in biomedicine: Current status and future prospect. European Polymer Journal, 59, 302–325. https://doi.org/10.1016/j. eurpolymj.2014.07.025

Lindman, B. R., Clavel, M. A., Mathieu, P., Iung, B., Lancellotti, P., Otto, C. M., & Pibarot, P. (2016). Calcific aortic stenosis. Nature Reviews Disease Primers, 2, 16006. https:// doi.org/10.1038/nrdp.2016.6

Lu, T., Li, Q., Chen, W., & Yu, H. (2014). Composite aerogels based on dialdehyde nanocellulose and collagen for potential applications as wound dressing and tissue engineering scaffold. Composites Science and Technology, 94, 132–138. https://doi.org/10.1016/j.compscitech.2014.01.020

Ludwicka, K., Kolodziejczyk, M., Gendaszewska-Darmach, E., Chrzanowski, M., Jedrzejczak-Krzepkowska, M., Rytczak, P., & Bielecki, S. (2019). Stable composite of bacterial nanocellulose and perforated polypropylene mesh for biomedical applications. Journal of Biomedical Materials Research - Part B Applied Biomaterials, 107(4), 978–987. https://doi.org/10.1002/jbm.b.34191

Luo, X., Zhang, H., Cao, Z., Cai, N., Xue, Y., & Yu, F. (2016). A simple route to develop transparent doxorubicin-loaded nanodiamonds/cellulose nanocomposite membranes as potential wound dressings. Carbohydrate Polymers, 143, 231–238. https://doi.org/10.1016/j.carbpol.2016.01.076

Maneerung, T., Tokura, S., Rujiravanit, R. (2008). Impregnation of silver nanoparticles into bacterial cellulose for antimicrobial wound dressing. Carbohydrate Polymers, 72(1), 43–51. https://doi.org/10.1016/J.CARBPOL.2007.07.025

Markstedt, K., Mantas, A., Tournier, I., Martínez Ávila, H., Hägg, D., & Gatenholm, P. (2015). 3D Bioprinting human chondrocytes with nanocellulose–Alginate bioink for cartilage tissue engineering applications. Biomacromolecules, 16(5), 1489–1496. https://doi.org/10.1021/acs.biomac.5b00188

Masoodi, R., El-Hajjar, R. F., Pillai, K. M., & Sabo, R. (2012). Mechanical characterization of cellulose nanofiber and bio-based epoxy composite. Materials & Design, 36, 570–576. https://doi.org/10.1016/j.matdes.2011.11.042

Masruchin, N., Park, B. D., & Causin, V. (2018). Dual-responsive composite hydrogels based on TEMPO-oxidized cellulose nanofibril and poly(N-isopropylacrylamide) for model drug release. Cellulose, 25(1), 485–502. https://doi.org/10.1007/s10570-017-1585-2

Millon, L. E., Guhados, G., & Wan, W. K. (2008). Anisotropic polyvinyl alcohol-bacterial cellulose nanocomposite for biomedical applications. Journal of Biomedical Materials Research - Part B Applied Biomaterials, 86(2), 444–452. https://doi.org/10.1002/jbm.b.31040

Mohamad Haafiz, M. K., Eichhorn, S. J., Hassan, A., & Jawaid, M. (2013). Isolation and characterization of microcrystalline cellulose from oil palm biomass residue. Carbohydrate Polymers, 93(2), 628–634. https://doi.org/10.1016/j.carbpol.2013.01.035

Mohamad, N., Loh, E. Y. X., Fauzi, M. B., Ng, M. H., & Mohd Amin, M. C. I. (2019). In vivo evaluation of bacterial cellulose/acrylic acid wound dressing hydrogel containing keratinocytes and fibroblasts for burn wounds. Drug Delivery and Translational Research, 9(2), 444–452. https://doi.org/10.1007/s13346-017-0475-3

Mohammad Kazemi, F., Doosthoseini, K., Ganjian, E., & Azin, M. (2015). Manufacturing of bacterial nano-cellulose reinforced fiber-cement composites. Construction and Building Materials, 101, 958–964. https://doi.org/10.1016/j.conbuildmat.2015.10.093

Mohd Amin, M. C. I., Ahmad, N., Halib, N., & Ahmad, I. (2012). Synthesis and characterization of thermo- and pH-responsive bacterial cellulose/acrylic acid hydrogels for drug delivery. Carbohydrate Polymers, 88(2), 465–473. https://doi.org/10.1016/j.carbpol.2011.12.022

Möller, T., Amoroso, M., Hägg, D., Brantsing, C., Rotter, N., Apelgren, P., Lindahl, A., Kölby, L., & Gatenholm, P. (2017). In vivo chondrogenesis in 3D bioprinted human cell-laden hydrogel constructs. Plastic and Reconstructive Surgery - Global Open, 5(2), e1227. https://doi.org/10.1097/gox.0000000000001227

Müller, A., Ni, Z., Hessler, N., Wesarg, F., Müller, F. A., Kralisch, D., & Fischer, D. (2013). The Biopolymer bacterial nanocellulose as drug delivery system: Investigation of drug loading and release using the model protein albumin. Journal of Pharmaceutical Sciences, 102(2), 579–592. https://doi.org/10.1002/jps.23385

Murugan, R., & Ramakrishna, S. (2005). Development of nanocomposites for bone grafting. Composites Science and Technology, 65(15–16), 2385–2406. https://doi.org/10.1016/J.COMPSCITECH.2005.07.022

Nandi, S. K., Roy, S., Mukherjee, P., Kundu, B., De, D. K., & Basu, D. (2010). Orthopaedic applications of bone graft & graft substitutes: A review. Indian Journal of Medical Research, 132(7), 15–30.

Nogi, M., Iwamoto, S., Nakagaito, A. N., & Yano, H. (2009). Optically transparent nanofiber paper. Advanced Materials, 21(16), 1595–1598. https://doi.org/10.1002/adma.200803174

Norizan, M. N., Abdan, K., Ilyas, R. A., & Biofibers, S. P. (2020). Effect of fiber orientation and fiber loading on the mechanical and thermal properties of sugar palm yarn fiber reinforced unsaturated polyester resin composites. Polimery, 65(2), 34–43. https://doi.org/10.14314/polimery.2020.2.5

Norrrahim, Mohd Nor Faiz, Ariffin, H., Yasim-Anuar, T. A. T., Ghaemi, F., Hassan, M. A., Ibrahim, N. A., Ngee, J. L. H., & Yunus, W. M. Z. W. (2018). Superheated steam pretreatment of cellulose affects its electrospinnability for microfibrillated cellulose production. Cellulose, 25(7), 3853–3859. https://doi.org/10.1007/s10570-018-1859-3

Norrrahim, M. N.F., Ariffin, H., Yasim-Anuar, T. A. T., Hassan, M. A., Nishida, H., & Tsukegi, T. (2018). One-pot nanofibrillation of cellulose and nanocomposite production in a twin-screw extruder. IOP Conference Series: Materials Science and Engineering, 368(1). https://doi.org/10.1088/1757-899X/368/1/012034

Nurazzi, N. M., Khalina, A., Sapuan, S. M., & Ilyas, R. A. (2019). Mechanical properties of sugar palm yarn/woven glass fiber reinforced unsaturated polyester composites : Effect of fiber loadings and alkaline treatment. Polimery, 64(10), 12–22. https://doi.org/10.14314/polimery.2019.10.3

Nurazzi, N. M., Khalina, A., Sapuan, S. M., Ilyas, R. A., Rafiqah, S. A., & Hanafee, Z. M. (2019). Thermal properties of treated sugar palm yarn/glass fiber reinforced unsaturated polyester hybrid composites. Journal of Materials Research and Technology, December. https://doi.org/10.1016/j.jmrt.2019.11.086

O'Donnell, K. L., Oporto-Velásquez, G. S., & Comolli, N. (2020). Evaluation of acetaminophen release from biodegradable poly (Vinyl Alcohol) (PVA) and nanocellulose films using a multiphase release mechanism. Nanomaterials, 10(2), 301. https://doi.org/10.3390/nano10020301

Goncalves, S., Padrao, J., Rodrigues, I. P., Silva, J. P., Sencadas, V., Lanceros-Mendez, S., Girão, H. Dourado, F., & Rodrigues, L. R. (2015). Bacterial Cellulose as a Support for the Growth of Retinal Pigment Epithelium. Biomacromolecules, 16(4), 1341–1351. https://doi.org/10.1021/acs.biomac.5b00129

Pandey, J. K., Nakagaito, A. N., & Takagi, H. (2013). Fabrication and applications of cellulose nanoparticle-based polymer composites. https://doi.org/10.1002/pen

Poonguzhali, R., Basha, S. K., & Kumari, V. S. (2017). Synthesis and characterization of chitosan-PVP-nanocellulose composites for in-vitro wound dressing application. International Journal of Biological Macromolecules, 105, 111–120. https://doi.org/10.1016/j.ijbiomac.2017.07.006

Qin, C., Soykeabkaew, N., Xiuyuan, Ni, & Peijs, Ton. (2008). The effect of fibre volume fraction and mercerization on the properties of all-cellulose composites. Carbohydrate Polymers, 71(3), 458–467. https://doi.org/10.1016/J.CARBPOL.2007.06.019

Ratner, B. D., Hoffman, A. S., Schoen, F. J., & Lemons, J. E. (2004). In Biomaterials Science : An Introduction to Materials in Medicine. Elsevier Academic Press.

Rezaie, H. R., Bakhtiari, L., & Öchsner, A. (2015). Application of Biomaterials. In Biomaterials and Their Applications (pp. 19–24). Springer International Publishing. https://doi.org/10.1007/978-3-319-17846-2_2

Sabir, M. I., Xu, X., & Li, L. (2009). A review on biodegradable polymeric materials for bone tissue engineering applications. Journal of Materials Science, 44(21), 5713–5724. https://doi.org/10.1007/s10853-009-3770-7

Sambhy, V., MacBride, M. M., Peterson, B. R., & Ayusman, S. (2006). Silver bromide nanoparticle/polymer composites: Dual action tunable antimicrobial materials. Journal of American Chemical Society, 128(30), 9798–9808. https://doi.org/10.1021/JA061442Z

Sanyang, M. L., Ilyas, R. A., Sapuan, S. M., & Jumaidin, R. (2018). Sugar Palm Starch-Based Composites for Packaging Applications. In Bionanocomposites for Packaging Applications (pp. 125–147). Springer International Publishing. https://doi.org/10.1007/978-3-319-67319-6_7

Saska, S, Barud, H. S., Gaspar, A. M. M., Marchetto, R., Ribeiro, S. J. L., & Messaddeq, Y. (2011). Bacterial cellulose-hydroxyapatite nanocomposites for bone regeneration. International Journal of Biomaterials, 2011, 175362. https://doi.org/10.1155/2011/175362

Saska, Sybele, Scarel-Caminaga, R. M., Teixeira, L. N., Franchi, L. P., dos Santos, R. A., Gaspar, A. M. M., de Oliveira, P. T., Rosa, A. L., Takahashi, C. S., Messaddeq, Y., Ribeiro, S. J. L., & Marchetto, R. (2012). Characterization and in vitro evaluation of bacterial cellulose membranes functionalized with osteogenic growth peptide for bone tissue engineering. Journal of Materials Science: Materials in Medicine, 23(9), 2253–2266. https://doi.org/10.1007/s10856-012-4676-5

Sharip, N. S., & Ariffin, H. (2019). Cellulose nanofibrils for biomaterial applications. Materials Today: Proceedings, 16, 1959–1968. https://doi.org/10.1016/j.matpr.2019.06.074

Sharpe, J. R., & Martin, Y. (2013). Strategies demonstrating efficacy in reducing wound contraction in vivo. Advances in Wound Care, 2(4), 167–175. https://doi.org/10.1089/wound.2012.0378

Sheikh, Z., Sima, C., & Glogauer, M. (2015). Bone replacement materials and techniques used for achieving vertical alveolar bone augmentation. Materials, 8(6), 2953–2993. https://doi.org/10.3390/ma8062953

Shi, Q., Zhou, C., Yue, Y., Guo, W., Wu, Y., & Wu, Q. (2012). Mechanical properties and in vitro degradation of electrospun bio-nanocomposite mats from PLA and cellulose nanocrystals. Carbohydrate Polymers, 90(1), 301–308. https://doi.org/10.1016/j.carbpol.2012.05.042

Shikinami, Y., Matsusue, Y., & Nakamura, T. (2005). The complete process of bioresorption and bone replacement using devices made of forged composites of raw hydroxyapatite particles/poly l-lactide (F-u-HA/PLLA). Biomaterials, 26(27), 5542–5551. https://doi.org/10.1016/j.biomaterials.2005.02.016

Silva, R. M., Pereira, F. V., Mota, F. A. P., Watanabe, E., Soares, S. M. C. S., & Helena, M. (2016). Dental glass ionomer cement reinforced by cellulose micro fibers and cellulose nanocrystals. Materials Science & Engineering C, 58, 389–395. https://doi.org/10.1016/j.msec.2015.08.041

Singh, B. N., Panda, N. N., Mund, R., & Pramanik, K. (2016). Carboxymethyl cellulose enables silk fibroin nanofibrous scaffold with enhanced biomimetic potential for bone tissue engineering application. Carbohydrate Polymers, 151, 335–347. https://doi.org/10.1016/J.CARBPOL.2016.05.088

Siró, I., & Plackett, D. (2010). Microfibrillated cellulose and new nanocomposite materials: A review. Cellulose, 17(3), 459–494. https://doi.org/10.1007/s10570-010-9405-y

Syafri, E., Sudirman, Mashadi, Yulianti, E., Deswita, Asrofi, M., Abral, H., Sapuan, S. M., Ilyas, R. A., & Fudholi, A. (2019). Effect of sonication time on the thermal stability, moisture absorption, and biodegradation of water hyacinth (*Eichhornia crassipes*) nanocellulose-filled bengkuang (*Pachyrhizus erosus*) starch biocomposites. Journal of Materials Research and Technology, 8(6), 6223–6231. https://doi.org/10.1016/j.jmrt.2019.10.016

Tang, J., Bao, L., Li, X., Chen, L., & Hong, F. F. (2015). Potential of PVA-doped bacterial nano-cellulose tubular composites for artificial blood vessels. Journal of Materials Chemistry B, 3(43), 8537–8547. https://doi.org/10.1039/c5tb01144b

Tummala, G. K., Joffre, T., Lopes, V. R., Liszka, A., Buznyk, O., Ferraz, N., Persson, C., Griffith, M., & Mihranyan, A. (2016). Hyperelastic nanocellulose-reinforced hydrogel of high water content for ophthalmic applications. ACS Biomaterials Science & Engineering, 2(11), 2072–2079. https://doi.org/10.1021/acsbiomaterials.6b00484

Tummala, G. K., Joffre, T., Rojas, R., Persson, C., & Mihranyan, A. (2017). Strain-induced stiffening of nanocellulose reinforced poly(vinyl alcohol) hydrogels mimicking collagenous soft tissues. Soft Matter, 13(21), 3936–3945. https://doi.org/10.1039/c7sm00677b

Tummala, G. K., Lopes, V. R., & Mihranyan, A. (2019). Biocompatibility of Nanocellulose-Reinforced PVA Hydrogel with Human Corneal Epithelial Cells for Ophthalmic Applications.

Uddin, N., & Kalyankar, R. R. (2011). Manufacturing and structural feasibility of natural fiber reinforced polymeric structural insulated panels for panelized construction. International Journal of Polymer Science, 2011. https://doi.org/10.1155/2011/963549

Wang, Y., Hua, H., Li, W., Wang, R., Jiang, X., & Zhu, M. (2019). Strong antibacterial dental resin composites containing cellulose nanocrystal/zinc oxide nanohybrids. Journal of Dentistry, 80(November 2018), 23–29. https://doi.org/10.1016/j.jdent.2018.11.002

Wippermann, J., Schumann, D., Klemm, D., Kosmehl, H., Salehi-Gelani, S., & Wahlers, T. (2009). Preliminary results of small arterial substitute performed with a new cylindrical biomaterial composed of bacterial cellulose. European Journal of Vascular and Endovascular Surgery, 37(5), 592–596. https://doi.org/10.1016/j.ejvs.2009.01.007

Xue, L., & Greisler, H. P. (2003). Biomaterials in the development and future of vascular grafts. Journal of Vascular Surgery, 37(2), 472–480. https://doi.org/10.1067/mva.2003.88

Yasim-Anuar, T. A. T., Ariffin, H., Norrrahim, M. N. F., Hassan, M. A., Tsukegi, T., & Nishida, H. (2019). Sustainable one-pot process for the production of cellulose nanofiber and polyethylene/cellulose nanofiber composites. Journal of Cleaner Production, 207, 590–599.

Zahedmanesh, H., Mackle, J. N., Sellborn, A., Drotz, K., Bodin, A., Gatenholm, P., & Lally, C. (2011). Bacterial cellulose as a potential vascular graft: Mechanical characterization and constitutive model development. Journal of Biomedical Materials Research Part B: Applied Biomaterials, 97B(1), 105–113. https://doi.org/10.1002/jbm.b.31791

Zhang, F., Wu, W., Zhang, X., Meng, X., Tong, G., & Deng, Y. (2016). Temperature-sensitive poly-NIPAm modified cellulose nanofibril cryogel microspheres for controlled drug release. Cellulose, 23(1), 415–425. https://doi.org/10.1007/s10570-015-0799-4

Zhang, H., Cui, S., Lv, H., Pei, X., Gao, M., Chen, S., Hu, J., Zhou, Y., & Liu, Y. (2019). A crosslinking strategy to make neutral polysaccharide nanofibers robust and biocompatible: With konjac glucomannan as an example. Carbohydrate Polymers, 215, 130–136. https://doi.org/10.1016/j.carbpol.2019.03.075

Zhang, R., & Ma, P. X. (1999). Poly(alpha-hydroxyl acids)/hydroxyapatite porous composites for bone-tissue engineering. I. Preparation and morphology. Journal of Biomedical Materials Research, 44, 446–455.

9 Medical Rubber Glove Waste As Potential Filler Materials in Polymer Composites

M. Nuzaimah,[1,2] *S. M. Sapuan,*[1,3]
R. Nadlene,[2] *M. Jawaid,*[3] *and R. A. Ilyas*[1,3]

[1]Advanced Engineering Materials and Composites
Research Centre (AEMC), Department of
Mechanical and Manufacturing Engineering,
Universiti Putra Malaysia (UPM), Malaysia

[2]Fakulti Teknologi Kejuruteraan Mekanikal
dan Pembuatan, Universiti Teknikal Malaysia
Melaka (UTeM), Melaka, Malaysia

[3]Institute of Tropical Forest and Forest Products
(INTROP), Universiti Putra Malaysia (UPM), Malaysia

CONTENTS

9.1 Introduction 191
    9.1.1 General Overview of Rubber Gloves.. 192
    9.1.2 Rubber Glove Types... 193
    9.1.3 Medical Rubber Glove Manufacturing.. 194
    9.1.4 Medical Rubber Glove Properties 197
9.2 Incorporation of Waste Rubber Products in the Composites Industry......... 198
9.3 Incorporation of Waste Medical Rubber Gloves in the Composite Industry 200
9.4 Conclusion 201
Acknowledgment 201
References.. 202

9.1 INTRODUCTION

Today's rubber products have evolved from old age's rubber balls, rubber human figurines, and rubber bands that had been developed by the ancient Mesoamerican people back in 1600 B.C. for their daily use and as a ritual element in their religious events (Hosler et al., 1999). A great history of rubber was discovered in 1839 by Charles Goodyear who invented the sulfur vulcanization

CH₃   CH₃   Sulphur   CH₃   CH₃
| | | |
– CH₂ – C = CH – CH₂ – CH₂ – C = CH – CH₂ –   – CH₂ – C – CH – CH₂ – CH₂ – C – CH – CH₂ –
              +   + | | | |
                                                                    S   S   Sulphur   S   S
CH₃   CH₃ | | cross-linkage | |
| | S   S   S   S
– CH₂ – C = CH – CH₂ – CH₂ – C = CH – CH₂ –   ⟶ | | | |
                                                             – CH₂ – C – CH – CH₂ – CH₂ – C – CH – CH₂ –
                                                                    | |
                                                                   CH3   CH3

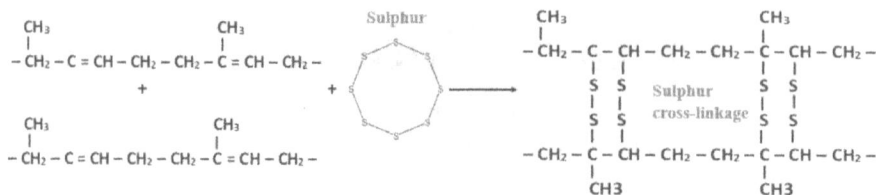

FIGURE 9.1 General scheme of rubber-sulfur vulcanization.

process (Isayev, 2013; Francis, 2016; Sienkiewicz et al., 2017). Sulfur vulcanization of rubber is considered a noteworthy achievement, bringing about an extensive variety of utilizations for rubber in various applications. Properties of rubber improved tremendously with sulfur vulcanization as it enhanced elasticity; strength; resistance to heat, chemicals, and electricity; abrasion resistance; as well as exceptional permeability to gases (Ali Shah et al., 2013; Francis, 2016; Nuzaimah et al., 2018).

Sulfur vulcanization is a process in which sulfur forms cross-links (bridges) between rubber polymer chains. Cross-linkage between sulfur and rubber polymer chains forms a three-dimensional (3D) chemical network that makes rubber a thermoset material, which is incredibly elastic, insoluble, and infusible with high strength (Martınez et al., 2013, Ikeda, 2014; Safia & Fajula, 2015; Kruželák et al., 2016; Nuzaimah et al., 2018). Figure 9.1 shows the general scheme of rubber-sulfur vulcanization.

The vulcanization process provides rubber with remarkable properties that allow countless usages; this rubber has been applied in items as simple as balloons for children to play with, up to more demanding and sophisticated industries like outer-space and marine. Advancement in modern society nowadays requires highly engineered rubbers in order to fulfill the desirable demand of technologies, indirectly leading to growing numbers of rubber products globally. Thus, they now progress rapidly in crucial industries, namely automotive, defense, power generation, medical, aerospace, and aviation. This rapid growth is due to many distinctive attributes of rubber, such as high resilience, resistance to creep, and low sensitivity to strain effects in dynamic loading (Ilyas, Sapuan, & Ishak, 2018; Ilyas, Sapuan, Ishak, et al. 2018; Noor Azammi et al., 2018; Azammi et al., 2020). Rubber blends for manufacturing rubber products not only produce highly cross-linked thermoset material, but also in combination with several additives such as reinforcing agents, curatives, accelerators, antiozonants, and antioxidants, they provide notable properties to the rubber products. The process turns rubber into a material extremely resistant toward degradation and decomposition (Imbernon & Norvez, 2015; Ramarad et al., 2015).

9.1.1 GENERAL OVERVIEW OF RUBBER GLOVES

Meanwhile, the development and rise of standards of life have dynamically created greater demand for rubber gloves as the population is more exposed to healthcare awareness. The yearly forecast for global rubber glove demand is 3.9%

between 2016 and 2020 (Nuzaimah et al., 2018; Gloves Industry: Malaysia hand-in-glove with the rubber glove market, 2019). Rubber gloves are the supreme protective barrier to harmful substances and hazardous risks such as viruses and bacteria, blood-borne pathogens, contagions, diseases, electric shocks, gaseous and chemical attack. For medical practitioners, the glove is their first line of protection as it prevents cross contaminations from both patients and medical personnel or vice versa (Yip & Cacioli, 2002). In food processing, services, and the car painting industry, gloves provide protection to products to avoid defects and contaminants.

The history of the rubber glove, specifically the medical glove, began in 1889 by Dr. William Stewart Halsted, the chief surgeon of Johns Hopkins Hospital. He needed some kind of protection or gear for his nurse, Caroline, who had skin so sensitive that she developed severe dermatitis when in contact with surgical disinfectants. Thus, the very first thin rubber gloves were invented back then in cooperation with the Goodyear Rubber Company. Caroline was the first user to wear rubber gloves in the operating room (MacCallum, 2004; Lathan, 2010; Dahham et al., 2015). Since this discovery, the medical glove industry has evolved and the functionality of the glove has been constantly developed to serve the needs of users.

9.1.2 RUBBER GLOVE TYPES

Rubber gloves generally can be divided into three different categories: medical, household and industrial gloves. Medical gloves are worn by healthcare workers during medical procedures and examinations to prevent cross contaminations between them and the patients. Medical gloves are further categorized into surgical gloves and examination gloves. Surgical gloves worn by surgeons are made with the strictest specifications to prevent rupture or tear during usage. Meanwhile, examination glove applications are more general and extensive, such as caregivers, healthcare examiners, laboratories, room cleaning, food handling, and processing. Figure 9.2 shows

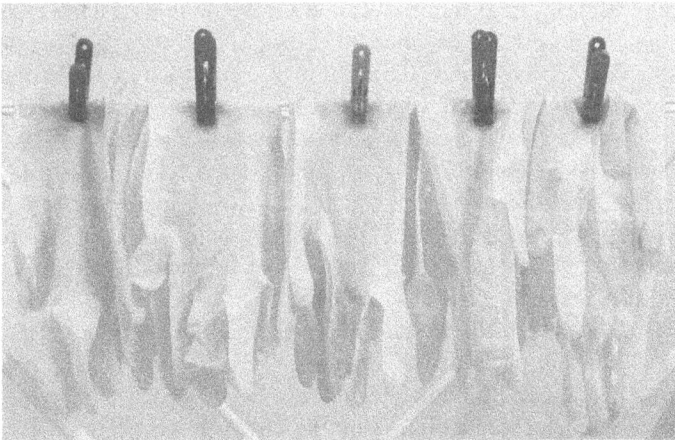

FIGURE 9.2 Examples of medical examination gloves.

typical medical examination gloves. Household gloves, on the other hand, are used for household chores, professional cleaners, and food processing industries. Household gloves normally are thicker with longer cuffs to provide excellent protection against abrasions, cuts, and tears. Industrial gloves are commonly used in food, electronics, painting and coating, printing, and dyeing industries. These gloves not only protect the hands of employees, but also protect their products from contamination or damage (Meleth, 2012).

9.1.3 Medical Rubber Glove Manufacturing

The basic raw material used in the manufacture of medical rubber gloves is latex, either natural rubber latex or synthetic latex. Natural rubber latex is runny, milky white sap from the *Hevea Brasiliensis* tree. The rubber tree is tapped by making an incision at the inner bark of the tree at a depth of a quarter inch to extract the sap. The latex sap is collected in cups and harvested several hours later and preserved with ammonia to stop it from premature coagulation. Because of its high water and non-rubber contents, about 70%, the latex is concentrated and purified by centrifugation to a 60% strength latex concentrate and stabilized using stabilizing agents such as lauric soap and ammonia to prevent coagulation and allow for long-term storage. Figure 9.3 shows how natural rubber latex is harvested from *Hevea Brasiliensis* trees. Major natural rubber producing countries are from South East Asia; Thailand, Indonesia, Vietnam, China, Malaysia, Myanmar, and Cambodia. In 2017, the major rubber consumer countries were China, US, India, Japan, Thailand, Malaysia, Indonesia, Brazil, Germany, and Russia. On the other hand, synthetic rubber is derived from petroleum through the polymerizing process of one or more monomers. Common synthetic latexes used to manufacture gloves are: nitrile rubber (NBR), styrene butadiene rubber (SBR), polychloroprene (CR), vinyl rubber, or polyvinyl chloride (PVC) (Yip & Cacioli, 2002; Hanhi et al., 2007; Rattanapan et al., 2012; Malaysian Rubber Export Promotion Council, 2018).

Besides latex, the other ingredients that are essential in producing rubber gloves are sulfur as a cross-linking or vulcanizing agent, activators, and accelerators to expedite the curing process, and additional ingredients such as a pH adjuster, antioxidants, fillers, pigments, stabilizers, surfactants, and anti-webbing agents. All the ingredients are mixed in a specific sequence and amounts and allowed to mature over a period of 24–36 hours. The pre-vulcanization process by sulfur is performed during this time and further vulcanization continues throughout the manufacturing process (Yip & Cacioli, 2002; Meleth, 2012; Joseph, 2013). Table 9.1 shows the basic ingredients and formulations of medical rubber gloves (Nocil Limited, 2010; Joseph, 2013; Akabane, 2016).

Figure 9.4 illustrates the manufacturing processes for common medical rubber gloves. The gloves are made with a dipping process, in which hand-shaped formers, normally porcelain, dip into solution tanks. The process starts with cleaning the former by removing contaminants on the former's surface in order to prevent glove defects. The cleaning agents are surfactants, alkaline, and acidic solutions. The formers are immersed in the cleaning solution and proceed for drying in the

FIGURE 9.3   Harvesting natural rubber latex from *Hevea Brasiliensis* trees.

TABLE 9.1

Basic Latex Ingredients and Formulations for Medical Rubber Gloves

Ingredient	Formulation (%)
Latex	95.0–98.0
Latex stabilizer	0.01–0.20
Surfactant	0.10–0.20
Sulfur	0.60–1.30
Activator such as zinc oxide (ZnO)	0.35–0.60
Accelerator such as zinc diethyldithiocarbamate (zdec)	0.20–0.80
Pigments such as titanium dioxide (TiO$_2$)	0.50–0.80
Antioxidant	0.60–1.00
Anti-webbing agent	0.05–0.10

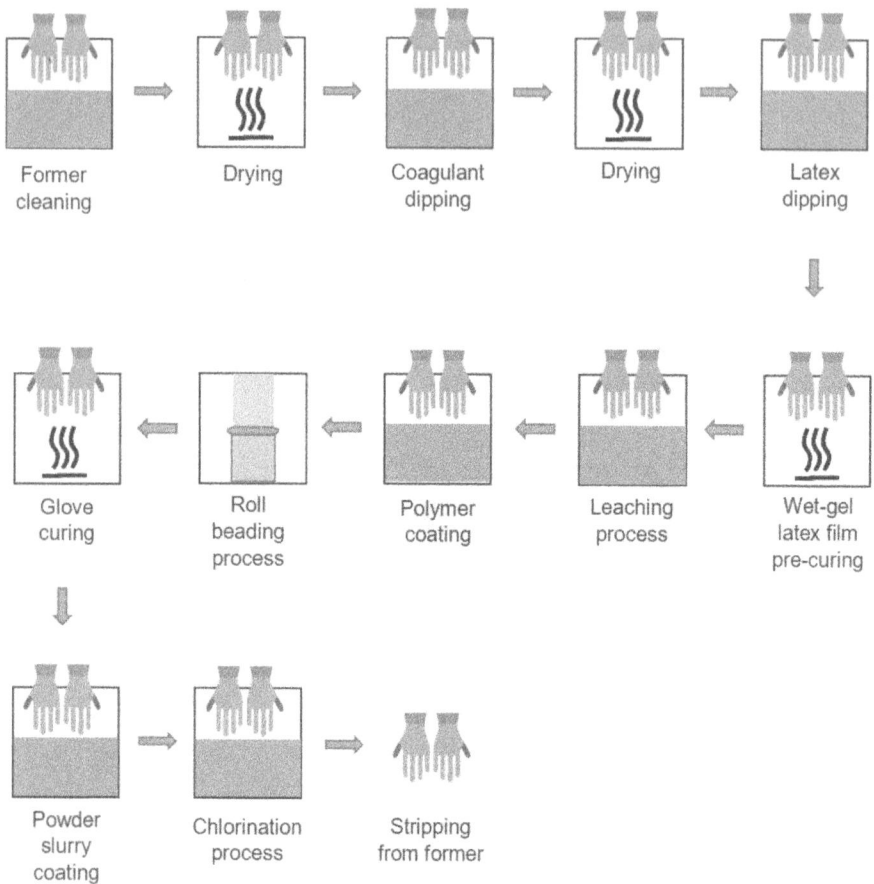

FIGURE 9.4   Schematic of glove manufacturing process.

drying oven. Figure 9.5 shows a schematic process of former cleaning. The cleaned and dried formers are next dipped into a coagulant slurry of calcium nitrate salt-based solution that will assist latex to deposit onto the former surface. Coagulant also contains surfactants and calcium carbonate that help to prevent the rubber from sticking to the former surface and allow the cured glove to be easily stripped off the former. Table 9.2 shows the typical coagulation formulations for rubber gloves (Meleth, 2012; Singh & Jagannath, 2016).

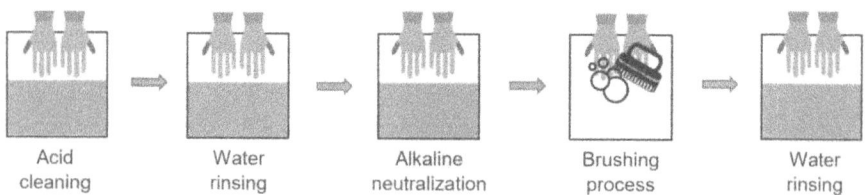

FIGURE 9.5   Schematic process of former cleaning.

TABLE 9.2

Coagulation Formulations

Ingredient	Formulation (%)
Calcium zitrate	15–30
Wetting agent	0.1–1
Surfactant	0.1–1
Anti-tack	2–4
Thickener	Adjust until get desired viscosity
Water	for concentration control

After dipping in coagulant solution, formers go through the drying oven and then are immersed in the compounded latex. The wet-gel latex film requires pre-curing in a gelling oven. The amount of latex deposited on the former surface will determine the thickness of the glove. The wet gel rubber deposit requires a pre-curing process to strengthen the latex article in the pre-curing oven. The semi-cured gloves are next leached in hot water to remove residual calcium nitrate, latex protein, and chemicals (Nurul et al., 2010; Meleth, 2012; Akabane, 2016; Singh and Jagannath, 2016; Chen et al., 2017).

The gloves are further dipped into polymer coating or powder slurry tanks to lubricate the inner surface of the gloves for easy stripping as well as to help for better donning on the hands. After that, the gloves go through a beading station to form a roll bead end at the gloves' cuff. The beaded cuff facilitates gripping when the user dons the gloves (Meleth, 2012; Akabane, 2016). Semi-wet gloves go to the curing oven; the vulcanization process continues to take place during this time. Fully vulcanized rubber provides the gloves with the essential mechanical and physical properties (Yip & Cacioli, 2002; Jirasukprasert et al., 2012; Meleth, 2012; Singh & Jagannath, 2016).

Upon exiting the curing oven, the cured gloves once again go through the powder slurry coating to assist in stripping gloves off from the former. The final dipping is into chlorine solution tanks to allow a chlorination process on the rubber surface. The chlorination hardens the surface of the inner gloves to lessen rubber tackiness and facilitate donning. Chlorination also significantly reduces the level of latex proteins. The gloves are then stripped off from the formers and proceed for drying and quality checking (Noorman & Yuen, 2002; Nurul et al., 2010; Ng et al., 2013; Akabane, 2016; Chen et al., 2017).

9.1.4 MEDICAL RUBBER GLOVE PROPERTIES

Medical rubber gloves must comply with strict specifications to ensure that they are capable of providing very effective barrier performance against harmful substances and blood pathogens. Thus, they are specially formulated and manufactured in order to minimize the possibility that gloves will rupture or tear under stress during use. The gloves must possess important in-use characteristics such as high strength, good elasticity, tactile sensitivity, comfort, good fit, high durability, high tear resistance, high chemical resistance, and ease of donning. For example, medical

TABLE 9.3

Typical Medical Examination Glove Properties (MREPC, 2017)

Property	Minimum Requirement	Unit
Tensile strength before accelerated aging	18	MPa
Tensile strength after accelerated aging	14	MPa
Elongation before accelerated aging	650	%
Elongation after accelerated aging	500	%
Force-at-break before accelerated aging	6	N
Force-at-break after accelerated aging	6	N
Elongation before accelerated aging	650	%
Elongation before accelerated aging	500	%

Note: Accelerated aging at 7 days/70±2°C.

examination gloves manufactured in Malaysia must comply to Standard Malaysian Glove (SMG). SMG is a product quality certification scheme of Malaysia in consultation with regulatory bodies such as the US Food and Drug Administration (FDA) and other relevant authorities. SMG establishes standards for barrier performance, protein, and powder contents in the gloves. Table 9.3 shows properties of typical medical examination glove according to SMG (Yip & Cacioli, 2002; MREPC, 2017; Nuzaimah et al., 2019).

9.2 INCORPORATION OF WASTE RUBBER PRODUCTS IN THE COMPOSITES INDUSTRY

Although natural rubber is biodegradable, vulcanized rubber is resistant to degradation (Ikram, 1999). Rubber obtains a very complex cross-linked structure as a result of the vulcanization process, making it very durable and resistant to many environmental agents; thus, biodegradation seems difficult. The increased demand for rubber products has an adverse effect when it becomes a threat to the environment. High usage of rubber gloves globally has led to high disposal of used products, which have accumulated in stockpiles over the years in landfills. For example, approximately 400–500 million discarded tires were dumped into landfills yearly, and the number is expected to increase in the future. Stockpiling of rubber products in landfills leads to serious ecological hazards when it becomes a breeding ground for disease vectors such as mosquitoes and rodents, as well as dangerous animals like snakes and scorpions. Furthermore, rubber stockpiles can leach harmful chemicals, such as zinc, into the environment, and pollute both soil and water. Also, some of the waste and discarded rubber products like tires are being burned and generate toxic products that are released to the atmosphere (Adhikari et al., 2000; Bailey et al., 2002; Ramarad et al., 2015; Thomas & Gupta, 2016; Nuzaimah et al., 2018).

Therefore, recycling rubber products is one of the important efforts to reduce the amount of rubber to be disposed of. The waste rubber has been discovered to be useful in many applications such as an energy and fuel source, gravel or aggregate substitute for civil works, filler in rubberized asphalt cement concrete, wastewater treatment filters, crash barriers, damping and insulation material, artificial reef, and breakwaters for oceans and rivers (Yang, 1999; Adhikari et al., 2000; Rajan et al., 2006; Yehia et al., 2012; Lo Presti, 2013; Shu & Huang, 2013; Forrest, 2014; Wang, 2014; Moustafa & ElGawady, 2015; Ramarad et al., 2015).

Recycling the waste rubber by making it an alternative for the new raw materials or substituting for the existing material is getting popular. The rubber is ground into smaller sizes and incorporated into matrices as the fillers. The history of using waste rubber as fillers begins with Charles Goodyear, the vulcanization inventor. He patented his work in 1853, when he melded polymer materials that were obtained from rubber granulate and natural rubber (Sienkiewicz et al., 2017). The works soon grew with many researchers exploring the mixing of waste rubber products into many other mediums such as cements and resins. Back in 1985, Pittolo and Burford (1985) studied the effects of adding waste rubber to polystyrene composites. According to their findings, the addition of rubber crumbs improved the impact strength of the composite, which resulted from the action of rubber as a toughening agent (Pittolo & Burford, 1985). Most composites have been fabricated to improve combinations of mechanical characteristics such as tensile, flexural, stiffness, impact, toughness, and ambient and high-temperature strength, as well as lower the density, weight, and cost, while increasing durability (Atikah et al., 2019; Ilyas, Sapuan, Ibrahim, Abral, et al., 2019; Mazani et al., 2019; Nurazzi, Khalina, Sapuan, Ilyas, et al., 2019; Nurazzi et al., 2019; Asyraf et al., 2020; Hazrol et al., 2020; Ilyas & Sapuan, 2020; Ilyas, Sapuan, Atiqah, et al., 2020; Ilyas, Sapuan, Ibrahim, et al., 2020; Norizan et al., 2020). This is due to enhancement of the interfacial bond between filler and matrix (Abral et al., 2019; Aisyah et al., 2019; Ilyas, Sapuan, Ibrahim, Atikah, et al., 2019; Jumaidin, Ilyas, et al., 2019; Jumaidin, Khiruddin, et al., 2019; Jumaidin, Saidi, et al., 2019; Abral, Ariksa, et al., 2020; Abral, Atmajaya, et al., 2020). Meanwhile, Adhikari et al. (2000) reported in their review paper that waste rubber products such as tires can be used as fillers in many civil works such as playgrounds and parking lot flooring. Also, because rubber has better thermal insulation and damping properties, it has been used as asphalt fillers in road construction. The rubber enhanced the durability, fatigue, and skid resistance of the asphalt (Adhikari et al., 2000).

Shu and Huang (2013) reported that rubber is being recycled as lightweight fillers, additives in Portland cement concrete, and modifiers in asphalt paving mixtures. Meanwhile, Moustafa and ElGawady (2015) highlighted that the incorporation of waste tires had improved the properties of the concrete. The ductility, damping, and impact resistance of the concrete improved with the addition of the waste rubber (Shu & Huang, 2013; Moustafa & ElGawady, 2015).

Ramarad et al. (2015) mentioned some other applications of waste rubber especially waste tires as tools for breakwater and floating purposes. The rubber can also be added into plastic material to produce mats, playground surfaces, and trails for athletic uses. Waste tires are used as a fuel source for industries that require high energy sources such as cement, pulp, paper, lime, steel, electrical, energy industry, and steam production (Ramarad et al., 2015).

Yang (1999) conducted an investigation of using scrap tires as filler in drainage structures such as for underground and horizontal drains and for ravine crossing. Lo Presti (2013) mentioned in his review that waste rubber tires have been used in flooring for playground and sport stadiums, in shock-absorbing mats, roofing materials, and also as an alternative to fossil fuel for energy recovery.

Yehia et al. (2012) used waste rubber as filler-extender, as an ion exchanger to clear industrial water, and as an asphalt modifier for road pavement. In their study, the waste rubber undergoes a reclamation process, and later is reused as filler-extender. The surface of waste rubber powder is treated with oxidizing agents to create a functional group such as a carbonyl group. This functional group acts as a filler extender in the asphalt to enhance interphase adhesion between aggregate and asphalt. The waste rubber was added to natural rubber formulations to replace a certain amount of virgin rubber. Besides, in order to make the rubber powder as an ion exchanger, the powder was first sulfonated and tested for its ability to eliminate heavy metals from industrial waste. On the other hand, the waste rubber in a powder form was chemically modified prior to mixing with asphalt in order to enhance interphase addition to improve the road pavement performance (Yehia et al., 2012).

9.3 INCORPORATION OF WASTE MEDICAL RUBBER GLOVES IN THE COMPOSITE INDUSTRY

The main reason many researchers choose to recycle waste rubber gloves as their substitute raw material is due to the gloves' unique properties such as having very large elongation behavior. The material has been especially designed with high strength and elasticity to reduce the risk of gloves rupturing or tearing during use. Also, rubber gloves are relatively cheap and easy to get, which leads to low processing costs. Thus, instead of being treated as useless garbage and pollutant waste rubber, gloves have become a valuable source of new functional materials (Ahmad et al., 2016).

Some of the research has proven that filler from waste gloves helps to give better properties to the composite. For example, blending of filler from waste acrylonitrile butadiene rubber (NBR) gloves with epoxidized natural rubber (ENR 50) and NBR rubber glove waste blended ENR 50 has shown improvement of tensile strength, modulus, and elongation at break (Ahmad et al., 2016; Salleh et al., 2016). One report says that waste rubber gloves were chosen as their filler in order to utilize the excellent puncture and tear resistance of the rubber gloves. These properties give additional value to the newly developed material or composite (Ahmad et al., 2016).

On top of that, rubber flexible properties play a major role in polymer-based composite in which commonly the matrix is a stiff material. Rubber provides flexibility or a soft phase in the composite, while matrix polymer such as polyester provides a hard phase that provides strength to the composite (Esmizadeh et al., 2017). The combination of the two distinct behaviors of these constituents gives the composite a remarkable property for the composite and can be applied in wide ranges of industry. Nuzaimah et al. (2019) have carried out a study by incorporating waste rubber gloves into unsaturated polyester. Their work showed that the incorporation of waste rubber improved a composite's toughness but did encounter a setback of experiencing lower tensile and flexural strength (Nuzaimah et al., 2019). Another interesting finding was by Riyajan et al. (2012), who developed a polymer composite using waste rubber gloves blended with waste polystyrene foam and sugar cane leaves. The composite with the addition of waste rubber gloves produced better composite mechanical properties (Riyajan et al., 2012).

9.4   CONCLUSION

Rubber gloves, especially medical gloves, have been widely used in various fields for barrier protection because they are extremely elastic, very resilient, durable, and resistant to many chemicals, gaseous, and environmental agents. The manufacturing process for medical rubber gloves is very strict and must comply with the specific requirements for gloves in order to ensure that the gloves are of the finest quality and are capable of providing full protection. Growing demand for rubber products, including rubber gloves, has a negative impact on the environment as the highly durable properties of rubber make it difficult to degrade. Hence, ideas for recycling the rubber products have been developed, and the works are evolving. Waste rubber products, rubber medical gloves included, were used as an alternative to new raw materials or to replace existing material. Additionally, the waste rubber gloves are relatively inexpensive and easy to obtain, which results in low processing costs. Many studies have found that the use of the waste rubber in composites has improved the composites' properties. Rubber provides composites with better toughness, damping, and fatigue properties as well as enhanced durability and flexibility. In conclusion, the recycling of waste rubber and waste medical rubber gloves into composites has a great potential to benefit many areas while at the same time helping to save the environment.

ACKNOWLEDGMENT

The authors would like to thank Universiti Putra Malaysia for the financial support provided through the Putra Grant IPS (9607000), Universiti Teknikal Malaysia Melaka, and Ministry of Education Malaysia for providing scholarship to the principal author to conduct this research project and the facilities support by Institute of Tropical Forestry and Forest Products (INTROP), Department of Mechanical and Manufacturing Engineering and Department of Chemical and Environmental Engineering, Faculty of Engineering, Universiti Putra Malaysia.

REFERENCES

Abral, H., Ariksa, J., Mahardika, M., Handayani, D., Aminah, I., Sandrawati, N., Pratama, A.B., Fajri, N., Sapuan, S.M., and Ilyas, R.A., 2020. Transparent and antimicrobial cellulose film from ginger nanofiber. *Food Hydrocolloids*, 98 (August 2019), 105266.

Abral, H., Ariksa, J., Mahardika, M., Handayani, D., Aminah, I., Sandrawati, N., Sapuan, S.M., and Ilyas, R.A., 2019. Highly transparent and antimicrobial PVA based bionano-composites reinforced by ginger nanofiber. *Polymer Testing*, 106186.

Abral, H., Atmajaya, A., Mahardika, M., Hafizulhaq, F., Kadriadi, Handayani, D., Sapuan, S.M., and Ilyas, R.A., 2020. Effect of ultrasonication duration of polyvinyl alcohol (PVA) gel on characterizations of PVA film. *Journal of Materials Research and Technology*, 9(2), 2477–2486.

Adhikari, B., De, D., and Maiti, S., 2000. Reclamation and recycling of waste rubber. *Progress in Polymer Science*, 25(7), 909–948.

Ahmad, H.S., Ismail, H., and Rashid, A.A., 2016. Tensile properties and morphology of epoxidized natural rubber/recycled acrylonitrile-butadiene rubber (enr 50/nbrr) blends. *Procedia Chemistry*, 19, 359–365.

Aisyah, H.A., Paridah, M.T., Sapuan, S.M., Khalina, A., Berkalp, O.B., Lee, S.H., Lee, C.H., Nurazzi, N.M., Ramli, N., Wahab, M.S., and Ilyas, R.A., 2019. Thermal properties of woven kenaf/carbon fibre-reinforced epoxy hybrid composite panels. *International Journal of Polymer Science*, 2019 (December), 1–8.

Akabane, T., 2016. Production method & market trend of rubber gloves. *International Polymer Science and Technology*, 43(5), 369–373.

Ali Shah, A., Hasan, F., Shah, Z., Kanwal, N., and Zeb, S., 2013. Biodegradation of natural and synthetic rubbers: A review. *International Biodeterioration and Biodegradation*, 83, 145–157.

Asyraf, M.R.M., Ishak, M.R., Sapuan, S.M., Yidris, N., and Ilyas, R.A., 2020. Woods and composites cantilever beam: A comprehensive review of experimental and numerical creep methodologies. *Journal of Materials Research and Technology*, (January).

Atikah, M.S.N., Ilyas, R.A., Sapuan, S.M., Ishak, M.R., Zainudin, E.S., Ibrahim, R., Atiqah, A., Ansari, M.N.M., and Jumaidin, R., 2019. Degradation and physical properties of sugar palm starch/sugar palm nanofibrillated cellulose bionanocomposite. *Polimery*, 64(10), 27–36.

Azammi, A.M.N., Ilyas, R.A., Sapuan, S.M., Ibrahim, R., Atikah, M.S.N., Asrofi, M., and Atiqah, A., 2020. Characterization studies of biopolymeric matrix and cellulose fibres based composites related to functionalized fibre-matrix interface. *In: Interfaces in Particle and Fibre Reinforced Composites*. London: Elsevier, 29–93.

Bailey, R.A., Clark, H.M., Ferris, J.P., Krause, S., and Strong, R.L., 2002. Solid waste disposal and recycling. *Chemistry of the Environment*, 769–792.

Chen, S.F., Wang, S., Wong, W.C., and Chong, C.S., 2017. Glove coating and manufacturing process. *U.S. Patent Application 15/409,983*.

Dahham, O., Noriman, N., Sam, S.T., Omar, M.F., and Alakrach, A., 2015. Cure characteristics, tensile and physical properties of recycled natural rubber latex glove (nrl-g) filled acrylonitrile butadiene rubber. *Applied Mechanics and Materials*, 754–755, 693–697.

Esmizadeh, E., Naderi, G., Bakhshandeh, G.R., Fasaie, M.R., and Ahmadi, S., 2017. Reactively compatibilized and dynamically vulcanized thermoplastic elastomers based on high-density polyethylene and reclaimed rubber. *Polymer Science, Series B*.

Forrest, M.J., 2014. *Recycling and Re-use of Waste Rubber*. Smithers Rapra.

Francis, R., 2016. *Recycling of Polymers: Methods, Characterization and Applications*. 1st ed. Weinheim, Germany: John Wiley & Sons.

Gloves Industry: Malaysia hand-in-glove with the rubber glove market [online], 2019. *Rubber Journal Asia.*

Hanhi, K., Poikelispaa, M., and Tirila, H.M., 2007. Elastometric Materials. *Tampere University of Technology.*

Hazrol, M.D., Sapuan, S.M., Ilyas, R.A., Othman, M.L., and Sherwani, S.F.K., 2020. Electrical properties of sugar palm nanocrystalline cellulose, reinforced sugar palm starch nanocomposites. *Polimery*, 55(5), 33–40.

Hosler, D., Burkett, S.L., and Tarkanian, M.J., 1999. Prehistoric polymers: Rubber processing in ancient Mesoamerica. *Science*, 284(5422), 1988–1991.

Ikeda, Y., 2014. Understanding Network Control by Vulcanization for Sulfur Cross-Linked Natural Rubber (NR). In: *Chemistry, Manufacture and Applications of Natural Rubber.* Kidlington, UK: Woodhead Publishing (Imprint of Elsevier), 119–134.

Ikram, A., 1999. *Environment – Friendly Natural Rubber Gloves.* Malaysian Rubber Board (MRB).

Ilyas, R.A. and Sapuan, S.M., 2020. The preparation methods and processing of natural fibre bio-polymer composites. *Current Organic Synthesis*, 16(8), 1068–1070.

Ilyas, R.A., Sapuan, S.M., Atiqah, A., Ibrahim, R., Abral, H., Ishak, M.R., Zainudin, E.S., Nurazzi, N.M., Atikah, M.S.N., Ansari, M.N.M., Asyraf, M.R.M., Supian, A.B.M., and Ya, H., 2020. Sugar palm (Arenga pinnata [Wurmb.] Merr) starch films containing sugar palm nanofibrillated cellulose as reinforcement: Water barrier properties. *Polymer Composites*, 41(2), 459–467.

Ilyas, R.A., Sapuan, S.M., Ibrahim, R., Abral, H., Ishak, M.R., Zainudin, E.S., Atikah, M.S.N., Mohd Nurazzi, N., Atiqah, A., Ansari, M.N.M., Syafri, E., Asrofi, M., Sari, N.H., and Jumaidin, R., 2019. Effect of sugar palm nanofibrillated concentrations on morphological, mechanical and physical properties of biodegradable films based on agro-waste sugar palm (Arenga pinnata [Wurmb.] Merr) starch. *Journal of Materials Research and Technology*, 8(5), 4819–4830.

Ilyas, R.A., Sapuan, S.M., Ibrahim, R., Abral, H., Ishak, M.R., Zainudin, E.S., Atiqah, A., Atikah, M.S.N., Syafri, E., Asrofi, M., and Jumaidin, R., 2020. Thermal, biodegradability and water barrier properties of bio-nanocomposites based on plasticised sugar palm starch and nanofibrillated celluloses from sugar palm fibres. *Journal of Biobased Materials and Bioenergy*, 14(2), 234–248.

Ilyas, R.A., Sapuan, S.M., Ibrahim, R., Atikah, M.S.N., Atiqah, A., Ansari, M.N.M., and Norrrahim, M.N.F., 2019. Production, Processes and Modification of Nanocrystalline Cellulose from Agro-Waste: A Review. In: *Nanocrystalline Materials.* IntechOpen, 3–32.

Ilyas, R.A., Sapuan, S.M., and Ishak, M.R., 2018. Isolation and characterization of nanocrystalline cellulose from sugar palm fibres (Arenga Pinnata). *Carbohydrate Polymers*, 181, 1038–1051.

Ilyas, R.A., Sapuan, S.M., Ishak, M.R., and Zainudin, E.S., 2018. Development and characterization of sugar palm nanocrystalline cellulose reinforced sugar palm starch bionanocomposites. *Carbohydrate Polymers*, 202, 186–202.

Imbernon, L. and Norvez, S., 2015. From landfilling to vitrimer chemistry in rubber life cycle. *European Polymer Journal*, 82, 347–376.

Isayev, A.I., 2013. Recycling of Rubbers. In: *The Science and Technology of Rubber.* Oxford UK: Academic Press (Imprint of Elsevier), 697–764.

Jirasukprasert, P., Garza-reyes, J.A., Soriano-meier, H., and Rocha-Iona, L., 2012. A Case Study of Defects Reduction in a Rubber Gloves Manufacturing Process by Applying Six Sigma Principles and DMAIC Problem Solving Methodology. In: *International Conference on Industrial Engineering and Operations Management.* Istanbul, Turkey, 472–481.

Joseph, R., 2013. Latex Compounding Ingredients. In: *Practical Guide to Latex Technology.* Smithers Rapra, 27–40.

Jumaidin, R., Ilyas, R.A., Saiful, M., Hussin, F., and Mastura, M.T., 2019. Water transport and physical properties of sugarcane bagasse fibre reinforced thermoplastic potato starch biocomposite. *Journal of Advanced Research in Fluid Mechanics and Thermal Sciences*, 61(2), 273–281.

Jumaidin, R., Khiruddin, M.A.A., Asyul Sutan Saidi, Z., Salit, M.S., and Ilyas, R.A., 2019. Effect of cogon grass fibre on the thermal, mechanical and biodegradation properties of thermoplastic cassava starch biocomposite. *International Journal of Biological Macromolecules*, 146, 746–755.

Jumaidin, R., Saidi, Z.A.S., Ilyas, R.A., Ahmad, M.N., Wahid, M.K., Yaakob, M.Y., Maidin, N.A., Rahman, M.H.A., and Osman, M.H., 2019. Characteristics of cogon grass fibre reinforced thermoplastic cassava starch biocomposite: Water absorption and physical properties. *Journal of Advanced Research in Fluid Mechanics and Thermal Sciences*, 62(1), 43–52.

Kruželák, J., Sýkora, R., and Hudec, I., 2016. Sulphur and peroxide vulcanisation of rubber compounds-overview. *Chemical Papers*, 70(12), 1533–1555.

Lathan, S.R., 2010. Caroline Hampton Halsted: The first to use rubber gloves in the operating room. *Proceedings (Baylor University. Medical Center)*, 23(4), 389–392.

Lo Presti, D., 2013. Recycled tyre rubber modified bitumens for road asphalt mixtures: A literature review. *Construction and Building Materials*, 49, 863–881.

MacCallum, W.G., 2004. William stewart halsted (1852-1922): Thyroid surgeon. *National Academy Biographical Memoirs*.

Malaysian Rubber Export Promotion Council, 2018. World Rubber Production, Consumption and Trade [online]. *Malaysian Rubber Export Promotion Council*.

Martınez, J.D., Puy, N., Murillo, R., Garcia, T., Navarro, M.V., and Mastral, A.M., 2013. Waste tyre pyrolysis – A review. *Renewable and Sustainable Energy Reviews*, 23, 179–213.

Mazani, N., Sapuan, S.M., Sanyang, M.L., Atiqah, A., and Ilyas, R.A., 2019. Design and Fabrication of a Shoe Shelf From Kenaf Fiber Reinforced Unsaturated Polyester Composites. *In: Lignocellulose for Future Bioeconomy.* Elsevier, 315–332.

Meleth, J.P., 2012. *An Introduction to Latex Gloves.* 1st ed. Deutschland, Germany: Lambert Academic Publishing.

Moustafa, A. and ElGawady, M.A., 2015. Mechanical properties of high strength concrete with scrap tire rubber. *Construction and Building Materials*, 93, 249–256.

MREPC, 2017. Standard Malaysian Glove (SMG) [online]. *Malaysian Rubber Export Promotion Council*.

Ng, M.C., Ab-samat, H., and Kamaruddin, S., 2013. Reduction of defects in latex dipping production: A case study in a Malaysian company for process improvement. *The International Journal of Engineering and Science (IJES)*, 2(6), 1–11.

Nocil Limited, 2010. *Starting Point Rubber Compounding Formulations.*

Noor Azammi, A.M., Sapuan, S.M., Ishak, M.R., and Sultan, M.T.H., 2018. Mechanical and thermal properties of kenaf reinforced thermoplastic polyurethane (TPU)-natural rubber (NR) composites. *Fibers and Polymers*, 19(2), 446–451.

Noorman, A.H. and Yuen, C.C., 2002. Powder-free medical gloves.

Norizan, M.N., Abdan, K., Ilyas, R.A., Zin, M., Muthukumar, C., Rafiqah, S., and Aisyah, H. (2020). Effect of fiber orientation and fiber loading on the mechanical and thermal properties of sugar palm yarn fiber reinforced unsaturated polyester resin composites. *Polimery*, 65(2), 34–43.

Nurazzi, N.M., Khalina, A., Sapuan, S.M., and Ilyas, R.A., 2019. Mechanical properties of sugar palm yarn/woven glass fiber reinforced unsaturated polyester composites : Effect of fiber loadings and alkaline treatment. *Polimery*, 64(10), 12–22.

Nurazzi, N.M., Khalina, A., Sapuan, S.M., Ilyas, R.A., Rafiqah, S.A., and Hanafee, Z.M., 2019. Thermal properties of treated sugar palm yarn/glass fiber reinforced unsaturated polyester hybrid composites. *Journal of Materials Research and Technology*, (December).

Nurul, H.Y., Hasma, H., Ma'Zam, M.D.S., and Amir, H.M.Y., 2010. Study on protein profiles in commercial examination glove production. *Journal of Rubber Research*, 13(4), 207–217.

Nuzaimah, M., Sapuan, S.M., Nadlene, R., and Jawaid, M., 2018. Recycling of waste rubber as fillers: A review. *IOP Conference Series: Materials Science and Engineering*, 368 (012016).

Nuzaimah, M., Sapuan, S.M., Nadlene, R., and Jawaid, M., 2019. Microstructure and mechanical properties of unsaturated polyester composites filled with waste rubber glove crumbs. *Fibers and Polymers*, 20(6), 1290–1300.

Pittolo, M. and Burford, R.P., 1985. Recycled Rubber Crumb as a Toughener of Polystyrene. *Rubber Chemistry and Technology*.

Rajan, V.V., Dierkes, W.K., Joseph, R., and Noordermeer, J.W.M., 2006. Science and technology of rubber reclamation with special attention to NR-based waste latex products. *Progress in Polymer Science (Oxford)*, 31(9), 811–834.

Ramarad, S., Khalid, M., Ratnam, C.T., Chuah, A.L., and Rashmi, W., 2015. Waste tire rubber in polymer blends: A review on the evolution, properties and future. *Progress in Materials Science*, 72, 100–140.

Rattanapan, C., Suksaroj, T.T., and Ounsaneha, W., 2012. Development of eco-efficiency indicators for rubber glove product by material flow analysis. *Procedia – Social and Behavioral Sciences*, 40, 99–106.

Riyajan, S.A., Intharit, I., and Tangboriboonrat, P., 2012. Physical properties of polymer composite: Natural rubber glove waste/polystyrene foam waste/cellulose. *Industrial Crops and Products*, 36(1), 376–382.

Safia, A. and Fajula, X. C., 2015. Manufacturing and Characterization of materials obtained by reactivated/devulcanized Ground Tire Rubber (GTR) blended with Styrene Butadiene Rubber (SBR). Chemical Engineering Department School of Engineering Universitat Politecnica de Catalunya Terrassa.

Salleh, S.Z., Ahmad, M.Z., and Ismail, H., 2016. Properties of natural rubber/recycled chloroprene rubber blend: Effects of blend ratio and matrix. *Procedia Chemistry*, 19, 346–350.

Shu, X. and Huang, B., 2013. Recycling of waste tire rubber in asphalt and Portland cement concrete: An overview. *Construction and Building Materials*, 67, 217–224.

Sienkiewicz, M., Janik, H., Borzędowska-Labuda, K., and Kucińska-Lipka, J., 2017. Environmentally friendly polymer-rubber composites obtained from waste tyres: A review. *Journal of Cleaner Production*, 147, 560–571.

Singh, K. and Jagannath, S., 2016. Defects reduction using root cause analysis approach in gloves manufacturing unit. *International Research Journal of Engineering and Technology*, 3(7), 173–183.

Thomas, B.S. and Gupta, R.C., 2016. A comprehensive review on the applications of waste tire rubber in cement concrete. *Renewable and Sustainable Energy Reviews*, 54, 1323–1333.

Wang, J., 2014. Saving Latex Gloves from Landfills : Evaluating Sustainable Methods of Waste Disposal such as Recycling, Composting, and Upcycling.

Yang, S., 1999. Use of scrap tires in civil engineering applications. Iowa State University.

Yehia, A., Abdelbary, E.M., Mull, M., Ismail, M.N., and Hefny, Y., 2012. New trends for utilization of rubber wastes. *Macromolecular Symposia*, 320(1), 5–14.

Yip, E. and Cacioli, P., 2002. The manufacture of gloves from natural rubber latex. *Journal of Allergy and Clinical Immunology*, 110 (2).

10 Fabrication and Properties of Polylactic Acid/Hydroxyapatite Biocomposites for Human Bone Substitute Materials

N. Bano,[1] S. S. Jikan,[2] H. Basri,[2]
S. Adzila,[3] N. A. Badarulzaman,[3]
Z. M. Yunus,[2] and S. Asman[2]

[1]Department of Chemistry, Government Postgraduate College for Women, Raiwind, Lahore, Pakistan

[2]Faculty of Applied Sciences and Technology, Universiti Tun Hussein Onn Malaysia, Pagoh Educational Hub, Pagoh, Johor, Malaysia

[3]Faculty of Mechanical and Manufacturing Engineering, Universiti Tun Hussein Onn Malaysia, Parit Raja, Johor, Malaysia

CONTENTS

10.1 INTRODUCTION

Due to the outstanding properties of bioactive compounds in biomaterials, they are broadly studied to accomplish osteointegration, the direct bonding of implant bio-material to bone tissue (Jeong et al., 2008). Generally, the implant biomaterial can be designed by the reinforcement of biofillers into the matrix and it is named a bio-composite. Composites contain two or more different phases of materials in which one phase is continuous, called a matrix, whereas the other is a discontinuous phase, such as a filler within a matrix. In biocomposites, polymers can function as a matrix that supports cell growth. Polymers have many characteristics such as porosity, bio-compatibility, biodegradability, mechanical strength, and hydrophobicity. Indeed, introducing many function groups, monitoring polymerization rates, and changing parts of monomers in different polymer ratios can improve a polymer's properties and introduce new desirable biocomposite materials.

Biocomposite materials are the topic of current research areas on the manufac-turing of implants for bone augmentation. Due to their biocompatible property, biocomposites have been used as bone reconstruction materials. Biocomposites are also highly used for flexible loading applications such as artificial ligaments and tendon replacements due to their good strength (Park & Lakes, 2007). The calcium phosphate family, for instance HAP and α- and β-TCP, has been used to produce biocomposites with high biocompatibility. In principle, the HAP, if sintered at high temperature, is an excellent ceramic material for biocomposite production because of its resemblance to other human bone ceramic components. The HAP that pro-duced bioactivity in bone restoration was found to be highly reactive. Because of the great bioactivity of HAP with body tissue, via this bioactivity behavior, it is an ideal biomaterial in bone repair and implant applications in which tissue attachment to transplant is necessary.

Polymer/HAP biocomposite, which is one type of biocomposite, is designed as bone grafts for numerous orthopedic applications (Ishaug et al., 1996; Furukawa et al., 2000; Ducheyne & Qiu, 1999). It has been reported in the literature that in the biomedical research field, polylactic acid (PLA) has been wildly used with hydroxy-apatite (HAP) or TCP to fabricate PLA/HAP composites for biodegradable bone grafts in orthopedic applications (Ishaug et al., 1996). Incorporation of HAP in a PLA matrix enhances the bending strength of the biocomposite leading to vigorous characteristics in new bone implant development. In fact, the new bone construction and bending strength of biocomposites have been enhanced by mixing HAP with PLA (Furukawa et al., 2000).

Human bone is a type of composite (organic/inorganic). It is composed of collagen and HAP arranged in a hierarchical structure (Liu & Ma, 2004). Most important calcium phosphate compounds are incorporated in polymer matrices producing biocomposites of HAP or calcium-deficient HAP, and α- and β-TCP. In this nature, many biocomposites provide mechanical properties, which are essen-tial as substitute materials in particular applications such as the teeth, skeleton, or cells of living organisms. These biocomposites have mineral components and an organic matrix (LeGeros et al., 2003; Murugan & Ramakrishna, 2005). The incor-poration of calcium phosphate in the polymer matrix producing biocomposites

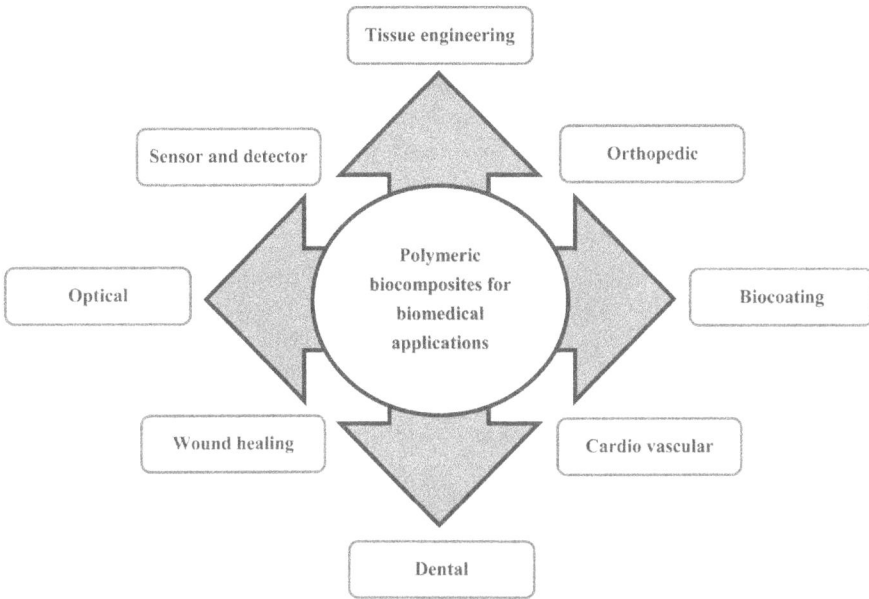

FIGURE 10.1 Different applications of polymeric biocomposites in the biomedical engineering field.

generates a highly biocompatible product by increasing bone tissue and biomaterial interactions when compared to that of the polymer used alone (Rizzi et al., 2001). Additionally, the continuous discharge of calcium and phosphate ions from those biocomposites act as substrates in the remodeling reactions of mineralized tissues, which is an extra benefit (Skrtic et al., 2003). Both solid and porous HAP implants have certain mechanical disadvantages, such as low tensile strength and high brittleness. Even though a number of biodegradable polymers are available, biodegradable polymer/calcium phosphate biocomposites were found to be the most important group. HAP is the best choice due to its resemblance to bone's own material. HAP, when sintered, is a famous bioceramic. It is chemically analog to inorganic hard tissues of humans and used in bone tissue engineering applications (Liu & Ma, 2004; Bouyer et al., 2000). Recent advances and awareness on nanotechnology have influenced the biomaterial industries in fabricating biocomposites with a nanostructured scale (bionanocomposite). There are numerous applications of polymeric biocomposites in the biomedical field such as dental and orthopedic implants, fabrication of biosensors, and cardiovascular stents, as shown in Figure 10.1.

10.2 FABRICATION TECHNIQUES OF PLA/HAP BIOCOMPOSITE

Biocomposite materials have been fabricated with different technologies such as solvent casting and particulate leaching (Nguyen et al., 2017; Karacan et al., 2017), thermally induced phase separation (TIPS) technique (Nishida et al., 2015;

Wang et al., 2010; Han et al., 2013), surface grafting and microemulsions (Sha et al., 2016), three-dimensional (3D) printing (Zhang et al., 2016), electrospinning (Wang et al., 2016), solvent-blending (Kaavessina et al., 2014), melt-mixing techniques (Tajbakhsh & Hajiali, 2016; Ferri et al., 2016), solid–liquid phase separation (Nejati et al., 2009), hot pressing (Kasuga et al., 2001), cold pressing (Ignjatovic et al., 2001), compression molding, extrusion and injection molding processes (Ferri et al., 2016; Park et al., 2018), emulsion freeze drying (Nagata et al., 2005), gas foaming, and solid freeform fabrication (SFF) (Puppi et al., 2010). These techniques have further improved the extensive acceptance of PLA-related biomaterials for a variety of applications as stated by Lim et al. (2008). Based on this acceptance, bioactive, bioreabsorbable, biocompatible, biodegradable, and osteoconductive synergistic biomaterials can be fabricated by taking advantage of the distinct properties of PLA and HAP to yield PLA/HAP biocomposites. The foremost advantage of these degradable biomaterials is their appropriate tunability for diverse applications. Therefore, these biomaterials can assist to decrease the need for second operating procedures frequently related to nondegradable transplant biomaterials (Yu et al., 2012).

Macroporous PLA/HAP composites were recently prepared by a 3D printing method with 60% porosity and 500 μm pore size (Zhang et al., 2016). Commercial PLA and HAP had been used in this experiment. The 3D-printed PLA/HAP scaffolds showed little inflammation response for *in vivo* study, but it has been revealed that these composite scaffolds have good biocompatibility, cell proliferation, and osteoinductive activity in both *in vivo* and *in vitro* studies. Therefore, they can be good candidates for bone substitute materials.

Recent developments in nanotechnology have increased the needs for producing PLA/n-HAP bionanocomposite by incorporating a nanohydroxyapatite (n-HAP) filler. PLA/n-HAP biocomposites modified by deoxyribonucleic acid (DNA) were successfully prepared by Nishida et al. (2015) with different porosity and morphology. The TIPS technique was used to fabricate this biocomposite. The double-stranded deoxyribonucleic acid (ds-DNA) adsorption behavior was examined for the purpose of its gene therapy applications. The biocomposites were characterized by dynamic light scattering (DLS), Fourier transforms infrared (FTIR), and field emission scanning electron microscopy (FESEM). The *in vitro* cell culture test was performed on mesenchymal stem cells (hMSCs) of human bone. It was shown that the cells on PLA/n-HA/ds-DNA scaffold indicated more substantial growths rather than neat-PLA and PLA/n-HA scaffolds.

There are few established techniques to produce dense PLA/HAP biocomposites as conducted by previous researchers (Ignjatović et al., 1999; Gay et al., 2009). One of the techniques is to fabricate PLA/HAP biocomposites by mixing granules of HAP with fully melted PLA. Next, the samples are compressed at 49–490 MPa pressure by cold and hot pressing at temperatures of 20–184°C. The temperature for hot pressing was very near to the melt temperature of PLA (184°C). Thus, biocomposites with 99.6% density, 0.4% porosity, and 93 MPa compressive strength were manufactured (Ignjatović et al., 1999). An alternative approach conducted by Gay and coworkers (Gay et al., 2009) to prepare dense PLA/HAP biocomposites can be established via a hot pressing technique. It has been demonstrated that superior mechanical performances of dense PLA/HAP composites like cortical bone were achieved by dispersing HAP nanoparticles into the PLA matrix. Microscopic analysis revealed that HAP nanoparticles keep equally spread at the nanoscale all throughout the fabrication stages. The elastic modulus and the strength of the biocomposites were considerably increased by increasing the HAP content, but higher inorganic matters have changed the composite from ductile to brittle as exposed by mechanical tests.

The PLA/n-HAP biocomposite for bone tissue engineering application was synthesized and characterized by Nejati et al. (2008) via the TIPS method. Few molecular interactions and chemical linkages among n-HAP particles and PLA matrix had been revealed by FTIR and XRD analyses. The cell affinity and biocompatibility of the nanocomposite were also found to be higher than those of pure PLA.

Consequently, the even distribution of HAP in the polymer matrix at the micron and nano level and the good interfacial connection among the inorganic HAP and organic polymers are the two most important factors that are responsible for fabricating biocomposites with bone-like qualities. In addition, the high surface area and enhanced bioactivity of HAP materials are responsible for more intensive interaction of HAP with the lowest structure level of bone. Otherwise, the resulting biocomposites might have poor mechanical strength. The bone reestablishment and regeneration, both synthetic and natural types of biodegradable polymeric materials, have been used as the polymer matrix in polymer/HAP biocomposites. Thereby, in the last era, huge progress has been made to develop a polymer/HAP system that resembles bone.

10.3 PROPERTIES OF PLA/HAP BIOCOMPOSITE

Even though PLA/HAP biocomposites have emerged as technologically applicable and significant biomaterials in recent years, this area still largely lacks the general description of structure–property relations. The main reason for this information deficit is the lack of capacity to characterize precisely and quantitatively the structure and morphology of PLA/HAP biocomposites. Polymeric composites containing HAP have traditionally been characterized qualitatively, and therefore the establishment of the relationship between structure and macroscopic properties has never been final. In addition, the quantitative structural analyses of these biomaterials are not an easy task. It takes a longer time and needs a number of complementary techniques.

In the literature, different characterization techniques were employed to understand the physicochemical, morphological, and mechanical properties of PLA/HAP biocomposites. The nanomaterial characterizations at nanoscale are very important to understand the physical, structural, and thermal behavior of the PLA/HAP biocomposite. The main physicochemical approaches used for the characterizations of PLA/HAP biocomposites were microscopic techniques such as X-ray diffraction (XRD), FTIR, FESEM, melt flow index (MFI), mechanical testing, and thermal analysis methods.

10.4 THERMAL PROPERTIES

Thermal behaviors of PLA/HAP biocomposites produced by hot pressing for 5, 15, 30, 45, and 60 minutes were studied by Ignjatovic et al. (2004). The patterns of wide-angle X-ray scattering (WAXS) and DSC thermograms of the PLA/HAP bio-composites fabricated at different times exposed the reduction in melting temperature (T_m), glass transition temperature (T_g), and crystallinity % of PLA constituent through extending hot pressing time. These observations were due to the reshuffle of PLA chains on the HA surfaces during the biocomposite fabrication. Furthermore, gel permeation chromatography (GPC) outcomes confirmed the fragmentation of polymer chains due to the thermo-mechanical influences. Thermal gravimetric analyses (TGA) indicated that thermal stability of the achieved biocomposites reduced with increasing hot pressing time.

Deng et al. (2007) studied the fabrication of PLA/HAP hybrid nanofibrous composites using the electrospinning technique. From DSC results, melting enthalpies of the PLA/HAP composite were lower (34.8 J/g) than those of the PLA composite (44.6 J/g). The reduction in melting enthalpy of the PLA/HAP composite showed that hybrid composite contained microcrystals between PLAs (Shikinami & Okuno, 2001).

Zheng et al. (2006) evaluated the shape memory properties of poly(D, L-lactide) (PDLLA)/HAP biocomposites. It was found that the HAP amount greatly impacts the fluctuating T_g values of the biocomposites. DSC results exposed that the T_g of neat-PDLLA and PDLLA/HAP biocomposites at weight ratios of HAP of 25%, 30%, 45%, and 50% were 53.7°C, 55.6°C, 56.8°C, 57.2°C, and 59.0°C, respectively. When the HAP amount increases, the T_g of the biocomposite increases attributable to the interfacial interaction of PDLLA and HAP existing in the biocomposite.

Gay et al. (2009) studied the thermal characteristics of PLA and PLA/HAP bio-composites using DSC. The influences of HAP content on melting temperatures and crystallization of the composites were evaluated. In the second heating scan, crystal-lization, glass transition (T_g), and melting points of PLA and PLA/HAP biocompos-ite were detected. The DSC results showed that the glass transition temperature of the biocomposites increases with the addition of HAP content into PLA, but there was no noteworthy alteration in the crystallization and melting temperatures of the biocomposites as compared to the neat PLA.

The effects of different n-HAP loadings on thermal properties of the PLA/n-HAP bionanocomposite are shown in Table 10.1 (Bano, 2019). As shown in Table 10.1,

TABLE 10.1

DSC Results of Neat-PLA and PLA/n-HAP Bionanocomposites with Different wt% of n-HAP Loading

Sample	1.1 neat PLA	1.2 PLA/1 wt% n-HAP	1.3 PLA/3 wt% n-HAP	1.4 PLA/5 wt%n-HAP
Glass transition temperature, T_g (°C)	66	66.3	66.8	67
Melting temperature, T_m (°C)	172	174	176	179

there is a slight increase in the T_g of the bionanocomposite to a higher temperature when n-HAP loading increased. The change in T_g of the bionanocomposite can be attributed to restriction caused by the n-HAP loading on the molecular chain mobility of PLA (Damadzadeh et al., 2010). The resultant effect of this is also seen in the increase of the endothermic T_m peak. Hence, the higher T_m of the bionanocomposite is attributed to the presence of more imperfect crystals within the bionanocomposite due to the heterogeneous crystallization induced on the PLA matrix by the n-HAP biofillers (Bano, 2019).

10.5   MECHANICAL PROPERTIES

Todo et al. (2006) studied the correlation among fracture behavior and microstructure of bioabsorbable PLA/HAP biocomposites. Effects of particle shape and particle size of HAP on the fracture surface of PLA/HAP biocomposites were assessed by scanning electron microscopy (SEM). The SEM micrographs of fracture surfaces of PLA/HAP biocomposites revealed that the HAP with plate-like shape displayed comparatively rough surfaces. This behavior was due to the elastic deformation of the PLA matrix during the debonding of the matrix and HAP interfaces. In contrast, the HAP with nanoscale and spherical shape displayed even surfaces. Moreover, this nanoscale interaction between the PLA fibrils and the HAP particles corresponded to brittle fracture behavior of the PLA/HAP biocomposite.

The effects of HAP particle size on fracture energy and the mechanical properties of PLA/HAP biocomposites were studied by Takayama et al. (2009). SEM micrographs of cryofracture surfaces of PLA/HAP biocomposites indicated good distribution of microscale HAP particles in the PLA matrix. In contrast, SEM micrographs of PLA biocomposites comprising nanoscale HAP displayed additional accumulation of HAP particles with numerous sizes of accumulated particles. This behavior presented that the nanoscale HAP particles effortlessly agglomerated as related to that of microscale HAP particles. The reduction of the fracture energy and bending strength was caused by the existence of these agglomerations and brittle fracture behavior of HAP led to fracture due to weak bonding among particles. Such localized fracture becomes the initiator of the fracture of the whole biocomposite system. The micrograph of nanoscale PLA/HAP exhibited a very smooth fracture surface, demonstrating very little degenerate energy as compared to the fracture surface of the microscale PLA/HAP. The microscale PLA/HAP also showed localized plastic deformation at the surroundings of the debonded particles and a rough surface with interfacial failure at the microscale HAP and PLA interface.

Rakmae et al. (2011) studied the effect of mixing techniques along with filler content on the physical and thermal properties of PLA and bovine bone-derived carbonated HAP biocomposites. Melt-mixing or solution-mixing procedures were used to prepare PLA/HAP biocomposites at different contents of HAP. The morphologies, thermal properties, as well as mechanical properties of the PLA/HAP biocomposites together with the reduction in molecular weight of PLA matrices were examined. In the biocomposites, average molecular weights of PLA decreased with increasing HAP content, prepared by both techniques, while their molecular weight distributions (MWDs) increased. On the other hand, tensile strength, elongation at break, and impact strength of the composites were decreased with increasing HAP content, whereas the tensile moduli of the biocomposites were increased. In assessment between these two mixing procedures, the melt-mixing dispersed and distributed HAP into PLA matrix is more successful as compared to the solution-mixing method. Consequently, the melt-mixed biocomposites showed higher tensile strength, tensile moduli, and impact strength than those of the solution-mixed biocomposites of the corresponding HAP content. Furthermore, the % crystallinity and decomposition temperatures of the melt-mixed biocomposites were greater than those of the solution-mixed biocomposites.

The effects of HAP particle size on mechanical properties of PLA/HAP biocomposite were studied by Nejati et al. (2009). The tensile properties of PLA/HAP biocomposites having nanoscale HAP were compared with those of PLA biocomposites having microscale HAP. The compressive modulus and compressive strength of the micro- and nanobiocomposites were higher than those of the pure PLA. The compressive strength of 50 wt% nanoscale HAP fillers in the PLA/HAP biocomposite was remarkably greater as compared to that of the microscale HAP fillers in the PLA/HAP biocomposite. This fact is due to the uniform dispersal of the nanoscale HAP particles in PLA matrix and larger surface area of the nanoscale HAP as compared to those of microscale HAP particles. Nevertheless, the compressive modulus values of the micro- and nanobiocomposites were not statistically different.

The tensile properties and preparation of PLA with HAP (Ca-deficient) nanocrystals (d-HAP) nanocomposites were studied by Deng et al. (2001). The tensile modulus values of d-HAP/PLA composite were significantly increased when d-HAP was added into the PLA matrix, as shown in tensile test results. The yield stress of the biocomposite to some extent was varied with d-HAP loading due to effective adhesion among the filler and the polymer matrix. Similarly, Zhang et al. (2010) found that when the HAP content increased from 0 to 20 wt% in PLA, the young modulus and the compressive strength were increased gradually from 1.2 GPa to 3.6 GPa and from 53 MPa to 155 MPa, respectively.

Nejati et al. (2008) stated that PLA/HAP microcomposites (50 wt% HAP) and PLA/n-HAP nanocomposites (50 wt% HAP) had considerably greater average compressive strengths and elastic moduli, at 4.61 MPa and 13.68 MPa correspondingly for the microcomposites and for the nanocomposite at 8.67 MPa and 14.9 MPa, respectively, compared to 2.4 MPa and 1.79 MPa respectively, for the pure PLLA. These results revealed that the higher specific surface area of the nanocomposites resulted in more enhancement of the tensile properties when compared with the microcomposites.

On the other hand, the increasing of n-HAP loading decreases the tensile strength and elongation at break, whereas the tensile modulus of the system increases (Bano, 2019). This finding indicates the weak interaction between n-HAP and PLA matrix at higher filler loading. The tensile strength for pure PLA is 59.23 MPa. This value reduces to 56.78 MPa, 52.65 MPa, and 48.25 MPa after incorporation of 1 wt% n-HAP, 3 wt% n-HAP, and 5 wt% n-HAP, respectively, into the PLA matrix. At 1 wt% HAP content, the tensile strength of the bionanocomposite is 56.78 MPa, which is a reduction of about 2.59% when compared with neat PLA. With further increase in n-HAP loading to 3 wt%, the tensile strength is reduced to 52.65 MPa.

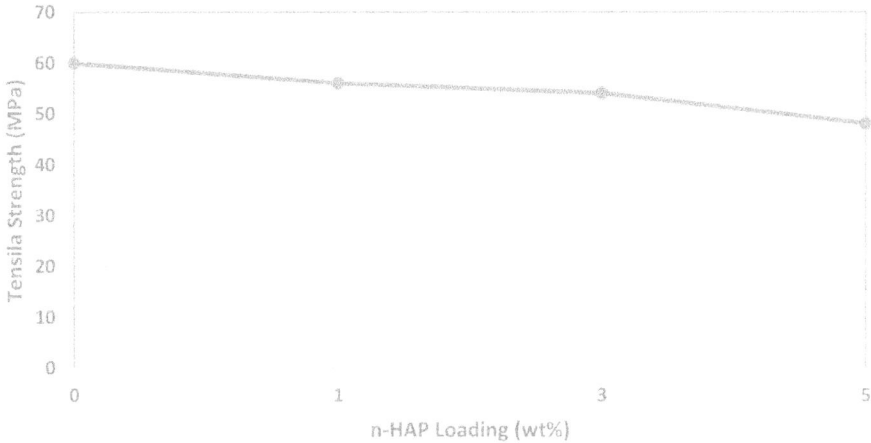

FIGURE 10.2 A typical tensile strength of neat-PLA and PLA/n-HAP bionanocomposites with different n-HAP loadings.

It is a reduction of about 11.10% if compared to that of neat PLA. Additionally, relating to 1% n-HAP, the tensile strength of the bionanocomposite having 5 wt% n-HAP biofiller decreased by 15.02%. This decrease can be correlated with the low strength of n-HAP, conceivably due to the elimination of organic components during extraction (Bano, 2019). Higher filler wt% changed the bionanocomposite from ductile to brittle. A typical tensile strength of neat-PLA and PLA/HAP bionanocomposites with different n-HAP loadings is displayed in Figure 10.2.

10.6   MELT FLOW PROPERTIES

The last topic for discussion of important PLA/HAP properties in this chapter is melt flow property measured by PLA/HAP biocomposites' index values. MFI is a quantity of the mass of polymer that is extruded via a capillary die at a definite temperature and force. Takayama et al. (2013) reported the effects of particle size distribution on physical properties of injection-molded HAP/PLA composites. They found that the properties of HAP as fillers that allow these effects are particle size distribution, high aspect ratio, and rod-like particle shape. Altogether, these factors are assumed to affect the flow rates and increase the MFI value of the composite. At this stage, a higher amount of filler loading leads to higher filler attraction. This is the main reason responsible for the increase in the MFI value.

Another research work conducted by Carrasco et al. (2010) stated that the injection and extrusion methods resulted in a lower viscosity, which is related to a reduction in molecular weight due to the degradation of PLA. Generally, incorporation of particles increases viscosity, which results in lower MFI value, though this result shows the opposite propensity. It is believed that during the melt-mixing process, the hydrolysis degradation of PLA that occurred was enhanced by HAP particle distribution because n-HAP has the hydroxyl group as a functional group. The influence of hydrolysis degradation on the decrease of viscosity is greater than the effect of

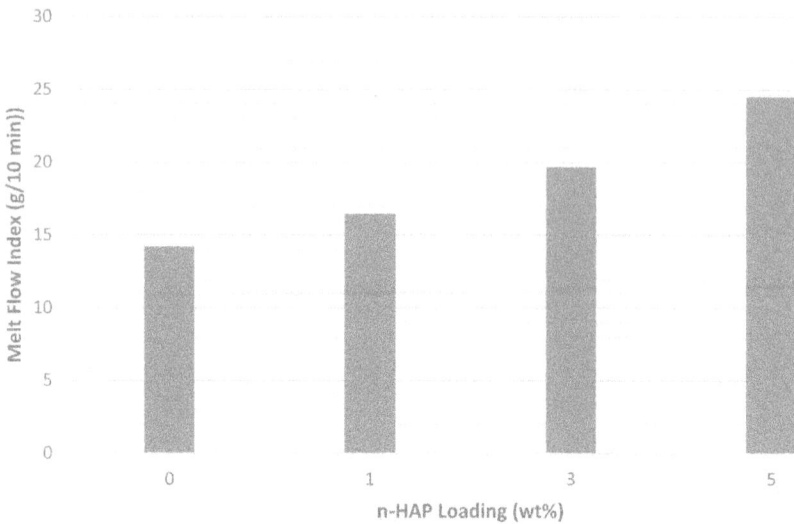

FIGURE 10.3 Typical MFI values of PLA/HAP bionanocomposite obtained at different n-HAP loadings.

incorporation of particles. In addition, the MFI values were increased concurrently with the increased filler loading.

PLA/n-HAP bionanocomposites, which were fabricated using a melt compounding and injection moulding, have shown that the MFIs of all bionanocomposites are higher than those of neat PLA (Bano, 2019). The MFI values of the bionanocomposite indicated that as the amount of n-HAP loading is increased, the MFI value increases gradually. The mineral properties of the filler have therefore considerably affected the flow properties of the bionanocomposite. Figure 10.3 presents a typical MFI value of PLA/HAP bionanocomposites obtained at different n-HAP loadings.

As discussed in this chapter, MFI is one of the studies on flow behavior of PLA/HAP biocomposite melt. It is very significant as it can identify the properties of the end product desired by any biomedical industry. The properties of the filler tend to affect the MFI. Moreover, the method of preparation, nature, and concentration of filler could define the efficacy of the nanocomposite melt flow behavior.

10.7 CONCLUSION

PLA/HAP biocomposite materials have fascinated much interest as bone substitute biomaterials for many biomedical industries. The bioactivity, mechanical properties, degradability, and architecture of biocomposite materials vastly depend on the interfacial interaction between fillers and matrix, the fabrication techniques, various properties, and compositions of the raw materials. It has been shown that PLA/HAP biocomposites can be tailored to meet numerous bone substitute requirements. The requirements of biocomposite material for bone substitute material are complicated. A diversity of distinctive parameters together with degradation rate, pore microstructures, mechanical strength, porosity, and surface chemistry should be prudently measured and controlled for the design and production of biocomposite materials to meet the needs of bone implant applications. In order to accomplish the abovementioned issues, PLA/HAP biocomposite biomaterials offer a multistage approach and consequently superior potential to control their physical and biological assets rather than fabricating biomaterials produced from HAP and PLA alone. This chapter accumulates the recent literature in PLA/HAP biocomposites towards potential applications in numerous biomedical fields. It highlights the information available on different properties of PLA/HAP biocomposites. Therefore, a basic understanding of the correlations between nanoparticle content, scattering, and flow properties (viscosity) and cell nucleation and growth on PLA/HAP biocomposites' properties such as morphological, thermal, mechanical, and MFI is very crucial. Excellent understanding of different biocomposite systems can help optimize and control the fabrication processes. This has the attention of researchers and consultants in the study of biomaterials. However, a number of issues such as mechanical strength, long-term degradation, and inflammatory responses have to be improved for wider applications of PLA/HAP biocomposites as human bone substitute materials.

ACKNOWLEDGMENT

The authors would like to acknowledge Universiti Tun Hussein Onn Malaysia and Ministry of Higher Education Malaysia under Fundamental Research Grant Scheme (FRGS Vote K199).

REFERENCES

Bano, N. (2019). Hydrothermally extracted nanohydroxyapatite from bovine bone as bioceramic and biofiller in bionanocomposite, *PhD Thesis*, Universiti Tun Hussein Onn Malaysia.

Bouyer, E., Gitzhofer, F., and Boulos, M. I. (2000). Morphological study of hydroxyapatite nanocrystal suspension. *Journal of Materials Science: Materials in Medicine*, *11*(8), 523–531.

Carrasco, F., Pages, P., Gamez-Perez, J., Santana, O. O., and Maspoch, M. L. (2010). Processing of poly(lactic acid): Characterization of chemical structure, thermal stability and mechanical properties. *Polymer Degradation and Stability*, *95*(2), 116–125.

Damadzadeh, B., Jabari, H., Skrifvars, M., Airola, K., Moritz, N., and Vallittu, P. K. (2010). Effect of ceramic filler content on the mechanical and thermal behaviour of poly-L-lactic acid and poly-L-lactic-co-glycolic acid composites for medical applications. *Journal of Materials Science: Materials in Medicine*, *21*(9), 2523–2531.

Deng, X. L., Sui, G., Zhao, M. L., Chen, G. Q., and Yang, X. P. (2007). Poly(L-lactic acid)/hydroxyapatite hybrid nanofibrous scaffolds prepared by electrospinning. *Journal of Biomaterials Science, Polymer Edition*, *18*(1), 117–130.

Deng, X., Hao, J., and Wang, C. (2001). Preparation and mechanical properties of nanocomposites of poly(D,L-lactide) with Ca-deficient hydroxyapatite nanocrystals. *Biomaterials*, *22*(21), 2867–2873.

Ducheyne, P. and Qiu, Q. (1999). Bioactive ceramics: The effect of surface reactivity on bone formation and bone cell function. *Biomaterials*, *20*(23–24), 2287–2303.

Ferri, J., Gisbert, I., Garcia-Sanoguera, D., Reig, M., and Balart, R. (2016). The effect of beta-tricalcium phosphate on mechanical and thermal performances of poly(lactic acid). *Journal of Composite Materials*, *50*(30), 4189–4198.

Furukawa, T., Matsusue, Y., Yasunaga, T., Shikinami, Y., Okuno, M., and Nakamura, T. (2000). Biodegradation behavior of ultra-high-strength hydroxyapatite/poly (l-lactide) composite rods for internal fixation of bone fractures. *Biomaterials*, *21*(9), 889–898.

Gay, S., Arostegui, S., and Lemaitre, J. (2009). Preparation and characterization of dense nanohydroxyapatite/PLLA composites. *Materials Science and Engineering C*, *29*(1), 172–177.

Han, W., Zhao, J., Tu, M., Zeng, R., Zha, Z., and Zhou, C. (2013). Preparation and characterization of nanohydroxyapatite strengthening nanofibrous poly(L-lactide) scaffold for bone tissue engineering. *Journal of Applied Polymer Science*, *128*(3), 1332–1338.

Ignjatovic, N., Savic, V., Najman, S., Plavsic, M., and Uskokovic, D. (2001). A study of HAp/PLLA composite as a substitute for bone powder, using FT-IR spectroscopy. *Biomaterials*, *22*(6), 571–575.

Ignjatovic, N., Suljovrujic, E., Budinski-Simendic, J., Krakovsky, I, and Uskokovic, D. (2004). Evaluation of hot-pressed hydroxyapatite/poly-L-lactide composite biomaterial characteristics. *Journal of Biomedical Materials Research – Part B Applied Biomaterials*, *71*(2), 284–294.

Ignjatović, N., Tomić, S., Dakić, M., Miljković, M., Plavšić, M., and Uskoković, D. (1999). Synthesis and properties of hydroxyapatite/poly-L-lactide Composite Biomaterials. *Biomaterials*, *20*(9), 809–816.

Ishaug, S. L., Payne, R. G., Yaszemski, M. J., Aufdemorte, T. B., Bizios, R., and Mikos, A. G. (1996). Osteoblast migration on poly(α-hydroxy esters). *Biotechnology and Bioengineering*, *50*(4), 443–451.

Jeong, S. I., Ko, E. K., Yum, J., Jung, C. H., Lee, Y. M., and Shin, H. (2008). Nanofibrous poly(lactic acid)/hydroxyapatite composite scaffolds for guided tissue regeneration. *Macromolecular Bioscience*, *8*(4), 328–338.

Kaavessina, M., Chafidz, A., Ali, I., and Al-Zahrani, S. M. (2014). Characterization of poly(lactic acid)/hydroxyapatite prepared by a solvent-blending technique: Viscoelasticity and in vitro hydrolytic degradation. *Journal of Elastomers and Plastics*, *47*(8), 753–768.

Karacan, I., Macha, I. J., Choi, G., Cazalbou, S., and Ben-Nissan, B. (2017). Antibiotic containing poly lactic acid/hydroxyapatite biocomposite coatings for dental implant applications. *Key Engineering Materials*, *758*, 120–125.

Kasuga, T., Ota, Y., Nogami, M., and Abe, Y. (2001). Preparation and mechanical properties of polylactic acid composites containing hydroxyapatite fibers. *Biomaterials*, *22*(1), 19–23.

LeGeros, R. Z., Lin, S., Rohanizadeh, R., Mijares, D., and LeGeros, J. P. (2003). Biphasic calcium phosphate bioceramics: Preparation, properties and applications. *Journal of Materials Science: Materials in Medicine*, *14*(3), 201–209.

Lim, L. T., Aurasb, R., and Rubinob, M. (2008). Processing technology for poly(lactic acid). *Progress in Polymer Science*, *33*(8), 820–852.

Liu, X. and Ma, P. X. (2004). Polymeric scaffolds for bone tissue engineering. *Annals of Biomedical Engineering*, *32*(3), 477–486.

Murugan, R. and Ramakrishna, S. (2005). Aqueous mediated synthesis of bioresorbable nanocrystalline hydroxyapatite. *Journal of Crystal Growth*, *274*(1–2), 209–213.

Nagata, F., Miyajima, T., Teraoka, K., and Yokogawa, Y., (2005). Preparation of porous poly(lactic acid)/hydroxyapatite microspheres intended for injectable bone substitutes. *Key Engineering Materials*, *284–286*, 819–822.

Nejati, E., Firouzdor, V., Eslaminejad, M. B., and Bagheri, F. (2009). Needle-like nano hydroxyapatite/poly(l-lactide acid) composite scaffold for bone tissue engineering application. *Materials Science and Engineering C*, *29*(3), 942–949.

Nejati, E., Mirzadeh, H., and Zandi, M. (2008). Synthesis and characterization of nano-hydroxyapatite rods/poly(l-lactide acid) composite scaffolds for bone tissue engineering. *Composites Part A: Applied Science and Manufacturing*, *39*(10), 1589–1596.

Nguyen, T. T., Hoang, T., Can, V. M., Ho, A. S., Nguyen, S. H., Nguyen, T. T. T., Pham, T. N., Nguyen, T. P., Nguyen, T. L. H., and Thi, M. T. D. (2017). In vitro and in vivo tests of PLA/d-HAp nanocomposite. *Advances in Natural Sciences: Nanoscience and Nanotechnology*, *8*(4), 045013.

Nishida, Y., Domura, R., Sakai, R., Okamoto, M., Arakawa, S., Ishiki, R., Salick, M. R., Turng, L. S. (2015). Fabrication of PLLA/HA composite scaffolds modified by DNA. *Polymer*, *56*, 73–81.

Park, J., Kim, B., Hwang, J., Yoon, Y., Cho, H., Kim, D., Lee, J. K., and Yoon, S. (2018). In-vitro mechanical performance study of biodegradable polylactic acid/hydroxyapatite nanocomposites for fixation medical devices. *Journal of Nanoscience and Nanotechnology, 18*(2), 837–841.

Park, J. and Lakes, R. S. (2007). *Biomaterials an Introduction.* Madison: Springer. 152–155.

Puppi, D., Chiellini, F., Piras, A. M., and Chiellini, E. (2010). Polymeric materials for bone and cartilage repair. *Progress in Polymer Science*, *35*(4), 403–440.

Rakmae, S., Ruksakulpiwat, Y., Sutapun, W., and Suppakarn, N. (2011). Effects of mixing technique and filler content on physical properties of bovine bone-based CHA/PLA composites. *Journal of Applied Polymer Science*, *122*(4), 2433–2441.

Rizzi, S. C., Heath, D. J., Coombes, A. G. A., Bock, N., Textor, M., and Downes, S. (2001). Biodegradable polymer/hydroxyapatite composites: Surface analysis and initial attachment of human osteoblasts. *Journal of Biomedical Materials Research*, *55*(4), 475–486.

Sha, L., Chen, Z., Chen, Z., Zhang, A., and Yang, Z. (2016). Polylactic acid based nanocomposites: Promising safe and biodegradable materials in biomedical field. *International Journal of Polymer Science*, *2016*, 1–11.

Shikinami, Y. and Okuno, M. (2001). Bioresorbable devices made of forged composites of hydroxyapatite (HA) particles and poly l-lactide (PLLA). Part II: Practical properties of miniscrews and miniplates. *Biomaterials*, *22*(23), 3197–3211.

Skrtic, D., Antonucci, J. M., and Eanes, E. D. (2003). Amorphous calcium phosphate-based bioactive polymeric composites for mineralized tissue regeneration. *Journal of Research of the National Institute of Standards and Technology*, *108*(3), 167–182.

Tajbakhsh, S. and Hajiali F. (2016). Polylactic acid/bioactive ceramic biocomposite scaffolds for bone tissue engineering: A brief overview. *International Journal of Research in Applied, Natural and Social Sciences, 4*(7), 165–174.

Takayama, T., Todo, M., and Takano, A. (2009). The effect of bimodal distribution on the mechanical properties of hydroxyapatite particle filled poly(L-lactide) composites. *Journal of the Mechanical Behavior of Biomedical Materials, 2*(1), 105–112.

Takayama, T., Uchiumi, K., Ito, H., and Kawai, T. (2013). Particle size distribution effects on physical properties of injection molded HA/PLA composites. *Advanced Composite Materials*, *22*(5), 327–337.

Todo, M., Park, S. D., Arakawa, K., and Takenoshita, Y. (2006). Relationship between micro-structure and fracture behavior of bioabsorbable HA/PLLA *composites. Composites Part A: Applied Science and Manufacturing*, 37(12), 2221–2225.

Wang, X., Song, G., and Lou T. (2010). Fabrication and characterization of nano composite scaffold of poly(l-lactic acid)/hydroxyapatite. *Journal of Materials Science: Materials in Medicine*, *21*(1), 183–188.

Wang, Z., Wang, Y., Ito, Y., Zhang, P., and Chen, X. (2016). A comparative study on the in vivo degradation of poly(L-lactide) based composite implants for bone fracture fixation. *Scientific Reports, 6,* 20770.

Yu, T., Wang, Y. Y., Yang, M., Schneider, C., Zhong, W., Pulicare, S., Choi, W., Mert, O., Fu, J., Lai, S. K., and Hanes, J. (2012). Biodegradable mucus-penetrating nanoparticles composed of diblock copolymers of polyethylene glycol and poly(lactic-co-glycolic acid). *Drug Delivery and Translational Research, 2*(2), 124–128.

Zhang, C. Y., Lu, H., Zhuang, Z., Wang, X. P., and Fang, Q. F. (2010). Nano-hydroxyapatite/poly(L-lactic acid) composite synthesized by a modified in situ precipitation: preparation and properties. *Journal of Materials Science: Materials in Medicine, 21,* 3077–3083.

Zhang, H., Mao, X., Du, Z., Jiang, W., Han, X., Zhao, D., Han, D., and Li, Q. (2016). Three dimensional printed macroporous polylactic acid/hydroxyapatite composite scaffolds for promoting bone formation in a critical-size rat calvarial defect model. *Science and Technology of Advanced Materials, 17*(1), 136–148.

Zheng, X., Zhou, S., Li, X., and Weng, J. (2006). Shape memory properties of poly(d,l-lactide)/hydroxyapatite composites. *Biomaterials, 27*(24), 4288–4295.

11 Hydrogel-Based Composites in Perfusion Cell Culture/Test Device

Drug Delivery through Diffusion

W. Zhao
School of Mechanical and Electronic Engineering,
Wuhan University of Technology, P. R. China

CONTENTS

11.1 INTRODUCTION

11.1.1 Background

Cell culture techniques, developed at the turn of the 20th century, are still today an enabling and fundamental tool in the study of cell function, tissue engineering, and pharmacology. The use of novel materials, such as hydrogel-based composites, is opening new interesting perspectives, especially for what concerns controlled release, tissue engineering, and *in vitro* cell culture applications. In these perspectives, hydrogel-based composites are very promising candidates for biological applications because they show appealing mechanical properties, such as ease of molding, and, for most of them, high biocompatibility. Poly(2-hydroxyethyl methacrylate) (PHEMA), one of the toughest artificial hydrogels with acceptable biocompatibility, was used in this study as the base material (e.g., matrix) for the composite membrane. When it is synthesized with carbon nanotubes (CNTs), electrical conductivity was endowed to the produced hydrogel-based composite material without changing its properties in terms of diffusion properties and biocompatibility.

The use of this PHEMA-CNT composite as thin membranes (with thickness ranging from tenths of nanometers to hundreds of microns) has the advantage of not only mimicking the physiological environment that cells experience *in vivo*, but also opens the possibility of controlling the absorption, diffusion, and release processes of proteins, drugs, and other biomolecules throughout their thickness (Majumda, 2014; Zhao et al., 2018). The hydrogel-based composite membranes have been realized by means of a number of techniques, including *in situ* fabrication processes (e.g., spin and dip coating) as well as *ex situ* molding and replication (Tokarev & Minko, 2009, 2010).

The needs of a microfluidic device for automatic culturing cells over long-term periods, inputting testing drugs simply and individually, were considered in this study to develop a multifunctional perfusion device. By using this device, it will highly reduce the difficulties and complexities that may be present to the end users who are investigating the effects of various drugs on living cells.

This chapter focuses on the investigation of the diffusion properties of nutrient concentration through the hydrogel-based composite membrane, and optimal combinations of parameters to enable maximal cell viability for the developed device.

11.1.2 Perfusion Cell Culture/Test Device

A multilayer structured hydrogel-based microfluidic platform for cell culture is proposed and shown in Figure 11.1. Two microfluidic chips with culture chamber and drug delivery reservoir, which are made of thermoplastic (poly(methyl methacrylate) [PMMA]), are assembled with an inserted hydrogel-based composite membrane (PHEMA-CNT). The packaging technology is required to hold the assembly together and provide a reliable fluid seal.

Such a designed device is able to fulfill the following tasks: (1) Cells will be kept and cultured in the culture chamber (upper chamber) at the thermoplastic top layer,

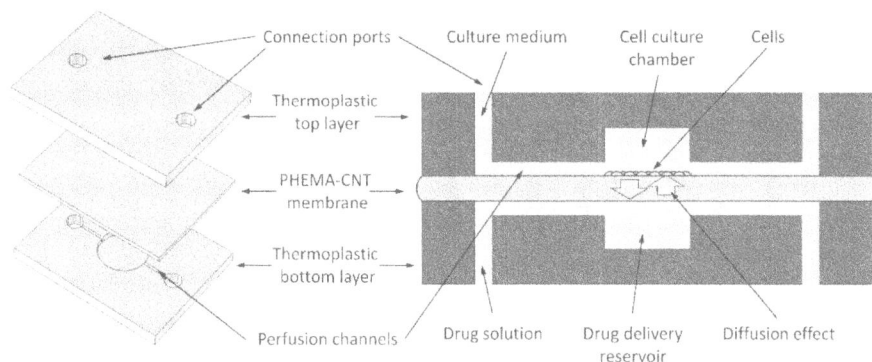

FIGURE 11.1 Schematic diagram of the microdevice used to culture cells. Three layers, two thermoplastic layers and one PHEMA-CNT composite membrane, make up the device as a two closed microfluidic system, culture chamber at the top and drug delivery reservoir at the bottom. Cells are cultured in the culture chamber. The exchange of molecules between culture chamber and drug delivery reservoir is taking place all the time, through diffusion inside the hydrogel-based composite membrane material.

surrounded with dynamic culture medium, which is carrying oxygen, glucose, and other necessary nutrients. (2) The culturing of cells will be achieved by controlling and adjusting the fluid flow or perfusion time of the culture system in the upper chamber, independent to the aqueous environment chamber (lower chamber) in the bottom layer. (3) Solution containing drugs or other small molecules, for the purposes of testing the reaction of cells to these molecules, is perfused into the drug delivery reservoir through microchannels. (4) These drugs will be delivered into the culture chamber through the hydrogel-based composite membrane, thanks to the diffusion characteristics of hydrogel materials.

Therefore, the characteristics of such device design can be summarized as follows:

- The system can treat culturing of cells and testing of cells individually, without any interruption, which offers a single variable environment for research on cells.
- Based on microfluidic structure, the usage of solutions is at a small amount, thus highly reducing the consumption of culture medium and other expensive growing factors for cells.
- The diffusion properties of the hydrogel-based composite are used to deliver drugs into the culture chamber and deliver the excreta of cells out of the culture chamber.
- The hydrogel-based composite's rubber-like mechanical property is used for sealing the device to achieve a reliable bonding and closed microfluidic environment when the system operates.
- The packaging method enables disassembly and reassembly bonding at room temperature, without the assistance of heating sources or adhesives.

11.2 THEORETICAL BASIS

11.2.1 Kinetic Theories of Diffusion in Composite Membrane

When a swollen hydrogel sinks into buffer solution, the exchange of molecules between the hydrogel and its external environments starts immediately. In this study, the cylindrical PHEMA-CNT composite sample, which is fully swollen by glucose solution (200 g/L), is placed into PBS buffer (phosphate buffered saline) to investigate the kinetics of diffusion of the PHEMA-CNT composite to glucose, including diffusion speed, diffusion mechanism, and capacity of absorption. PBS buffer is a commonly used buffer solution in biological research, which can mimic the environment of inside the human body, because the pH value, osmolarity, and ion concentrations of PBS buffer are identical to the isotonic environment of the human body.

To analyze the characteristics of molecule diffusion in the PHEMA-CNT composite, a time-dependent model, which is based on an empirical power equation, was developed by Peppas et al. (Lustig & Peppas, 1988):

$$M_t /M_\infty = kt^n \tag{11.1}$$

where M_t and M_∞ denote the absolute cumulative amounts of molecule released from the hydrogel at time t and at the equilibrium, respectively; k is a constant related to the properties of the hydrogel network, for example, structure and degree of cross-linker. And n is the diffusion exponent. According to the literature (Li et al., 2006), three models, which are dependent on the numerical range of the diffusion exponent n, can indicate the characteristics of releasing molecules from hydrogel materials: (1) Fickian diffusion ($n = 0.5$), diffusion-controlled release process; (2) zero-order model ($n = 1$), swelling-controlled release process; (3) Ritger–Peppas's empirical model ($0.5 < n < 1$), anomalous release with respect to the applicability. According to Equation (11.1), it is easy to note that the relationship between $\ln(M_t/M_\infty)$ and $\ln(t)$ is linear, with slope n and intercept k. Meanwhile, the values of M_t/M_∞ and t can be obtained from the cumulative release experiment. The value of diffusion exponent n can be determined based on the results from the cumulative release experiment to indicate which kind of kinetics model can best describe the diffusion characteristics of glucose in the PHEMA-CNT composite.

11.2.2 Basic Fluid Dynamics in Perfusion Channels

To determine the characteristics of fluidic flow in terms of laminar or turbulent forms, a preliminary calculation of the Reynolds constant (Re), which is correlated to the dimensions of the microchannel, is made (Batchelor, 1967). It indicates that the maximum Reynolds number is located at the smallest cross-section when fluid flows in a tube or closed channel. Comparing with the fluidic port and culture chamber, the smallest cross-section in fluid passage in this microfluidic device appears at the microchannel. Thus, the maximum Reynolds number occurs at the microchannel. According to the dimensions of the trapezoidal cross-section in the present study: widths $w_1 = 0.48$ mm, $w_2 = 0.1$ mm, height $h_c = 0.7$ mm, and dip angle $\theta = 15°$ (see Figure 11.1), the maximum Reynolds number (Re) of this microfluidic device can be calculated based on the equivalent dimensions of the microchannel:

$$Re = \frac{\rho v_{in} D_H}{\mu} = \frac{2\rho h_c (w_1 + w_2)}{\mu(w_1 - w_2) + \mu \sin\theta (w_1 + w_2)} \cdot v_{in} \qquad (11.2)$$

where μ is dynamic viscosity of fluid (Pa·s or kg/[m·s]), ρ is the density of fluid (kg/m^3), and v_{in} is the mean velocity of the fluid in the channel (mm/s). The hydraulic diameter of the trapezoidal duct D_H is given by $D_H = 4*$Cross-sectional area (mm^2)/Wetted perimeter (mm). The wetted perimeter in this case is the total perimeter of cross section of channel, because the channel is fully filled with fluid, thus all channel walls are in contact with the fluid.

According to the initial flow rate (0.5 mL/hour), which is controlled by the micropump, and the area of the cross section of microchannel ($w_1 = 0.48$ mm, $w_2 = 0.1$ mm, and $h_c = 0.7$ mm), the value of v_{in} is calculated, which is 0.68 mm/s. Thus, with the parameters that are utilized in the simulation, that is, $\mu = 8.9*10^{-4}$ Pa·s, $\rho = 1000$ kg/m^3, and $v_{in} = 0.68$ mm/s, Equation (11.2) yields a Reynolds number of 0.304. It indicates that the characteristic of fluidic flow in the microdevice is laminar flow, because the value of the maximum Reynolds number in the fluidic passage is much less than 1,000 (Liu et al., 2004). Therefore, the theory on the momentum convection equation based on Navier–Stokes equations at steady state (Temam, 2001) can be implemented in the simulation, which is used to represent the laminar fluid flow in the microdevice.

11.3 MATERIALS AND EXPERIMENTAL APPROACHES

11.3.1 Preparation of Hydrogel-Based Composite Membrane

The monomer HEMA (70 wt%), water (30 wt%), the cross-linker ethylene glycol dimethacrylate (EGDMA, 0.1 mol% of HEMA), and the catalyst tetramethylethylenediamine (TEMED) were mixed to form a homogeneous aqueous solution (HEMA solution). Single-walled CNTs (purity $\geq$ 90 wt%, length = 1–3 μm, from Chengdu Organic Chemicals Co. Ltd.) were added into HEMA solution to form suspension with 1–2 mg/mL of CNTs. This suspension was firstly cooled to 4°C and then mixed quickly with 20 wt% of ammonium persulfate (APS) solution (1 vol% of suspension) under stirring, followed by casting in molds (for both cylindrical sample and thin membrane) and left to stand for 24 hours at 4°C. The reaction of polymerization was completed until the initiator APS was fully consumed. The suspended CNT particles were bonded by both physical and chemical bonds with the base material, PHEMA hydrogel (Kharismadewi et al., 2016; Massoumi et al., 2016). This synthetic process and the content of chemicals used in the process were chosen based on that, it can give the composite the best mechanical and electrical characteristics (Ansari et al., 2016).

11.3.2   Purification Treatment of Hydrogel-Based Composite

After the synthesis of the hydrogel-based composites, swelling and cleansing processes have to be conducted to release the residual stress between polymer chains and remove the water-soluble contaminants (Zhao et al., 2015). In this work, the samples of hydrogel composites after synthesis were immersed in phosphate-buffered saline (PBS) to absorb ions. PBS was used as an electrolyte for the purpose of mimicking a biocompatible environment for any potential bioelectric applications (Huang et al., 2017). During this stage, ions and solvent molecules entered into the gaps between molecular chains and pushed them apart. Meanwhile, the solvent molecules attached to the chains and occupied the space between the hydrogel chains, which made the volume of swollen hydrogel composites larger than those samples before the swelling. The PBS was renewed every 4 hours to remove the residuals inside the composites. Such a renewal process was conducted six times before conducting experimental tests on the samples.

11.3.3   Experimental Approaches for Tests

Two cumulative release experiments were conducted, with the purposes of (i) investigating the diffusion characteristics of PHEMA-CNT composite, when it is interacted with glucose molecules; (ii) studying the capacity of absorption of PHEMA-CNT composite hydrogel to glucose.

For the first type of cumulative release experiment (Purpose I), the experimental set-ups, conditions, and descriptions are illustrated in Figure 11.2. Steps 1–3 are the preparation procedures, aiming to fabricate the PHEMA-CNT composite, which is filled with glucose molecules. Three days of stirring ensure that the glucose concentration inside the PHEMA-CNT composite sample saturates completely. Step 4 aims to obtain the speed that glucose molecules diffuse out from the PHEMA-CNT composite sample, by monitoring the glucose concentration in the PBS buffer over time. The constant environment for the diffusion experiment under 37°C and 100 r/min stirring is maintained in a water bath and using a magnetic stirring system.

According to Step 4 in Figure 11.2, 3 mL solution is withdrawn from the PBS buffer solution outside the PHEMA-CNT composite sample and immediately replaced by 3 mL fresh 37°C PBS buffer solution at the designated time interval points (t_x). The samples of such withdrawn solution are stored in a −5°C fridge in order to avoid the glucose diffusing out from the PHEMA-CNT composite sample to be consumed by germs, which may affect the accuracy of the experimental results. Thus, an ICB SBA-90 laboratory glucose meter (±0.01 g/L) is employed to measure the glucose concentration in the withdrawn solution samples. However, in Step 4, a small amount of glucose exists in 3 mL solution, which is extracted from the PBS buffer solution every time at t_x, but the 3 mL replaced PBS buffer solution does not contain any glucose. This process causes a glucose loss, thus underestimating the amount of glucose that is released from the PHEMA-CNT composite sample. Aiming to eliminate such glucose

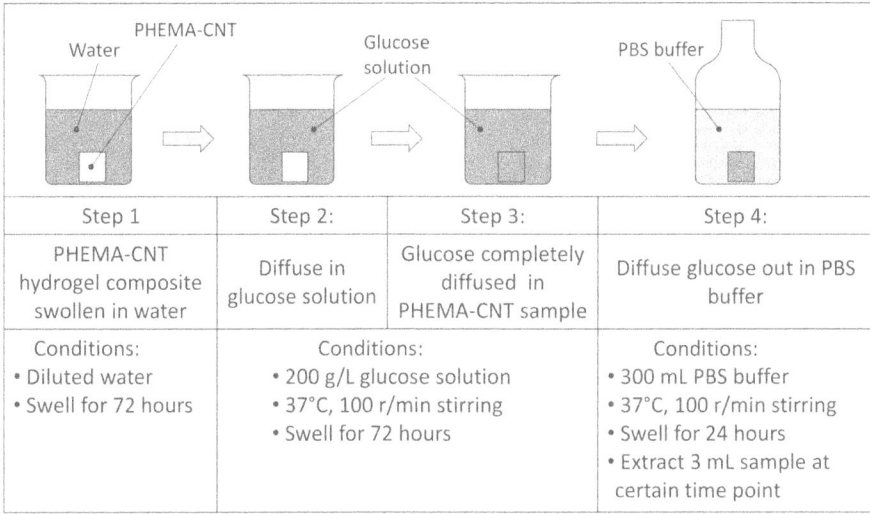

Step 1	Step 2:	Step 3:	Step 4:
PHEMA-CNT hydrogel composite swollen in water	Diffuse in glucose solution	Glucose completely diffused in PHEMA-CNT sample	Diffuse glucose out in PBS buffer
Conditions: • Diluted water • Swell for 72 hours	Conditions: • 200 g/L glucose solution • 37°C, 100 r/min stirring • Swell for 72 hours		Conditions: • 300 mL PBS buffer • 37°C, 100 r/min stirring • Swell for 24 hours • Extract 3 mL sample at certain time point

FIGURE 11.2 Process of the cumulative release experiment to determine diffusion characteristics of glucose in PHEMA-CNT composite.

loss, the amount of glucose released from PHEMA-CNT composite sample has to be calculated:

$$M(t_x) = M_m(t_x) + \frac{3[mL]}{300[mL]} \times \sum_{x=1}^{x-1} M_m(t_x) \qquad (11.3)$$

where $M(t_x)$ and $M_m(t_x)$ are the actual amount and measuring amount of glucose at the designated time interval points t_x, respectively; x is the order number, which indicates the withdraw time of the 3 mL PBS buffer solution. The equilibrium value of $M(t_x)$ equals the value of M_∞ from Equation (11.3).

The experimental set-ups, conditions, and descriptions for the other type of cumulative release experiment (Purpose II), which aims to study the capacity of absorption of PHEMA-CNT composite to glucose, are summarized in Figure 11.3. Step 1 in Figure 11.3 is the same as Step 3 in Figure 11.2, ensuring the glucose is completely saturated inside the PHEMA-CNT composite sample. The experimental steps of such a cumulative release experiment (Purpose II) are described as follows: (1) Place the PHEMA-CNT composite sample into PBS buffer solution to diffuse the glucose out from the sample. (2) Keep the environment under 37°C, 100 r/min stirring for 24 hours, which allows the release of the glucose to reach the equilibrium. (3) Extract 3 mL of the PBS buffer, mark as a sample number y, and store in fridge to keep temperature of −5°C, to avoid consumption of the glucose by germs. (4) Then place the PHEMA-CNT composite sample into 300 mL fresh PBS buffer solution, and repeat Steps 2 and 3. This repeated process is terminated if the concentration of glucose in the PBS buffer is zero (Step 4). To guarantee the zero concentration of glucose, many repeats of Steps 2 and 3 have to be implemented. In this study, seven

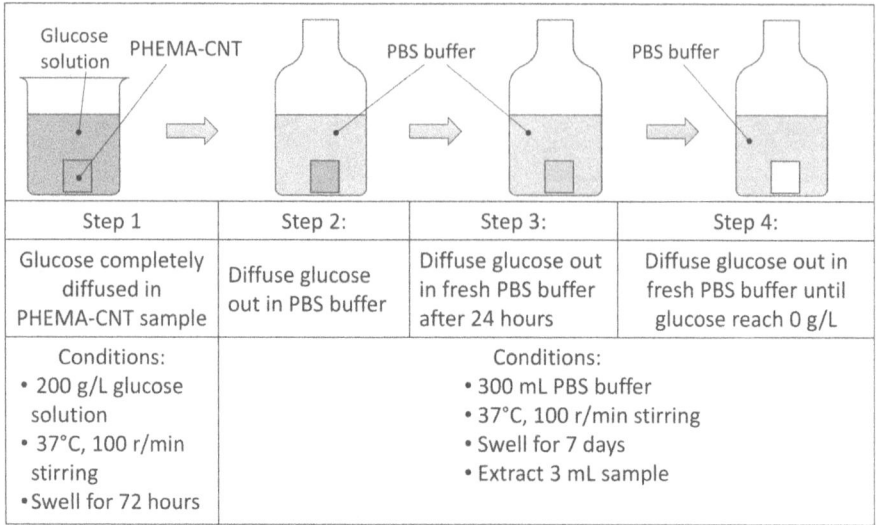

Step 1	Step 2:	Step 3:	Step 4:
Glucose completely diffused in PHEMA-CNT sample	Diffuse glucose out in PBS buffer	Diffuse glucose out in fresh PBS buffer after 24 hours	Diffuse glucose out in fresh PBS buffer until glucose reach 0 g/L
Conditions: • 200 g/L glucose solution • 37°C, 100 r/min stirring • Swell for 72 hours		Conditions: • 300 mL PBS buffer • 37°C, 100 r/min stirring • Swell for 7 days • Extract 3 mL sample	

FIGURE 11.3 Process of the cumulative release experiment to determine the capacity of absorption of PHEMA-CNT composite on glucose molecules.

times of repeat were carried out, which means samples marked with y (y = 1, 2, 3, …,7) have been obtained. Therefore, the amount of glucose that is absorbed by the PHEMA-CNT composite sample ($M_{glucose}$ [mol]) is given by:

$$M_{glucose}[mol] = \sum_{y=1}^{7} M_y[g] \times \frac{1}{180}[mol/g]$$

$$= 300[mL] \times \sum_{y=1}^{7} C_y[g/mL] \times \frac{1}{180}[mol/g] \qquad (11.4)$$

where M_y denotes the mass of glucose (g) that diffused out from the PHEMA-CNT composite sample at the time corresponding to the number y. C_y is the measured concentration of glucose (g/mL) from the withdrawn solution sample, which is marked as number y. As previously, the concentration of glucose was measured utilizing the ICB SBA-90 laboratory glucose meter (±0.01 g/L). Thus, the capacity of absorption (Γ) of the PHEMA-CNT composite sample to glucose can be given by:

$$\Gamma [mol/kg]$$

$$= \frac{\text{Amount of glucose absorbed by PHEMA} - \text{CNT sample} \left(M_{glucose}[mol] \right)}{\text{Mass of PHEMA} - \text{CNT sample} \left(M_{composite}[kg] \right)} \qquad (11.5)$$

11.4   EXPERIMENTAL RESULTS

11.4.1   GLUCOSE RELEASE RATE

According to Equation (11.1), the experimental results from the cumulative release experiment (Purpose I) are presented in Figure 11.4a in terms of the amount of glucose releasing from the PHEMA-CNT composite sample over time. It indicates that the diffusion speed of glucose from the PHEMA-CNT composite decreases with time and approaches the equilibrium after 24 hours. The high diffusion speed at the beginning is caused by the initial large difference of glucose concentration between inside and outside environments of the PHEMA-CNT composite sample. With the increase in time, glucose molecules diffuse out from the PHEMA-CNT composite sample and spread into the PBS buffer solution, thus reducing the difference of glucose concentration between inside the composite sample and the PBS buffer environment. To determine the value of the diffusion exponent n in Equation (11.1), the linear equation of the positive correlation between $\ln(M_t/M_\infty)$ versus $\ln(t)$ is plotted in Figure 11.4b. According to the intercept and slope of the linear fit, the values of k and n are calculated by k = 0.273, n = 0.456. Based on the literature (Korsmeyer et al., 1983), the values of n ~ 0.5/0.45/0.43 for the sample geometry of slab/cylinder/sphere, respectively, indicate a diffusion-controlled drug release process, in other words, Fickian diffusion mechanism. Thus, the diffusion characteristics of the PHEMA-CNT composite to the water-soluble molecules can be described by Fick's law.

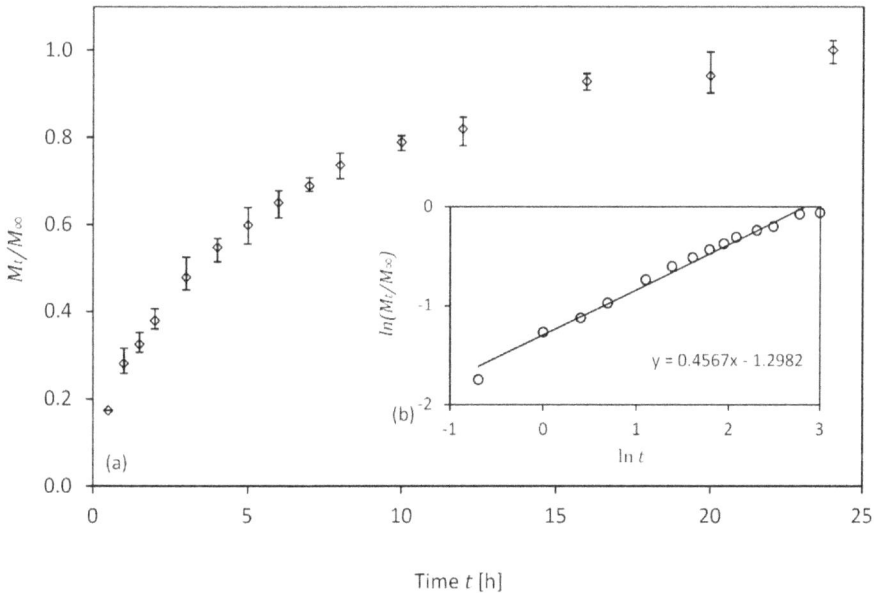

FIGURE 11.4   (a) Release profile of glucose from PHEMA hydrogel in PBS buffer solution at 37°C for 24 hours. (b) Linear plot of $\ln(Mt/M\infty)$ versus $\ln(t)$ to determine the diffusion exponent. The equation of the linear fit of the data is shown. Each set of data was the value averaged from four parallel experiments.

This result offers an experimental validation for Fick's law being applicable to describe the behavior of swelling of the PHEMA-CNT hydrogel-based composite. Thus, the Fickian diffusion for one-dimensional molecule transport, the equation for determining the diffusion coefficient D for initial stage, is given by (Tomic et al., 2007):

$$\frac{M_t}{M_\infty} = 4\sqrt{\frac{Dt}{\pi \delta^2}} \tag{11.6}$$

where δ is the height of the PHEMA-CNT composite sample, which is measured as 26 mm. Thus, the approximate value of the diffusion coefficient of PHEMA-CNT composite for early time (t = 1 s) is calculated as D = 0.099 (cm²/s), by employing Equation (11.6).

11.4.2 CAPACITY OF ABSORPTION

The relationship between the mass of glucose that diffused out from the PHEMA-CNT composite at the time corresponding to the number y is illustrated in the bar graph of Figure 11.5. Two parallel experiments on the same PHEMA-CNT composite sample (Samples 1 and 2) were conducted in this study. As is shown in Figure 11.5, the PBS buffer solution from the first time refresh (y = 1) contains the most glucose releasing from the PHEMA-CNT composite sample, which is caused by large diffusion speed induced by the large difference of glucose concentration between the PHEMA-CNT composite sample inside and in the PBS buffer environment.

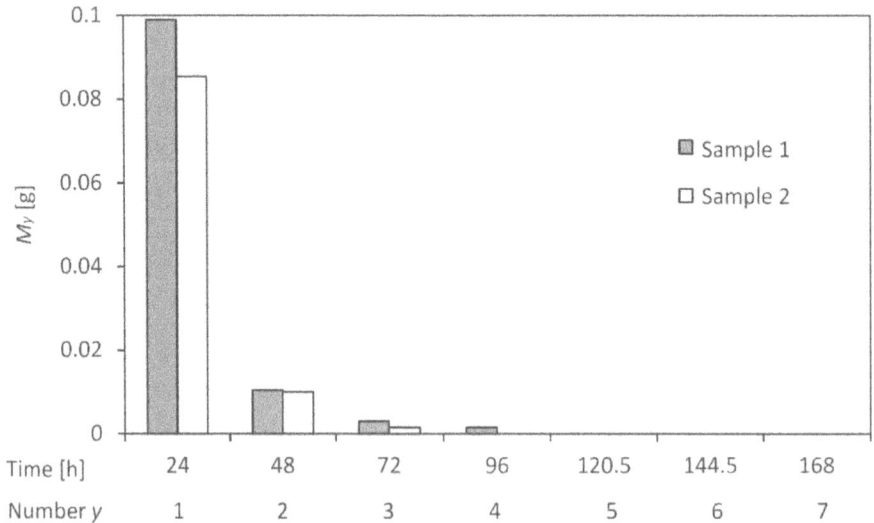

FIGURE 11.5 Bar graph of the amount of glucose detected from the PBS buffer solution versus the time. By summing the amount of glucose from all the refresh PBS buffer solutions, the amount of glucose absorbed by the PHEMA-CNT hydrogel sample can be obtained.

TABLE 11.1

Geometric Information of the Samples Used in the Cumulative Release Experiment (Purpose II) and the Capacity of Absorption

	Section 1.01 Sample 1	Section 1.02 Sample 2
$M_{glucose}$ (mol)	$6.33*10^{-4}$	$5.39*10^{-4}$
Dimensions (mm)	Ø29, h26	Ø29, h24
Volume (cm³)	17.16	15.84
Mass (kg)	0.016	0.015
Capacity of absorption (mol/kg)	0.039	0.037

Ø: Diameter. h: Height.

After the wash by refreshing of PBS buffer (y = 2, 3, 4), the glucose inside material is removed from the PHEMA-CNT composite; thereby fewer and fewer glucose molecules are detected from such PBS buffer samples. To ensure enough times of the refreshing process have been conducted (Steps 2 and 3 in Figure 11.3), the value of M_y (y = 7) has to be identical to zero, thus guaranteeing all the glucose molecules have been removed from the PHEMA-CNT composite sample, and the amount of such glucose molecules is considered in the calculation of the capacity of absorption (Γ). Therefore, according to Equations (11.4) and (11.5), the total amount of glucose $M_{glucose}$ (mol) and the capacity of absorption Γ (mol/kg) are calculated and listed in Table 11.1.

In summary, through the glucose release experiments, the diffusion characteristics of the PHEMA-CNT composite to glucose molecules can be described by Fickian diffusion, which is also known as the diffusion-controlled release process (Zhokh & Strizhak, 2019). The diffusion coefficient of glucose molecules in the PHEMA-CNT composite has been determined as D = 0.099 (cm²/s). Meanwhile, the capacity of absorption of PHEMA-CNT composite in terms of the amount of glucose absorbed by 1 kg swollen PHEMA-CNT composite sample that has been experimentally determined is 0.038 mol approximately. These two parameters, D = 0.099 (cm²/s) and Γ = 0.038 (mol/kg), will be retrieved and used in drug delivery simulations.

11.5 DRUG DELIVERY THROUGH DIFFUSION

When the cell culture/test microdevice is assembled, it consists of a culture chamber, hydrogel-based composite membrane, and the drug delivery reservoir. The glucose molecules exist not only in the culture chamber, but also in the drug delivery reservoir due to diffusion through hydrogel-based composite membrane. Likewise, due to the diffusion property of the hydrogel-based composite membrane, the testing drug that is stored in the drug delivery reservoir can be transported to the culture chamber through hydrogel-based composite membrane to react with cells. Distribution of glucose in the drug delivery reservoir and distribution of testing drug in the culture chamber can be investigated and visualized using numerical modeling. Overall, the modeling in this study verified the feasibility of using hydrogel-based composite membrane as an interlayer for transportation of molecules to connect the culture chamber and the drug delivery reservoir.

11.5.1   GLUCOSE FROM CULTURE CHAMBER TO DRUG DELIVERY RESERVOIR

To investigate the characteristics of the transport behavior of glucose molecules, the numerical simulation that consists of two chambers embracing a semipermeable membrane has been established. According to the input parameters listed in Table 11.2, the glucose concentration at inflow entrance (Inlet 1 in Figure 11.6) is assigned to be 25 mol/m³. The glucose concentration at Inlet 2 is assigned as 0, because the reservoir chamber is designed to diffuse testing drug only through hydrogel-based composite membrane to the culture chamber. The fluid flow velocity (v_1) of the culture medium is assumed to be 0.1 mm/s. Meanwhile,

---

TABLE 11.2
Summary of the Parameters in Simulation

Section 1.03 Category	Section 1.04 Description	Section 1.05 Name	Section 1.06 Value
Geometrical dimensions	Radius of culture chamber	r	2.5 (mm)
	Height of culture chamber	h	1.5 (mm)
	Widths of channel	w_1, w_2	0.48, 0.1 (mm)
	Height of channel	h_c	0.7 (mm)
Properties of fluid	Dynamic viscosity	μ	$8.9*10^{-4}$ (Pa·s)
	Density	ρ	1000 (kg/m³)
Concentration of glucose	At inlet of culture chamber	c_{in}	25 (mol/m³)
	At inlet of drug delivery reservoir	c_{in_res}	0 (mol/m³)
	In cells under 37°C	c_t	8 (mol/m³)
	Half maximum response	$C_{Hf,g}$	12.5 (mol/m³)
Diffusion coefficients	Glucose in culture medium	D_g	$9.45*10^{-10}$ (m²/s)
	Glucose in tissue (cell)	$D_{g,t}$	$4.0*10^{-11}$ (m²/s)
	Glucose in PHEMA hydrogel	D	$9.9*10^{-6}$ (m²/s)
Other	Langmuir equilibrium constant	k	0.0024 (m³/mol)
	Maximum glucose consumption rate	$R_{max,g}$	$2.42*10^{-4}$ (mol/ m³·s)
Variables to be solved	Inflow velocity of culture medium	v_{in}	(mm/s)
	Glucose concentration around cells	c_g	(mol/m³)
	Maximum shear stress around cells	τ_{max_cell}	(dyn/cm²)

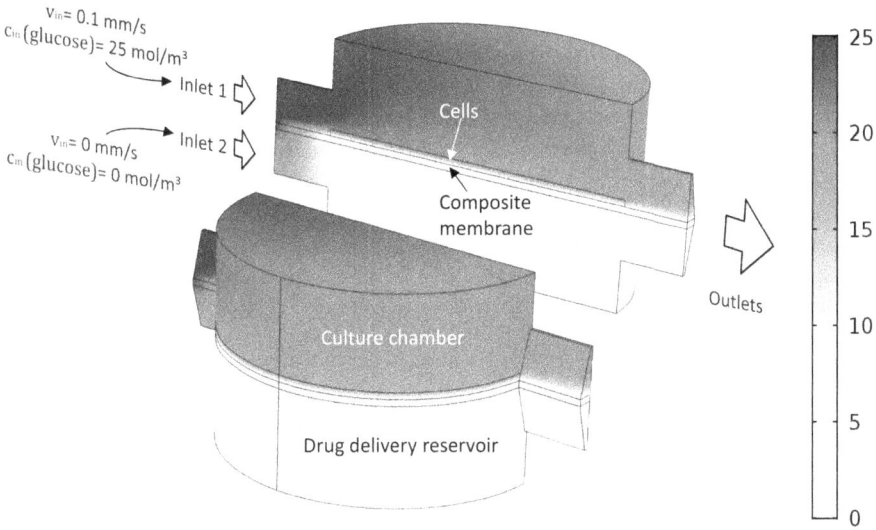

FIGURE 11.6   Sectional view of all the fluidic domains, including culture chamber, membrane, and drug delivery reservoir. The scale bar illustrates the distribution of glucose concentration in the whole system under the equilibrium. The initial constraints at inlets are indicated at the corresponding places. The glucose concentration in cells is 8 mol/m^3. Glucose transports from the culture chamber to the drug delivery reservoir through diffusion by PHEMA hydrogel-based composite membrane.

the fluid velocity of Inlet 2 (v_2) is assigned as 0. The cells are built as a film occupies the area of the culture chamber to replicate the cell groups. The diffusion of glucose in the hydrogel-based composite membrane between two chambers is simulated based on the Langmuir adsorption equation (Masel, 1996) for absorption of glucose molecules and the Fickian diffusion equation for molecule transportation.

The simulation results in terms of the glucose distribution and glucose concentration in the device, including the culture chamber, hydrogel-based composite membrane, and drug delivery reservoir, are illustrated in Figure 11.6. According to the gray scale in Figure 11.6, the maximum glucose concentration in the drug delivery reservoir is located at the entrance of Inlet 2 (see Figure 11.6). Due to the difference of glucose concentration between Inlet 1 (25 mol/m^3) and Inlet 2 (0 mol/m^3), glucose concentration is gradually decreased from Inlet 1 to Inlet 2 through the hydrogel-based composite membrane. As a glucose consumer, cells occupying the bottom of the culture chamber maintain the intracellular glucose concentration (c_I) at 8 mol/m^3 (Foley et al., 1980), and consume the glucose around them, including the glucose molecules flowing in the culture chamber and the glucose molecules diffusing up from the delivery reservoir through the hydrogel-based composite membrane. Thus, the consumption of cells causes dual results: (i) a homogeneous distribution of glucose at approximately 8 mol/m^3 in the drug delivery reservoir; and (ii) the difference of glucose concentration between Inlet 1 (25 mol/m^3) and outlet of the culture chamber (approximately at 18 mol/m^3).

According to the glucose concentration gradient of Figure 11.6, the glucose molecules that originate from Inlet 1 with the highest glucose concentration are transported through the culture chamber towards the cells, and contact the cells for reaction. Some glucose molecules that are not consumed by cells may (i) directly flow away above the cells to the outlets due to fluid flowing in the culture chamber; (ii) further travel through the hydrogel-based composite membrane to the drug delivery reservoir because of diffusion. Such results from numerical simulation demonstrated that glucose from the culture medium can be found in the drug delivery reservoir, and the concentration of such glucose in the drug delivery reservoir is approximately 8 mol/m³ under the working conditions.

11.5.2 Testing Drug from Delivery Reservoir to Culture Chamber

The transport characteristics of the drug molecules in the extracellular environment, including the culture chamber, drug delivery reservoir, and hydrogel-based composite membrane, are also numerically investigated. As designated, testing drugs are supplied from the drug delivery reservoir and transported through diffusion to the culture chamber for reaction with cells. The distribution of the drug concentration can be visualized in Figure 11.7 based on simulation results. The minimum

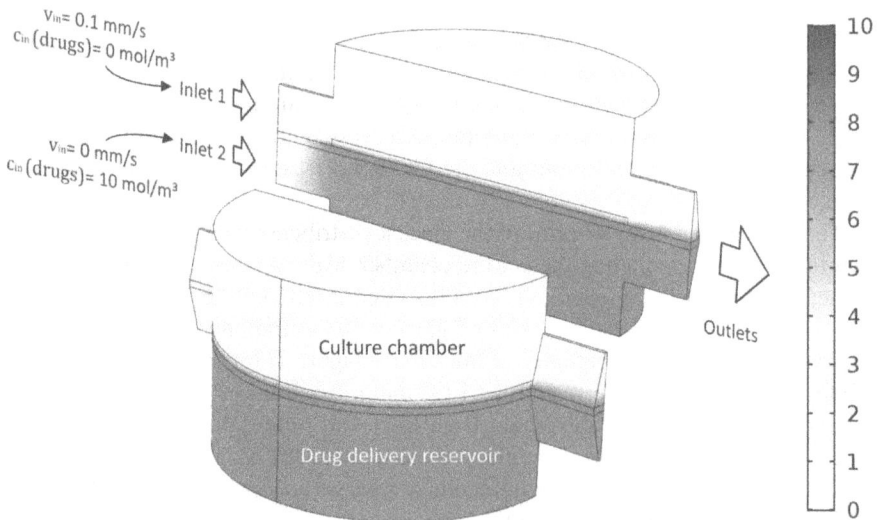

FIGURE 11.7 Sectional view of all the fluidic domains, including culture chamber, membrane, and drug delivery reservoir. The scale bar illustrates the distribution of testing drug concentration in the whole system under equilibrium. The initial constraints at inlets are indicated at the corresponding places. Drug molecules are transported from the drug delivery reservoir to the culture chamber through diffusion by the hydrogel-based composite membrane.

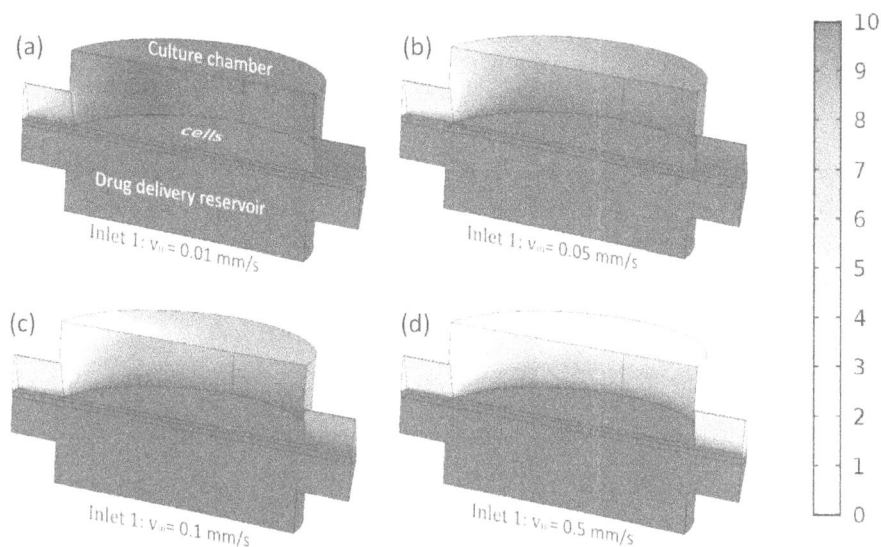

FIGURE 11.8 Sectional inside view of the fluidic domains. Gray scale shows the distribution of testing drug concentration with respect to the various inflow velocities at Inlet 1.

drug concentration can be found at Inlet 1, because only nutrients are perfused into the culture chamber without testing drug. The drug molecules will be transported through hydrogel-based composite membrane to reach the cells and react with cells, due to the difference of the testing drug concentration between the culture chamber and delivery reservoir. This has qualitatively demonstrated that the drug molecules from the drug delivery reservoir are able to reach the culture chamber and react with cells through diffusion.

To understand the interaction of testing drug and cells at the bottom of the culture chamber, quantitative evaluation of the testing drug concentration around cells has been carried out by simulation. Figure 11.8 shows the gray-scale map in terms of the distribution of testing drug concentration with respect to the inflow velocity (v_{in}) at Inlet 1. As shown in this figure, different inflow velocity causes various distributions of drug concentration in the culture chamber, and larger inflow velocity (e.g., Figure 11.8d) leads to inhomogeneous drug concentration in the culture chamber. However, the drug concentration at the bottom of the culture chamber, where the cells are located, does not change significantly with the inflow velocity (v_{in}) at Inlet 1. The average values of the concentration of testing drug (c_{drug_avg}) at the bottom of the culture chamber are given in Table 11.3 with respect to the inflow velocity (v_{in}) at Inlet 1. As shown in the table, though the inflow velocity increases from its minimum ($v_{in} = 0$) to its maximum limit ($v_{in} \sim 0.5$ mm/s, if $v_{in} > 0.5$ mm/s, cells will be detached from the substrate), the value of c_{drug_avg} is changed from 10 to 9.99781 mol/m^3. The variation of c_{drug_avg} is less than 0.3%. Similar results can also be found according to the work done by Shah et al. (2011). Thus, it is evident that

TABLE 11.3

Average Concentration of Testing Drug with Respect to Culture Chamber Inflow Velocity

v_{in} at Inlet 1 (mm/s)	0	0.01	0.05	0.1	0.5
Average concentration of testing drug (c_{drug_avg}) (mol/m³)	10	9.99934	9.99867	9.99803	9.99781

the supply of testing drug from the drug delivery reservoir for cells through diffusion in hydrogel is stable and can provide homogeneous concentrations of drug to react with cells.

11.6 CONCLUSION

Two parameters, that is, glucose release rate and capacity of absorption, which are related to the diffusion characteristics of PHEMA-CNT hydrogel-based composite material, have been experimentally examined. The glucose was chosen to be the only type of molecule for investigating the diffusion characteristics of PHEMA-CNT hydrogel-based composite, because it is the prime energy source for cells during incubation of cells. The diffusivity of glucose molecules in the PHEMA-CNT composite has been experimentally proven to obey the Fickian diffusion behavior in this study. Accordingly, the diffusion coefficient, which indicates the transport speed of glucose in PHEMA-CNT composite, calculated as $D = 0.099$ cm²/s, is based on the formula of Fickian diffusion for one-dimensional molecule transport. Meanwhile, the capacity of absorption (Γ) has also been determined from the experiment, with the value of $\Gamma = 0.038$ mol/kg.

Numerical simulations have been established to visualize the transport of testing drugs from delivery reservoir to culture chamber, and the transport of glucose from culture chamber to delivery reservoir. The studies demonstrated that molecules of testing drugs are able to reach cells by crossing the hydrogel-based composite membrane through diffusion, and forming a homogeneous distribution on the bottom of culture chamber to react with cells. The simulation results showed that being independent of the inflow velocity (v_{in}), the glucose concentration around cells (c_g) varies in the range of 22.5–24.51 mol/m³, which induces the glucose consumption rate of cells larger than the minimum allowed glucose consumption rate ($R_{min,single}$) (Zhang et al., 2011). This can ensure sufficient glucose for cells to perform their proliferation and differentiation during the perfusion culture. Thus, the range of allowed inflow velocity, which reflects the cell culture capacity of the microfluidic device, is therefore identified as 0–0.49 mm/s. Utilizing this method, the controlling variables of the microfluidic device (e.g., inflow rate, type of culture medium, time of perfusion culture) can be directly derived for a given type of cell.

ACKNOWLEDGMENT

The original research was supported by the National Natural Science Foundation of China (51703176, 91848102), and the Fundamental Research Funds for the Central Universities (WUT2018IVB006).

REFERENCES

Ansari, R., Pourashraf, T., Gholami, R., and Shahabodini, A. (2016). Analytical solution for nonlinear postbuckling of functionally graded carbon nanotube-reinforced composite shells with piezoelectric layers. *Compos Part B Eng*, 90, 267–277.

Batchelor, G.K. (1967). An Introduction to Fluid Dynamics. Cambridge University Press, Cambridge, UK, pp. 211–215.

Foley, J.E., Cushman, S.W., and Salans, L.B. (1980). Intracellular glucose concentration in small and large rat adipose cells. *Am J Physiol*, 238(2), E180–185.

Huang, W.F., Tsui, C.P., Tang, C.Y., Yang, M., and Gu, L. (2017). Surface charge switchable and pH-responsive chitosan/polymer core-shell composite nanoparticles for drug delivery application. *Compos Part B-Eng*, 121, 83–91.

Kharismadewi, D., Haldorai, Y., Nguyen, V.H., Tuma, D., and Shim, J.J. (2016). Synthesis of graphene oxide-poly(2-hydroxyethyl methacrylate) composite by dispersion polymerization in supercritical CO_2: Adsorption behaviour for the removal of organic dye. *Compos Interface*, 23(7), 719–739.

Korsmeyer, R.W., Gurny, R., Doelker, E., Buri, P., and Peppas, N.A. (1983). Mechanism of solute release from porous hydrophilic polymers. *Int J Pharm*, 15, 25–35.

Li, S., Shen, Y., Li, W., and Hao, X. (2006). A common profile for polymer-based controlled releases and its logical interpretation to general release process. *J Pham Phamaceut Sci*, 9(2), 238–244.

Liu, D., and Garimella, S.V. (2004). Investigation of liquid flow in microchannels. *J Thermophys Heat Tr*, 18(1), 65–72.

Lustig, S.R., and Peppas, N.A. (1988). Solute diffusion in swollen membranes. IX. Scaling laws for solute diffusion in gels. *J Appl Polym Sci*, 36(4), 735–747.

Majumda, P. (2014). Modelling and simulation of hydrogel growth mechanism by analysis of experimental data. *Chin J Polym Sci*, 32(3), 350–361

Masel, R. (1996). Principles of Adsorption and Reaction on Solid Surfaces. Wiley Interscience, New York, pp. 239–245. ISBN 0471303925.

Massoumi, B., Ghandomi, F., Abbasian, M., Eskandani, M., and Jaymand. M. (2016). Surface functionalization of graphene oxide with poly(2-hydroxyethyl methacrylate)-graft-poly(e-caprolactone) and its electrospun nanofibers with gelatin. *Appl Phys A*, 122, 1000.

Shah, P., Vedarethinam, I., Kwasyn, D., Andresen, L., Dimaki, M., Skov, S., and Svendsen, W.E. (2011). Microfluidic bioreactors for culture of non-adherent cells. *Sens Actuator B*, 156, 1002–1008.

Temam, R. (2001). Navier–Stokes Equations, in Theory and Numerical Analysis. American Mathematical Society, Chelsea, pp. 107–112.

Tokarev, I., and Minko, S. (2009). Stimuli-responsive hydrogel thin films. *Soft Matter*, 5(3), 511–524.

Tokarev, I., and Minko, S. (2010). Stimuli-responsive porous hydrogels at interfaces for molecular filtration, separation, controlled release, and gating in capsules and membranes. *Adv Mater*, 22(31), 3446–3462.

Tomic, S.L., Micic, M.M., Filipovic, J.M., and Suljovrujic, E.H. (2007). Swelling and drug release behaviour of poly(2-hydroxyethyl methacrylate/itaconic acid) copolymeric hydrogels obtained by gamma irradiation. *Radiat Phys Chem*, 76, 801–810.

Zhang, F., Tian, J., Wang, L., He, P., and Chen, Y. (2011). Correlation between cell growth rate and glucose consumption determined by electrochemical monitoring. *Sens Actuator B*, 156, 416–422.

Zhao, W., Shi, Z., Chen, X., Yang, G., Lenardi, C., and Liu, C. (2015). Microstructural and mechanical characteristics of PHEMA-based nanofibre-reinforced hydrogel under compression. *Compos B Eng*, 76, 292–299.

Zhao, W., Shi, Z., Hu, S., Yang, G., and Tian, H. (2018). Understanding piezoelectric characteristics of PHEMA-based hydrogel nanocomposites as soft self-powered electronics. *Adv Compos Hybrid Mater*, 1, 320–31.

Zhokh, A., and Strizhak, P. (2019). Crossover between Fickian and non-Fickian diffusion in a system with hierarchy. *Micropor Mesopor Mat*, 282, 22–28.

12 Nanocomposites for Human Body Tissue Repair

C. Wang,[1,2] *L. He,*[2] *and M. Wang*[1]

[1]Department of Mechanical Engineering, The University of Hong Kong, Pokfulam Road, Hong Kong

[2]School of Mechanical Engineering, Dongguan University of Technology, Songshan Lake, Dongguan, China

CONTENTS

## 12.1	INTRODUCTION

### 12.1.1	The Composite Approach

With the rapid growth of the world population and with an increasingly aging population in developed countries, more and more people suffer from human body tissue diseases such as osteoarthritis and osteoporosis and tissue defects arising from trauma, sport injury, tumor resection, and congenital disorders, which cause tissue dysfunction and diminish patients' quality of life. A variety of strategies have now been adopted to repair damaged human tissues, with the use of man-made materials on the increase. Different types of artificial prostheses made of metals, ceramics, polymers, and composites have been developed to replace diseased or damaged human body tissues (Hou et al., 2016; Toney et al., 2016; Zimel et al., 2016). Among these materials, nanocomposites, especially polymer-based particulate nanocomposites, which consist of dispersed nanosized particles (mainly bioceramics) and polymer matrices, have gained increasing attention for human tissue repair owing to a number of reasons: hierarchical similarity to bone tissue at the micro- and nanostructural level, resemblance in composition and structure to natural tissue, good biocompatibility, relative ease of manufacture, improved and well-matched mechanical properties, desirable biological performance, etc. (Wang & Zhao, 2019). Bioceramic-polymer composites with different bioceramic contents are promising composite materials for both soft and hard tissue substitution with excellent biological activity (e.g., osteoconductivity for bone tissue repair) and appropriate mechanical properties (Wang, 2003). From the materials point of view, human body tissues, particularly hard tissues such as bone, consist of an organic matrix (mainly collagen), water and nano-sized inorganic bone apatite. Therefore, blending bioactive bioceramics with suitable polymers to form nanocomposites for soft or hard tissue repair is a promising approach for new biomaterial development (Sarkar, 2015; Wang, 2003; Wang & Zhao, 2019). In this chapter, the design, manufacture and assessment of micro- and nanocomposites with specific features for soft or hard tissue repair will be introduced, and some of our research on bioceramic-polymer micro- and nanocomposites will also be presented. Throughout this chapter, human tissue repair means either tissue replacement or tissue regeneration.

### 12.1.2	Classification of Biomedical Composites

Composites for human tissue repair can be classified as metal matrix, ceramic matrix or polymer matrix composites. Metals are materials with excellent hardness, mechanical strength and elastic modulus, and hence are suitable for load-bearing medical implants. They are also thermally and electrically conductive. Metals used for human tissue repair are normally bioinert, that is, they do not form chemical bonding with the host tissue. For example, titanium, Co-based alloys and stainless steel have been used to produce metal prostheses for hard tissue replacement. To improve the biocompatibility and bioactivity, metal-based composites can be produced by coating biomedical polymers on the surface of metal products or coating bioceramic layers on the metal surface. Ceramics are inorganic, nonmetallic materials with high corrosion resistance. According to their biological properties, ceramics in medicine (i.e., bioceramics) can be bioinert ceramics such as Al_2O_3 and ZrO_2, which are non-biodegradable in the body, and bioactive ceramics such as hydroxyapatite (HA, $Ca_{10}(PO_4)_6(OH)_2$) and tricalcium phosphate (TCP, $Ca_3(PO_4)_2$), which are either biostable or biodegradable in the body. Bioceramics usually require high temperatures for sintering to form products with high strength and hardness. Also, brittleness is always an issue for bioceramic applications. To overcome the brittleness problem for bioceramics, a small amount of biocompatible polymer can be added into bioceramic powders as adhesives to form non-sintered bioceramic-based composites, while polymer melts can fill the pores of sintered bioceramic products to form toughened bioceramic composites. Polymers are organic materials and most of them have excellent ductility but low strength and stiffness. Polymers can be classified into synthetic polymers and natural polymers; and human body tissues are made of natural polymers. According to their degradability in the human body environment, polymers can also be classified into biodegradable polymers and non-biodegradable polymers. Non-biodegradable polymers such as ultra-high molecular weight polyethylene (UHMWPE) are suitable for constructing stable implants/devices in the body (e.g., acetabular cup in hip prosthesis), whereas biodegradable polymers, especially those producing nontoxic degradation products such as poly(lactic acid) (PLA), are increasingly used in biodegradable implants (e.g., biodegradable bone screw). To improve the mechanical strength and also bioactivity of polymer products, biodegradable metal powders such as magnesium and particulate bioceramics with high bioactivity such as Bioglass can be dispersed in polymers to form polymer-based bioactive composites.

12.2 MATRIX, REINFORCEMENT, AND INTERFACE IN BIOMEDICAL COMPOSITES

12.2.1 MATRIX MATERIALS

Bioinert metals can be made into prostheses for clinical use. These metal prostheses have excellent mechanical strength and very high Young's modulus, and hence are used for replacing damaged hard tissues. The most widely used metals include titanium and its alloys, Co-based alloys, and stainless steel. Titanium-based hip prostheses coated with HA have been marketed for decades to replace the diseased human hip joint, whereas titanium knee joint and non-biodegradable plastic meniscus have been simultaneously used to replace the damaged human knee joint. Owing to high temperature processing conditions, it is not feasible to make uniform polymer-metal composites. However, bioceramic particles can be dispersed in metals to form, at high temperatures, bioactive metal matrix composites for implants. Recently, due to their distinctive biodegradation characteristics, light metals of magnesium alloys have been investigated as raw materials to make implants for bone substitution and for interbody fusion.

Polymeric materials can be made into dense or macro-/microporous structures for medical applications. Nonporous structures are produced via extrusion, compression molding, injection molding, etc., while porous structures can be formed by using techniques such as solvent casting and salt leaching, emulsion freezing/freeze-drying, electrospinning and three-dimensional (3D) printing. For hard tissue repair, the sole use of polymers for implants is not adequate as they lack bioactivity and their mechanical strength is not sufficiently high. Therefore, bioceramic particles can be dispersed in polymers to form bioactive composites (Chen & Wang, 2002; Liu & Wang, 2007a). The first bioactive polymer matrix composite, HA reinforced high-density polyethylene (HDPE), has now been used clinically for bone substitution (Bonfield et al., 1998; Wang et al., 1994). Apart from HDPE, polymers such as polyetheretherketone (PEEK), polysulfone (PSU) and polypropylene (PP) are good candidates as matrices for composites. However, similar to most metal-based biomedical composites, all these composites are non-biodegradable and hence a second surgery may need to be performed to remove the implants or medical devices when they are no longer needed in the human body. Therefore, producing biodegradable bioceramic-reinforced polymer composites has become popular. Biodegradable polymers, including synthetic polymers such as poly(L-lactic acid) (PLLA), poly(glycolic acid) (PGA), poly(D,L-lactic acid) (PDLLA), poly(D,L-lactic-co-glycolic acid) (PLGA), poly(ε-carprolactone) (PCL) and polyvinyl alcohol (PVA), and natural polymers, such as collagen, chitosan, poly-β-hydroxybutyrate (PHB), and poly(3-hydroxybutyrate-co-3-hydroxyvalerate) (PHBV), are good candidates for matrices in composites developed for biodegradable implants. Many of these polymers have distinctive degradability, mechanical properties, hydrophilicity/hydrophobicity, cell affinity, and processability.

12.2.2 Reinforcements

Metal powders, metal dioxide powders and ceramic powders are common reinforcements for polymer-based composites. Bioceramics, including calcium phosphates such as HA, TCP, biphasic calcium phosphate (BCP) and amorphous calcium phosphate (Ca-P), Bioglass and A-W glass-ceramic, are excellent reinforcements for producing bioceramic-polymer composites for hard tissue or soft tissue repair. The Ca-P particles can act as nucleation centers and promote the deposition of calcium and phosphate ions *in vivo*, thus accelerating biomineralization and new bone formation. Bioactive glasses containing trace elements such as Mg, Mn, etc., can also contribute to improved osteogenesis. Different bioceramics have different degradability. For example, HA has a very low biodegradability (HA is perceived as "biostable" in the human body) while β-TCP can be gradually degraded and absorbed after implantation. In comparison, BCP, which consists of HA and β-TCP, has tunable degradability. Among all bioceramics, synthetic HA is the most widely used material for hard tissue repair due to its chemical similarity to bone apatite and its excellent biocompatibility and bioactivity (Cross et al., 2016; Duan et al., 2008; Liu & Webster, 2010). However, amorphous Ca-P particles are gaining popularity owing to their degradability and higher bioactivity. Regarding the size of reinforcements, particles of either micro-sizes or nano-sizes can be used. Microparticulate bioceramics have been used for decades to form bioceramic-polymer composites. In recent years, dense nanoparticle-polymer composites are increasingly investigated as new biomaterials for bone tissue repair due to improved mechanical properties as compared to conventional composites.

12.2.3 INTERFACE

Controlling the interfacial bonding strength is important in designing and fabricating nanocomposites. It is well recognized that there are several major factors that contribute to the mechanical properties of ceramic-polymer composites, including size and shape of ceramic particles, dispersion of particles in composites, composite composition (i.e., the ceramic to polymer ratio), physical or chemical interactions between the ceramic and polymer phases, and properties of polymer matrices (Wang, 2004; Wang & Zhao, 2019). He et al. (2012) produced HA/PLLA nanohybrids, where PLLA oligomers were grafted on HA nanoparticle surfaces in order to improve the dispersion of HA in PLLA/chloroform solution. Zou et al. (2011) prepared surface modified TCP reinforced PLLA composites through solvent-casting. The low molecular weight PLLA coating interacted with the TCP particles and formed a stable hindrance layer between TCP particles, which inhibited particle agglomeration and thus improved the dispersibility of TCP particles in the PLLA matrix greatly. Yang et al. (2009) fabricated surface-modified BCP/PLLA composite also through solvent casting. The modified BCP (mBCP) particles could be uniformly dispersed in PLLA. The improved mechanical strength of mBCP/PLLA composites was attributed to the enhanced adhesion between the inorganic BCP particles and PLLA matrix.

In the situation when the bioceramic particle sizes are different, compared to microparticle-polymer composites, the enhanced mechanical properties of nanocomposites may be attributed to the larger interfacial area between nanoparticles and the polymer matrix and also to the more uniform distribution of the nanoparticles in the composite. In some cases, bioceramic nanoparticles have a strong interface with polymers, which can be stronger than the polymer matrix itself. Microvoids will then unlikely initiate from the nanoparticles or the interface. Compared to microparticle-polymer composites, internal cavity formation or interfacial debonding hardly occur in these nanocomposites, and the connection between the surface of nanoparticles and the polymer chain is more intimate. Upon loading, stress concentrates around the nanoparticles. The stress fields overlap each other in the composite and highly constrain the development of the local crack dilatation in the matrix. Therefore, by adding nanoparticles

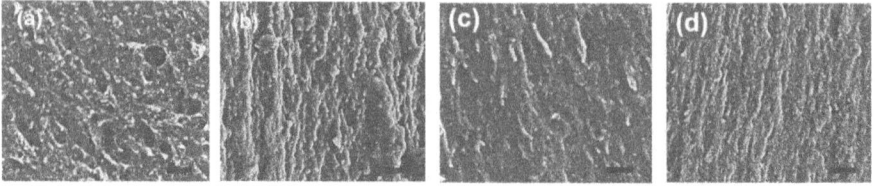

FIGURE 12.1 Fracture surface of bioceramic-polymer composites: (a) 5 wt% Ca-P/chitosan, (b) 15 wt% Ca-P/chitosan, (c) 5 wt% TCP/chitosan, (d) 15 wt% TCP/chitosan. (Scale bar: 2 μm)

instead of microparticles into the polymer matrix, the mechanical strength will be enhanced. In our studies, Ca-P nanoparticles were found to better improve the tensile strength and Young's modulus of the chitosan matrix than TCP microparticles (Duan et al., 2008; Liu & Wang, 2007a). The size and shape of bioceramic particles could significantly influence the mechanical properties of bioceramic-polymer composites (Wang et al., 1998). Figure 12.1 displays fracture surfaces of Ca-P/chitosan nanocomposites and TCP/chitosan microcomposites of different bioceramic contents. It can be seen that the Ca-P/chitosan nanocomposite had a rough fracture surface by exhibiting a great number of nanopores and microcracks, indicating a relatively ductile rupture locally. In comparison, the fracture surface of TCP/chitosan microcomposite was not as rough as that of Ca-P/chitosan nanocomposite and fewer nanopores were seen. Compared to TCP/chitosan microcomposite, the larger contact area between Ca-P nanoparticles and chitosan matrix in Ca-P/chitosan nanocomposite had led to a reduction in crack propagation. Higher energy was needed to force nanoparticles to detach from the polymer matrix and the crack may have to randomly pass more nanoparticles and go along a much longer path to propagate through the whole sample for the fracture to occur, as illustrated by Figure 12.2, resulting in significantly improved mechanical properties. Observations made by others (Ding et al., 2016) also support this theory.

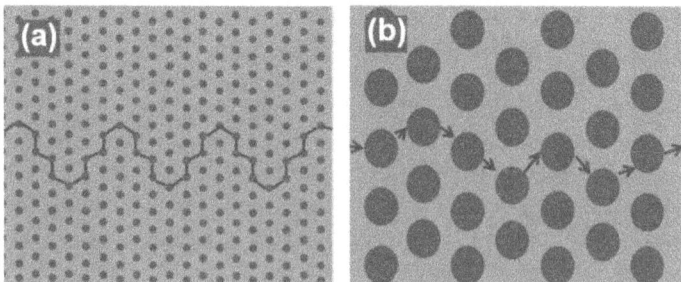

FIGURE 12.2 Schematic diagrams showing the crack propagation path for fracture in (a) a nanocomposite and (b) a microcomposite.

12.3 DESIGN AND FABRICATION OF NANOCOMPOSITES

12.3.1 DESIGN OF NANOCOMPOSITES

In recent years, the research emphasis on biomedical composites for hard tissue repair has moved to nanocomposites, which are based on biodegradable polymers. Such biodegradable nanocomposites can degrade gradually in the human body and are eventually replaced by the newly formed tissue (García-Gareta et al., 2015; Quinlan et al., 2015). When designing nanocomposites, it is highly important to select appropriate matrix material and reinforcement. For instance, if a nanocomposite aims at repairing a soft tissue such as skin, the nanocomposite should have a low compressive strength but a high tensile strength. In comparison, for the repair of bone tissue, nanocomposites with high compressive strength and bending strength are more suitable. It should be noted that a high stiffness can result in "stress shielding" that is undesirable in human tissue repair. Secondly, the degradability of the matrix and reinforcement should match the repair rate for the tissue (i.e., the new tissue formation rate). If the nanocomposite fully degrades before the formation of the new tissue, the nonunion of tissue will occur. However, nanocomposite degradation that is too slow will hinder the remodeling of the tissue under regeneration. In addition, the bioactivity of the designed nanocomposite is of great importance as favorable cellular responses can be achieved if biologically active agents can be incorporated in the matrix. For example, the incorporation of a sufficiently large amount of Ca-P nanoparticles in a biodegradable polymer not only improves the mechanical properties of nanocomposite but also enhances the osteogenesis and hence bone formation due to the excellent osteoconductivity of Ca-P nanoparticles. The further incorporation of osteogenic peptide can significantly accelerate bone tissue repair.

Apart from nonporous composites, which have a dense structure, porous composites have been increasingly made and studied for tissue engineering applications. Due to the presence of micropores and the required very high porosity, compared to the dense composites, much lower mechanical strength and modulus are obtained for porous composites. However, with the high porosity, excellent cellular responses may be achieved with many types of tissue engineering scaffolds for the target tissues. The design of porous composites (i.e., scaffolds) for tissue engineering should fulfill the following requirements: (1) the porous composites should be biocompatible and biodegradable; (2) the porous composites should have optimized pore size and porosity, in order to best regenerate the target tissue; (3) the porous composites should possess desirable surface chemistry and topography; and (4) the porous composites should still have adequate mechanical strength with matching modulus to provide certain mechanical support during implantation and tissue regeneration. In addition, it is highly desirable that the porous nanocomposites are capable of controlled delivery of bioactive biomolecules such as bone morphogenetic protein (BMP).

12.3.2 FABRICATION TECHNIQUES

For fabricating nonporous biomedical composites, a number of techniques can be employed. Solvent casting is the simplest way to produce uniform bioceramic-polymer composites. Polymer solutions used for solvent casting had a low viscosity (Porras et al., 2016; Tong et al., 2017). Therefore, with the assistance of ultrasonication, which provides both energy and entropy to disperse bioceramic particles in the polymer solution through an intense oscillation, an excellent dispersion of TCP microparticles, instead of their aggregates in polymer solutions, could be achieved (Liu & Wang, 2007b). During the solvent evaporation process, the particle dispersed state could be maintained to a certain extent, thereby forming composites with homogeneous particle distribution in the resultant solid composite products. Ikumi et al. (2018) fabricated HA/PLLA membranes through solvent casting, followed by UV irradiation for improving the hydrophilicity of composite membranes. The HA/PLLA composite induced the highest bone mineral density, bone mineral content, and relative bone growth area in rat cranial defects. In our research, TCP/chitosan microcomposite and Ca-P/chitosan nanocomposite were produced also using the solvent casting technique. Injection molding is another common technique for composite fabrication. Sadeghi-Avalshahr et al. (2018) produced HA/PLLA interference screws using a Brabender mixer, followed by a two-stage injection molding. Kobayashi and Sakamoto (2009) mixed PLLA pellets and β-TCP powder in the dry condition and fabricated nonporous composites through injection molding.

For producing porous bioceramic-polymer composites for tissue engineering applications, the hot pressing and salt leaching technique can be used. Yang et al. (2016) made nano-BCP/PLLA porous scaffolds by hot pressing of non-porous nano-BCP/PLLA composite, followed by salt leaching. In another study, Wang et al. (2010) produced HA/PLLA fibrous scaffolds via thermal-induced phase separation. Their scaffolds degraded gradually and exhibited excellent protein adsorption ability. In our studies, porous Ca-P/PLGA and carbonated HA/PHBV nanofibrous composites were produced through electrospinning of mixtures of bioceramic nanoparticles and polymer solutions, in which as high as 30 wt% bioceramic nanoparticles could be incorporated (Tong et al., 2010; Wang & Wang, 2017). He et al. (2014) fabricated PLLA nanofibrous scaffolds, followed by electrodeposition to form mineralized scaffolds, in which nanosized HA were formed on the surface of PLLA scaffolds. Yang et al. (2016) produced PLLA coated BCP composite particles first, followed by hot pressing of uniformly blended BCP/PLLA composite particles and salt particles. After cooling and then salt leaching in distilled water, porous composite samples were obtained. 3D printing is now increasingly used to produce porous composite scaffolds. For instance, bioceramic powders and polymer powders can be dry-mixed and subjected to selective laser sintering (SLS) to make bone tissue engineering scaffolds. Gao et al. (2013) used PLLA as a binder to coat BCP particles and produced BCP scaffolds through SLS. Rectangular porous scaffolds were eventually made using optimized SLS parameters. Nurqadar et al. (2013) prepared stereocomplex of TCP/PCL using supercritical fluid technology. The composite containing PLLA grafted TCP had higher stereocomplex degree and more homogeneous TCP distribution when compared to the composite containing pristine TCP only. In our study, Ca-P/PLLA porous composite scaffolds with a Ca-P content of 30% were produced through novel cryogenic 3D printing, in which water-in-oil composite emulsions containing Ca-P and PLLA were printed into 3D porous patterns in a cryogenic environment (Wang et al., 2017).

12.4 CHARACTERIZATION, PERFORMANCE, AND APPLICATIONS OF NANOCOMPOSITES

12.4.1 Characterization of Biomedical Nanocomposites

The morphology (shape and surface features) of nanoparticles and microparticles as reinforcements and the surface of micro-/nanocomposites are normally examined using scanning electron microscopy (SEM). X-ray diffraction (XRD) analysis is conducted to investigate the crystallographic structure of particles and composites. The particle size of nanoparticles and microparticles can be measured using a particle analyzer. An energy dispersive X-ray spectrometer (EDX) attached to an SEM is often used to determine the elements and atomic ratio between calcium to phosphorus for bioceramic particles. Fourier transform infrared (FTIR) spectroscopy is commonly used to investigate the chemical structure of particles, polymer matrices, and resultant micro-/nanocomposites. The wettability of biomedical nanocomposites is closely related to the cell affinity to the surface of nanocomposites (Mooyen et al., 2017). The distribution of micro- or nanoparticles in composites is investigated by examining both the top and bottom side of the composites as well as the cross-sections which are produced by brittle fracture in liquid nitrogen. In our studies, Ca-P/chitosan and Ca-P/PLGA nanocomposite and TCP/chitosan microcomposite were made via solvent casting. Figure 12.3 shows the characteristics of Ca-P nanoparticles produced in-house

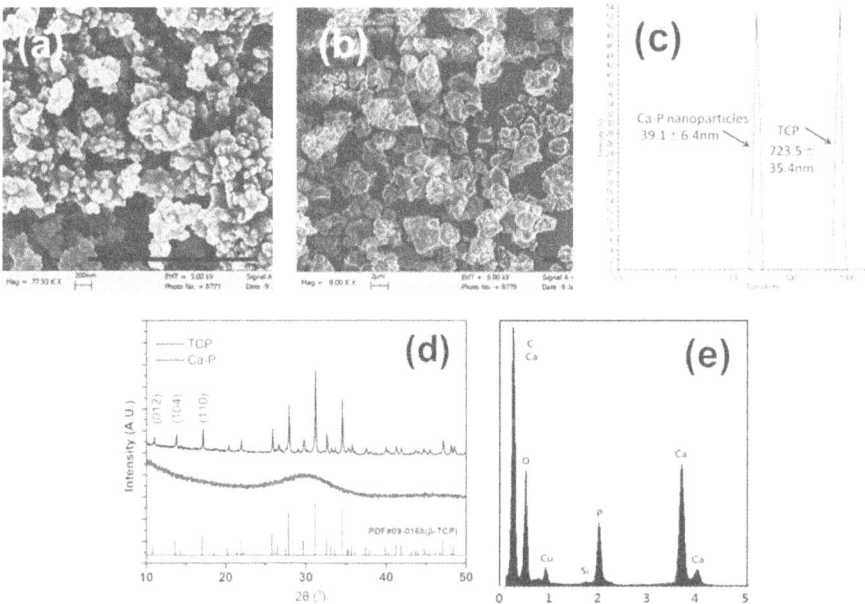

FIGURE 12.3 Morphology and structure of Ca-P nanoparticles and TCP microparticles: (a) SEM image of Ca-P nanoparticles; (b) SEM image of TCP microparticles; (Scale bar: 2 μm); (c) particle size distributions; (d) XRD patterns; (e) EDX spectrum for Ca-P.

and also commercial TCP microparticles. Ca-P nanoparticles were spherical in shape and had a diameter of approximately 30–60 nm (Figure 12.3a), whereas TCP microparticles exhibited an irregular shape with a diameter around 2 μm (Figure 12.3b). The sizes of Ca-P nanoparticles and TCP particles were measured using a Particle Analyzer (Bruker Tensor 2, Bruker Optics, Germany), which gave an average diameter of 39.1 ± 6.4 nm for Ca-P particles and 723.5 ± 35.4 nm for TCP particles (Figure 12.3c). The XRD patterns of Ca-P particles and TCP particles are displayed in Figure 12.3d. For Ca-P nanoparticles, only one single hump was observed between 20 and 40 degrees, indicating that Ca-P nanoparticles were amorphous. In comparison, TCP particles exhibited characteristic peaks from 10.9 to 48.5 degrees, which are consistent with the peaks shown in the PDF for TCP. The EDX spectrum of Ca-P particles obtained from SEM analysis is shown in Figure 12.3e. The atomic ratio between Ca and P in Ca-P nanoparticles was measured to be 1.48, which is similar to the Ca:P ratio of TCP (1.5 for $Ca_3(PO_4)_2$).

Figures 12.4a and 12.4b show FTIR spectra obtained from Ca-P/chitosan nanocomposite and TCP/chitosan microcomposite, while Figure 12.4c gives FTIR

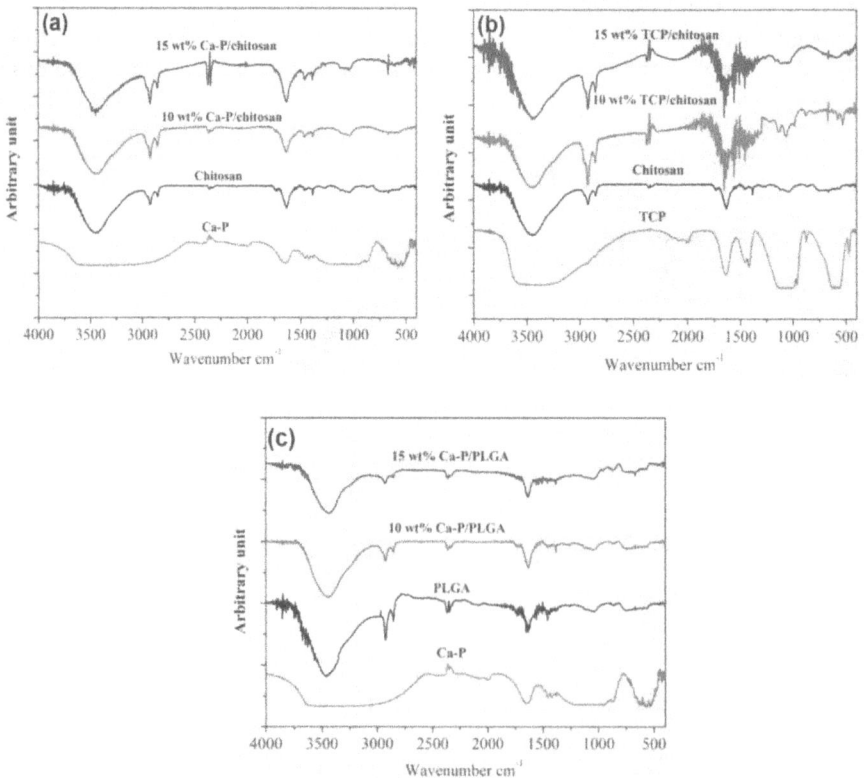

FIGURE 12.4 FTIR spectra obtained for (a) Ca-P/chitosan nanocomposite, (b) TCP/chitosan microcomposite, and (c) Ca-P/PLGA nanocomposite.

spectra of Ca-P/PLGA nanocomposite. For neat chitosan, the peaks at approximately 3,500 cm^{-1} could be attributed to the N-H and O-H stretching vibrations, and the distinctive amide bands were around 1500 cm^{-1}. The peaks around 2,800 cm^{-1} could be ascribed to the presence of C-H stretching of chitosan polymer. The spectrum for 15 wt% Ca-P/chitosan composite exhibited distinctive combined peaks for both chitosan polymer matrix and Ca-P nanoparticles. The absorption peaks at approximately 1,000 and 600 cm^{-1} could be assigned to the phosphate group, indicating that Ca-P nanoparticles were well incorporated in the composites. The spectrum of TCP microparticles was similar to the spectrum of Ca-P nanoparticles since both types of particles had very similar chemical compositions. For Ca-P/PLGA composite, absorption bands for the phosphate group at approximately 572 and 1435 cm^{-1} were present in the FTIR spectra of Ca-P nanoparticles and Ca-P/PLGA composite with different Ca-P contents.

The distribution of Ca-P nanoparticles or TCP microparticles in composites could be examined using SEM. It can be seen from Figure 12.5 that for chitosan-based composites, when a larger amount (15 wt%) of Ca-P nanoparticles was incorporated, no significant difference in Ca-P distribution could be observed on both sides of composite plates. However, the distribution of TCP microparticles on the two sides of 15 wt% TCP/chitosan composite plates showed a discernable difference, with a smooth top surface and a rough bottom surface, which was covered with TCP particles owing to particle sedimentation during composite manufacture. When Ca-P nanoparticles were incorporated in PLGA, no significant difference in particle distribution on two sides of the composite was found when the Ca-P particle content was increased from 5 wt% to 15 wt%. In order

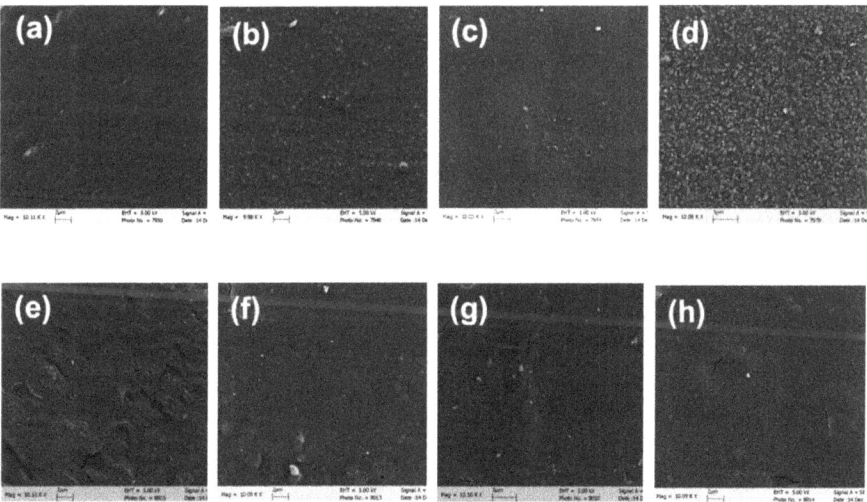

FIGURE 12.5 Particle distribution on the top and bottom surfaces of composites: (a) and (b) 15 wt% Ca-P/chitosan, (c) and (d) 15 wt% TCP/chitosan, (e) and (f) 15 wt% Ca-P/PLGA, and (g) and (h) 5 wt% Ca-P/PLGA. [(a), (c), (e), (g): top surface; (b), (d), (f), (h): bottom surface.]

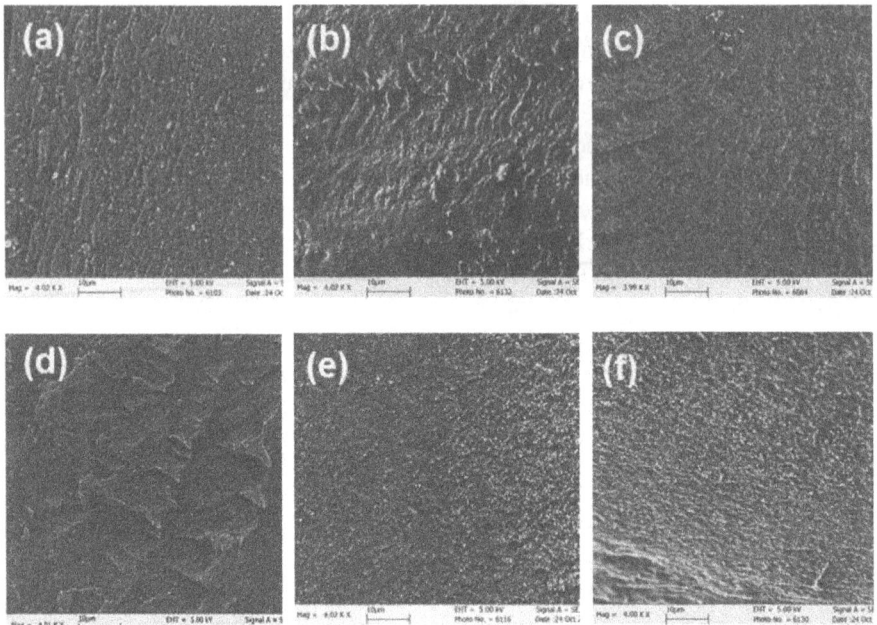

FIGURE 12.6 Cross-sectional views of composites: (a) 5 wt% Ca-P/chitosan, (b) 10 wt% Ca-P/chitosan, (c) 15 wt% Ca-P/chitosan, (d) 5 wt% TCP/chitosan, (e) 10 wt% TCP/chitosan, and (f) 15 wt% TCP/chitosan. [Left side of each micrograph: toward top surface; right side of each micrograph: toward bottom surface.]

to investigate the particle distribution across the sample thickness, the cross-sections of composites obtained by low-temperature brittle failure were observed under SEM. Figure 12.6 shows that Ca-P nanoparticles were well dispersed in composites incorporated with different amounts (5–15 wt%) of Ca-P nanoparticles and had a homogeneous distribution in the polymer matrix (chitosan or PLGA), whereas TCP microparticles exhibited a less even distribution in the chitosan matrix, especially when the TCP content was high (10–15 wt%). It can be seen from Figure 12.6e that more TCP microparticles were present on the right side of the micrograph, indicating TCP particle segregation on the bottom side of TCP/chitosan microcomposite.

12.4.2 Physical and Mechanical Properties of Biomedical Nanocomposites

12.4.2.1 Tensile Testing

Nanocomposites for repairing human body tissues such as skin, muscle, tendon, and long bone are frequently subjected to tensile loads and hence should have sufficient tensile strength and strain-at-fracture to avoid fracture and have Young's modulus matching or similar to that of the target tissue to avoid stress shielding.

These mechanical properties can be measured using tensile tests (Wang & Wang, 2019). For tensile testing, a test standard such as ASTM D638 should be followed. Normally, an electromechanical testing machine is used for conducting tensile tests. The test standard specifies the shape and size of test samples and also testing conditions (testing speed, temperature, etc.). During tensile testing, the sample is gripped at the two ends and a tensile force is applied until the sample fractures. Properties such as ultimate tensile strength, elongation at break, and area reduction can be directly measured, and other properties such as Young's modulus, yield strength, and Poisson's ratio can be subsequently determined. Sadeghi-Avalshahr et al. (2018) fabricated HA/PLLA interference screws and tested them. The HA/PLLA screws showed both increased tensile strength and improved cell attachment when they were compared to the pure PLLA screws.

In our studies, tensile testing is frequently used to assess biomedical composites. Figure 12.7 gives typical stress-strain curves of both Ca-P/chitosan nanocomposite and TCP/chitosan microcomposite of different bioceramic particle contents. The tensile strength, strain at fracture, and Young's modulus of the composites with the incorporation of either Ca-P or TCP particles are summarized in Figure 12.8. For both types of composites, the incorporation of a small amount of Ca-P or TCP particles increased both tensile strength and Young's modulus, and a drastic increase was observed when the content of Ca-P or TCP particles reached 5 wt%. When the Ca-P content was further increased to 10 wt% and 15 wt%, the tensile strength of Ca-P/chitosan nanocomposite decreased whereas the Young's modulus further increased. For TCP/chitosan microcomposite, the incorporation of 10 wt% or 15 wt% TCP microparticles resulted in not only a reduced tensile strength but also a reduced Young's modulus. However, the Young's modulus was still higher than that of the neat chitosan samples. The strain at fracture of both Ca-P/chitosan nanocomposite and TCP/chitosan microcomposite was found to decrease with increasing bioceramic particle content. Figure 12.9 shows typical tensile stress-strain curves of the Ca-P/PLGA composite. When 5 wt% Ca-P particles were incorporated, the Ca-P/PLGA nanocomposite had an enhanced tensile strength, whereas the nanocomposite with 10 wt% or 15 wt% Ca-P particles

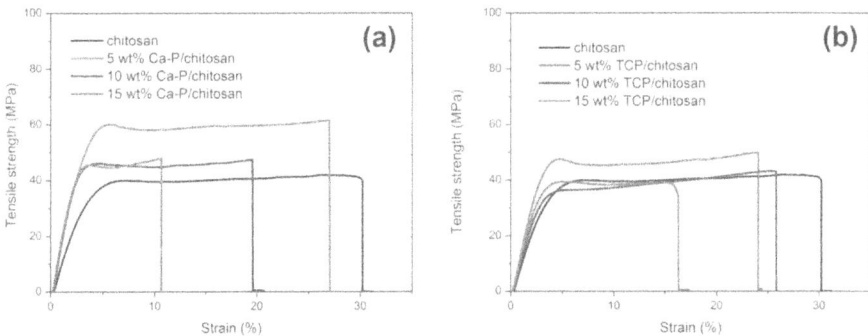

FIGURE 12.7 Tensile stress-strain curves of (a) Ca-P/chitosan nanocomposite and (b) TCP/chitosan microcomposite.

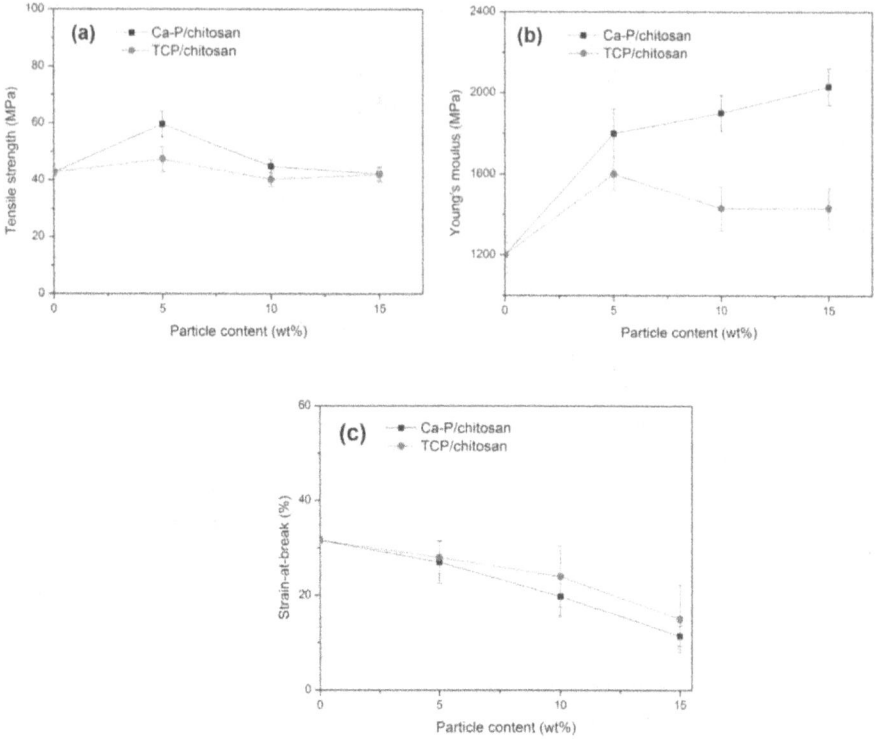

FIGURE 12.8 Mechanical properties of Ca-P/chitosan nanocomposite and TCP/chitosan microcomposite as a function of bioceramic particle content: (a) tensile strength, (b) Young's modulus, (c) strain at fracture.

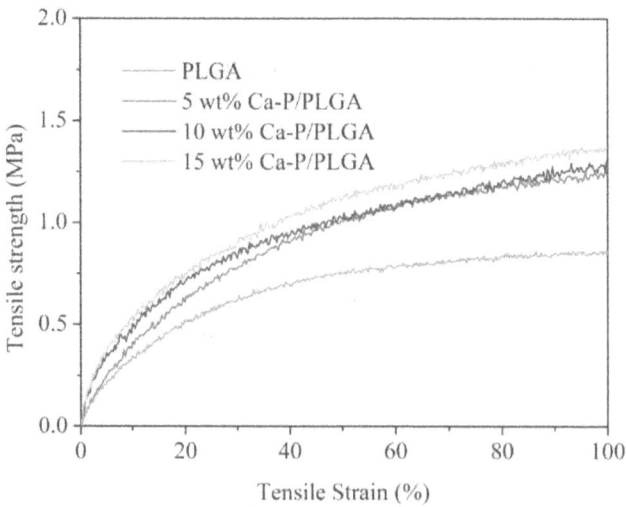

FIGURE 12.9 Tensile stress-stain curves of Ca-P/PLGA nanocomposite.

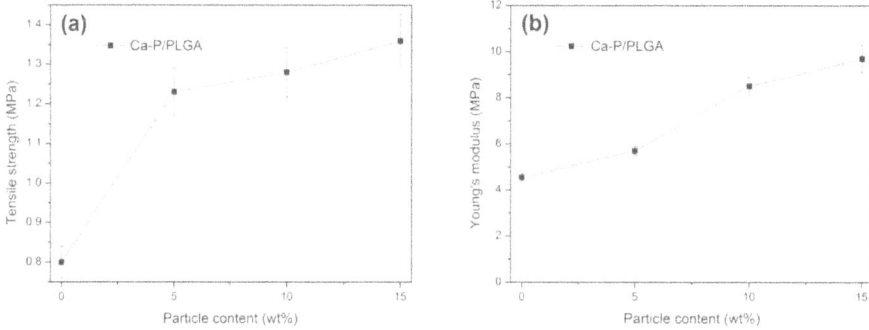

FIGURE 12.10 Mechanical properties of Ca-P/PLGA nanocomposite as a function of Ca-P nanoparticle content: (a) tensile strength, and (b) Young's modulus.

showed a further increased tensile strength (Figure 12.10a). Similarly, the increase of Ca-P content in the nanocomposite led to a continuous increase in the Young's modulus (Figure 12.10b), suggesting again that increasing the amount of Ca-P nanoparticles in composites could improve the elastic modulus of composites. As the Ca-P/PLGA nanocomposite did not fracture when 100% tensile strain was reached, no strain-at-fracture data was collected for comparison here. In comparison with Ca-P/chitosan nanocomposite, the values of both tensile strength and Young's modulus of Ca-P/PLGA nanocomposite were much lower while the strain at fracture appeared to be much larger.

12.4.2.2 Compression Testing

Biomaterials that are designed to repair human body tissues under compression should have sufficient compressive strength. Similarly, the compressive modulus and compressive fracture strain of these biomaterials also need to be determined (Wang, 2006). These compressive properties can be obtained using compression tests according to a test standard such as ASTM D695. For compression testing, an electromechanical testing machine is also used and upon the application of a compressive load, compressive stress-strain curves are recorded. Properties such as compressive modulus and compressive strength can then be measured. Gay et al. (2009) prepared dense HA/PLLA nanocomposites through extrusion-mixing and solvent casting. Increased Young's modulus and compressive strength were observed with increasing nano-HA content in the nanocomposite; but the failure mode of the nanocomposite changed from ductile to brittle. Yang et al. (2016) produced nano-BCP/PLLA porous scaffolds. The compressive strength of nano-BCP/PLLA scaffolds reached 10.25 MPa and 12.73 MPa at porosities of 60% and 50%, respectively, which were significantly higher than that of micro-BCP/PLLA composite scaffolds. Yan et al. (2016) fabricated spindle HA/PLLA nanocomposite for bone repair through co-precipitation and hot pressing. The compressive strength of this composite increased to 150 MPa when the HA content was raised to 20%.

12.4.2.3 Flexural Testing

Human hard tissues such as long bone must withstand large bending forces when heavy objects are held. Therefore, nanocomposites aiming at replacing body tissues, which are either occasionally or frequently subjected to bending forces, should have sufficient flexural strength and stiffness, which can be measured using either 3-point bending or 4-point bending tests (Wang et al., 2004). The 3-point bending testing is simpler and hence more often used, giving values for modulus of elasticity in bending, flexural stress, flexural strain, and the flexural stress-strain response of biomaterials. Zou et al. (2010) fabricated biodegradable bone-like TCP/PLLA composite. The addition of TCP reduced the bending strength of the composite, while the bending modulus increased with increasing TCP content. Kobayashi and Sakamoto (2009) prepared β-TCP particle reinforced PLLA composite by injection molding. The PLLA crystallinity increased with increasing heat treatment temperature. The bending and compressive moduli of the composite increased with increasing β-TCP content and crystallinity, whereas the bending strength decreased with increasing β-TCP content.

12.4.2.4 Microhardness Testing

Microhardness testing is used to determine the hardness of a material surface at the microscopic level. In microhardness testing, a diamond indenter of a specific geometry is impressed into the surface of a specimen with a test load and loading time and produces an indentation with the length of a side being tens of microns (Foxman et al., 2019). Microhardness testing can be conducted using a Vickers or Knoop indenter. In the Vickers microhardness tests, both diagonals of the indent are measured and the average value is used to compute the Vickers hardness number. Using Vickers microhardness testing, it was shown that with an increase in the bioceramic content, the microhardness of bioceramic-polymer composites increased (Wang, 2006).

12.4.2.5 Dynamic Mechanical Analysis

Human body tissues such as articular cartilage exhibit distinctive viscoelastic characteristics with substantial loading rate dependency. Therefore, dynamic mechanical analysis (DMA) of nanocomposites developed for repairing such tissues should be conducted (Nazhat et al., 2000). In a DMA test, a sinusoidal stress (tensile or bending) is normally applied to the specimen and the strain of the specimen is measured to determine the complex modulus. The temperature of the test chamber or the frequency of the applied stress can be varied during a DMA test for investigating its effect on the complex modulus. For a totally elastic solid, the resulting strain is in phase with the applied stress. For a purely viscous fluid, there is a 90-degree phase lag of strain with respect to applied stress. For a viscoelastic material, there is a phase lag of less than 90 degrees for the strain. Using DMA tests, the viscoelastic properties of a nanocomposite can be studied (Nazhat et al., 2000). DMA tests were used in our *in vitro* degradation studies on biomedical composites (Ni & Wang, 2002). It was observed in our studies that the storage modulus of HA/PHB composite increased initially with immersion time in a simulated body fluid (SBF) due to apatite formation on the composite surface and decreased after prolonged immersion in SBF, indicating degradation of the composite in the simulated body environment.

12.4.3 BIOLOGICAL EVALUATION OF BIOMEDICAL NANOCOMPOSITES

12.4.3.1 Cell Viability

The cytocompatibility of nanocomposites is critically important for their successful use for human tissue repair, and this can be determined by studying cell viability and cell proliferation through *in vitro* cell culture experiments (Wang & Wang, 2017). Live and dead staining is the most commonly used method to visualize live and dead cells cultured on nanocomposites under a fluorescence microscope. After incubating cell-composite constructs in phosphate buffer saline (PBS) containing calcein AM and EthD-1, green fluorescence signals can be observed from live cells whereas red color by dead cells can be seen (if there are dead cells). The cell proliferation is normally investigated through the use of 3-(4,5-Dimethylthiazol-2-yl)-2,5-diphenyltetrazolium bromide (MTT) assay. After incubating cell-composite constructs with MTT solution for a certain period of time, water or dimethyl sulfoxide (DMSO) is added to dissolve formazan and the absorbance is measured by using a microplate reader. Our studies showed that composite scaffolds for bone tissue engineering made by either electrospinning (Wang & Wang, 2017) or SLS (Duan et al., 2010) had excellent cytocompatibility and that different types of cells, including mesenchymal stem cells (MSCs), proliferated well on the scaffolds.

12.4.3.2 Cell Adhesion and Cell Morphology

Cell attachment to a nanocomposite, nonporous or porous, is critical to all subsequent cellular activities. Therefore, adhesion between cells and nanocomposites should be examined. Confocal laser scanning microscopy (CLSM) is very useful for evaluating cell adhesion. Through immunofluorescence staining, vinculin antibody could be conjugated to the adhesive plaques and fluorescence signals can be used to locate where the adhesive plaques are formed (Wang et al., 2017). Besides, phalloidin and 4-6-diamidino-2-phenylindole (DAPI) can also be used in staining for visualizing the F-actin and nucleus of cells, assisting to investigate cell activities. Morphology of cells cultured on nanocomposites can be observed using SEM. In SEM sample preparation, gradient dehydration should be conducted first for cell-nanocomposite constructs, followed by critical-point drying. Before SEM observation, a thin layer of gold is coated on the dried cell-nanocomposite constructs. With cell culture experiments going for weeks and by examining the cell morphology at different culture times, the morphological changes of cells on nanocomposites could be observed (Huang et al., 1997; Tong et al., 2010).

12.4.3.3 Cell Differentiation

After a period of culture (and induction, in the case of stem cells), differentiation of cells, particularly stem cells along different cell lineages, occurs, which can be determined using immunofluorescent staining and immunohistological staining.

For instance, the osteogenic differentiation of MSCs can be studied via alkaline phosphatase staining, alizarin red S staining, and alkaline phosphatase activity assay (Wang et al., 2017; Wang & Wang, 2017), whereas the chondrogenic differentiation of MSCs could be visualized via the immunofluorescent staining of chondrogenic molecular markers such as SOX9, aggrecan, and collagen II (Mahmoudifar and Doran, 2005).

12.4.3.4   *In Vivo* Animal Study

To investigate the efficacy and ability of nanocomposites for the substitution or regeneration of human body tissues, suitable animals with target defects can be used as *in vivo* models. With the assistance of surgery, implantation of micro-/nanocomposites (nonporous or porous) into the defect sites in the animal can be performed (Bonfield et al., 1998; Samadikuchaksaraei et al., 2016). After a period of implantation for new tissue formation (spanning from weeks to months), images by 3D reconstructed micro-computed tomography (Micro-CT) and magnetic resonance imaging (MRI) can be obtained to visualize the newly formed hard and soft tissues, respectively, and calculate the volume of tissues formed. Subsequently, thin specimens for histological analysis can be made from sections containing the implant-tissue interface, which are obtained from the animals. After histochemical staining, specimens can be examined using optical microscopy, SEM, or transmission electron microscopy (TEM). It should be noted that specimens for TEM examinations are the most difficult to prepare. SEM can be used to assess the overall mechanism of new tissue formation at the implant-tissue interface while TEM reveals ultrastructure and hence bonding mechanism at the interface. It was shown through SEM and TEM studies that new bone could form on an HA/PHB composite in 3 months after implantation (Luklinska & Bonfield, 1997). Micro-CT images of nanocomposite scaffolds implanted in the ilium of rabbits for 6 weeks revealed that surface-modified and BMP-2-loaded Ca-P/PHBV nanocomposite scaffolds promoted significantly new bone formation inside the scaffolds while the scaffolds degraded (Duan & Wang, 2011). Also, histological images of tissues after scaffold implantation in the ilium of rabbits for 6 weeks showed new bone formation adjacent to and inside surface-modified and BMP-2-loaded Ca-P/PHBV nanocomposite scaffolds (Duan & Wang, 2011). In our recent investigations, it was shown that the novel tricomponent fibrous scaffolds simultaneously incorporated with human vein endothelial growth factor (VEGF), BMP-2, and Ca-P nanoparticles had a balanced angiogenic and osteogenic capability (Wang et al., 2020). Eight weeks after scaffold implantation into the cranial defects of mice, obvious new bone regeneration and also newly formed capillaries were observed in the degraded tricomponent scaffolds.

12.4.4   Medical Applications of Nanocomposites

Nanocomposites have been widely investigated for human body tissue replacement or regeneration. Absorbable nanocomposite screws and plates are developed for internal fixation of bone fractures. Peng et al. (2015) found that at all time points following surgery, the degradation of HA/PLLA plates was significantly less than that of PLLA plates. Also, the lateral and longitudinal bending strengths of the surgically treated mandibles of beagle dogs in the HA/PLLA group were significantly greater than those of the PLLA group and reached almost the value of intact mandibles at 12 months postoperatively. Additionally, relatively rapid bone healing was observed in the HA/PLLA group with the formation of new lamellar bone tissues at 12 months after the surgery. Fibrous nanocomposite membranes made by electrospinning have been used as bone tissue engineering scaffolds to induce bone regeneration. Some nanocomposite membranes are employed as biomimetic artificial periosteum, which are wrapped on metal implants to induce bone regeneration, thus promoting rapid osteointegration of metal implants to bone (Takakuda et al., 2013). Nanocomposites are also used to repair damaged osteochondral tissue (Jeznach et al., 2018). As the osteochondral tissue can be divided into subchondral zone, calcified zone, deep zone, middle zone, and superficial zone, which contain different amounts of apatite (zero amount for several zones), well-designed multilayered nanocomposites with a gradient composition are desirable for repairing damaged osteochondral tissue. Tendon repair can also be achieved by using nanocomposites. Alshomer et al. (2017) showed that the cellular alignment, biological function of micropatterned nanocomposite polymer (POSS-PCU), and expression of various tendon-related proteins were significantly elevated on the micropatterned nanocomposite polymer surface as compared to flat, non-patterned samples. Nanocomposites comprising polymer matrix and nanoparticles with the photothermal effect are good candidates for photothermotherapy for cancer. After surgical removal of tumors, hydrogel-based nanocomposites with photothermal effect could be sprayed in the wound area and irradiated by a near-field infrared laser, which converts the light energy to heat to kill residual tumor cells postoperatively (Shao et al., 2018). There are many other examples for medical applications of nanocomposites, and it appears that there is still a large scope for the exploration.

12.5 CONCLUDING REMARKS

Compared to other types of biomedical materials (biomedical polymers, bioceramics, and implantable metals), biomedical composites have a relative short history of development. However, as biomedical composites have distinctive advantages over other materials used in medicine, there have been various efforts over the past 30 years to develop them into materials that can have real use for human body tissue repair. With the recent rapid advances in nanoscience and nanotechnology, increasing attention is now paid to the development of nanocomposite materials for biomedical applications. Biomedical nanocomposites, particularly the biodegradable porous structures, are highly promising materials that can be used for regenerating human body tissues. Compared to conventional microcomposites for medicine, there are particular issues for biomedical nanocomposites for human body tissue repair, in their design, manufacture, and assessment. Also, each biomedical nanocomposite is unique, as one change in one aspect of the nanocomposite (e.g., a processing parameter) can lead to a material having different properties. Therefore, great efforts are needed for the systematic study of nanocomposites in order to realize their great potential in medicine. Nevertheless, biomedical nanocomposites will play an increasingly important role in the repair of various human body tissues.

ACKNOWLEDGMENTS

Min Wang thanks UK's Engineering and Physical Sciences Research Council (EPSRC), Singapore's Ministry of Education, Hong Kong's Research Grants Council (RGC), and China's National Natural Science Foundation for providing research support for his R & D in biomedical composites. Chong Wang has been supported by the Dongguan University of Technology High-level Talents (Innovation Team) Research Project, Young Innovative Talent Project from the Department of Education of Guangdong Province, and Natural Science Foundation of Guangdong Province, China.

REFERENCES

Alshomer, F., Chaves, C., Serra, T., Ahmed, I., and Kalaskar, D. M. (2017). Micropatterning of Nanocomposite Polymer Scaffolds using Sacrificial Phosphate Glass Fibers for Tendon Tissue Engineering Applications. *Nanomedicine-Nanotechnology, Biology and Medicine*, *13*(3): 1267–1277.

Bonfield, W., Wang, M., and Tanner, K. E. (1998). Interfaces in Analogue Biomaterials. *Acta Materialia*, *46*(7), 2509–2518.

Chen, L. J., and Wang, M. (2002). Production and Evaluation of Biodegradable Composites Based on PHB-PHV Copolymer. *Biomaterials*, *23*(13), 2631–2639.

Cross, L. M., Thakur, A., Jalili, N. A., Detamore, M., and Gaharwar, A. K. (2016). Nanoengineered Biomaterials for Repair and Regeneration of Orthopedic Tissue Interfaces. *Acta Biomaterialia*, *42*, 2–17.

Ding, Y., Souza, M. T., Li, W., Schubert, D. W., Boccaccini, A. R., and Roether, J. A. (2016). Bioactive glass-biopolymer composites for applications in tissue engineering. In Handbook of Bioceramics and Biocomposites, pp. 325–356.

Duan, B., and Wang, M. (2011). Selective Laser Sintering and its Application in Biomedical Engineering. *MRS Bulletin*, *36*(12), 998–1005.

Duan, B., Wang, M., Zhou, W. Y., and Cheung, W. L. (2008). Synthesis of Ca–P nanoparticles and Fabrication of Ca–P/PHBV Nanocomposite Microspheres for Bone Tissue Engineering Applications. *Applied Surface Science*, *255*(2), 529–533.

Duan, B., Wang, M., Zhou, W. Y., Cheung, W. L., Li, Z. Y., and Lu, W. W. (2010). Three-Dimensional Nanocomposite Scaffolds Fabricated via Selective Laser Sintering for Bone Tissue Engineering. *Acta Biomaterialia*, *6*(12), 4495–4505.

Foxman, B., Kolderman, E., Salzman, E., Cronenwett, A., Gonzalez-Cabezas, C., Neiswanger, K., and Marazita, M. L. (2019). Primary Teeth Microhardness and Lead (Pb) Levels. *Heliyon*, *5*(4), e01551.

Gao, C., Yang, B., Hu, H., Liu, J., Shuai, C., and Peng, S. (2013). Enhanced Sintering Ability of Biphasic Calcium Phosphate by Polymers used for Bone Scaffold Fabrication. *Materials Science and Engineering: C*, *33*(7), 3802–3810.

García-Gareta, E., Coathup, M. J., and Blunn, G. W. (2015). Osteoinduction of Bone Grafting Materials for Bone Repair and Regeneration. *Bone*, *81*, 112–121.

Gay, S., Arostegui, S., and Lemaitre, J. (2009). Preparation and Characterization of Dense Nanohydroxyapatite/PLLA Composites. *Materials Science and Engineering: C*, *29*(1), 172–177.

He, C., Jin, X., and Ma, P. X. (2014). Calcium Phosphate Deposition Rate, Structure and Osteoconductivity on Electrospun Poly (L-lactic Acid) Matrix using Electrodeposition or Simulated Body Fluid Incubation. *Acta Biomaterialia*, *10*(1), 419–427.

He, J., Yang, X., Mao, J., Xu, F., and Cai, Q. (2012). Hydroxyapatite–Poly (l-Lactide) Nanohybrids via Surface-Initiated ATRP for Improving Bone-Like Apatite-Formation Abilities. *Applied Surface Science*, *258*(18), 6823–6830.

Hou, Y., Chen, C., Zhou, S., Li, Y., Wang, D., and Zhang, L. (2016). Fabrication of an Integrated Cartilage/Bone Joint Prosthesis and Its Potential Application in Joint Replacement. *Journal of the Mechanical Behavior of Biomedical Materials*, *59*, 265–271.

Huang, J., Di Silvio, L., Wang, M., Tanner, K. E., and Bonfield, W. (1997). *In Vitro* Mechanical and Biological Assessment of Hydroxyapatite-Reinforced Polyethylene Composite. *Journal of Materials Science: Materials in Medicine*, *8*(12), 775–779.

Ikumi, R., Miyahara, T., Akino, N., Tachikawa, N., and Kasugai, S. (2018). Guided Bone Regeneration using a Hydrophilic Membrane made of Unsintered Hydroxyapatite and Poly (L-Lactic Acid) in a Rat Bone-Defect Model. *Dental Materials Journal*, *37*(6), 912–918.

Jeznach, O., Kolbuk, D., and Sajkiewicz, P. (2018). Injectable Hydrogels and Nanocomposite Hydrogels for Cartilage Regeneration. *Journal of Biomedical Materials Research Part A*, *106*(10): 2762–2776.

Kobayashi, S., and Sakamoto, K. (2009). Bending and Compressive Properties of Crystallized TCP/PLLA Composites. *Advanced Composite Materials*, *18*(3), 287–295.

Liu, H., and Webster, T. J. (2010). Mechanical Properties of Dispersed Ceramic Nanoparticles in Polymer Composites for Orthopedic Applications. *International Journal of Nanomedicine*, *5*, 299–313.

Liu, Y., and Wang, M. (2007a). Developing a Composite Material for Bone Tissue Repair. *Current Applied Physics*, *7*(5), 547–554.

Liu, Y., and Wang, M. (2007b). Fabrication and Characteristics of Hydroxyapatite Reinforced Polypropylene as a Bone Analogue Biomaterial. *Journal of Applied Polymer Science*, *106*(4), 2780–2790.

Luklinska, Z. B., and Bonfield, W. (1997). Morphology and Ultrastructure of the Interface between Hydroxyapatite-Polyhydroxybutyrate Composite Implant and Bone. *Journal of Materials Science: Materials in Medicine*, *8*(6), 379–383.

Mahmoudifar, N., and Doran, P. M. (2005). Tissue Engineering of Human Cartilage in Bioreactors using single and Composite Cell-Seeded Scaffolds. *Biotechnology and Bioengineering*, *91*(3), 338–355.

Mooyen, S., Charoenphandhu, N., Teerapornpuntakit, J., Thongbunchoo, J., Suntornsaratoon, P., Krishnamra, N., and Pon-On, W. (2017). Physico-Chemical and in Vitro Cellular Properties of Different Calcium Phosphate-Bioactive Glass Composite Chitosan-Collagen (CaP@ ChiCol) for Bone Scaffolds. *Journal of Biomedical Materials Research Part B: Applied Biomaterials*, *105*(7), 1758–1766.

Nazhat, S. N., Joseph, R., Wang, M., Smith, R., Tanner, K. E., and Bonfield, W. (2000). Dynamic Mechanical Characterization of Hydroxyapatite Reinforced Polyethylene: Effect of Particle Size. *Journal of Materials Science: Materials in Medicine*, *11*(10), 621–628.

Ni, J., and Wang, M. (2002). *In Vitro* Evaluation of Hydroxyapatite Reinforced Polyhydroxybutyrate Composite. *Materials Science and Engineering: C*, *20*(1–2), 101–109.

Nurqadar, R. I., Purnama, P., and Kim, S. H. (2013). Preparation and Characterization of a Stereocomplex of Poly(Lactide-co-ε-caprolactone)/Tricalcium Phosphate Biocomposite Using Supercritical Fluid Technology. *eXPRESS Polymer Letters*, *7*(12), 974–983.

Peng, W. H., Zheng, W., Shi, K., Wang, W. S., Shao, Y., and Zhang, D. (2015). An *in vivo* Evaluation of PLLA/PLLA-gHA Nano-Composite for Internal Fixation of Mandibular Bone Fractures. *Biomedical Materials*, *10*(6), 065007.

Porras, R., Bavykin, D. V., Zekonyte, J., Walsh, F. C., and Wood, R. J. (2016). Titanate Nanotubes for Reinforcement of a Poly (Ethylene Oxide)/Chitosan Polymer Matrix. *Nanotechnology*, *27*(19), 195706.

Quinlan, E., Partap, S., Azevedo, M. M., Jell, G., Stevens, M. M., and O'Brien, F. J. (2015). Hypoxia-Mimicking Bioactive Glass/Collagen Glycosaminoglycan Composite Scaffolds to Enhance Angiogenesis and Bone Repair. *Biomaterials*, *52*, 358–366.

Sadeghi-Avalshahr, A., Khorsand-Ghayeni, M., Nokhasteh, S., Shahri, M. M., Molavi, A. M., and Sadeghi-Avalshahr, M. (2018). Effects of Hydroxyapatite (HA) Particles on the PLLA Polymeric Matrix for Fabrication of Absorbable Interference Screws. *Polymer Bulletin*, *75*(6), 2559–2574.

Samadikuchaksaraei, A., Gholipourmalekabadi, M., Erfani Ezadyar, E., Azami, M., Mozafari, M., Johari, B., Kargozar, S., Jameie, S. B., Korourian, A., and Seifalian, A. M. (2016). Fabrication and *in vivo* Evaluation of an Osteoblast-Conditioned Nano-Hydroxyapatite/Gelatin Composite Scaffold for Bone Tissue Regeneration. *Journal of Biomedical Materials Research A*, *104*(8), 2001–1010.

Sarkar, P. (2015). Properties Affected by Different Shape and Different Weight Percentages Nanoparticles, Embedded in Chitosan Polymer Thin Film. *European Journal of Advances in Engineering and Technology*, *2*(1), 91–100.

Shao, J. D., Ruan, C. S., Xie, H. H., Li, Z. B., Wang, H. Y., Chu, P. K., and Yu, X. F. (2018). Black-Phosphorus-Incorporated Hydrogel as a Sprayable and Biodegradable Photothermal Platform for Postsurgical Treatment of Cancer. *Advanced Science*, *5*(5), 1700848.

Takakuda, K., Kikuchi, M., Suzuki, S., and Moriyama, K. (2013). Hydroxyapatite/Collagen Nanocomposite-Coated Titanium Rod for Achieving Rapid Osseointegration onto Bone Surface. *Journal of Biomedical Materials Research Part B-Applied Biomaterials*, *101B*(6), 1031–1038.

Toney, C. B., Owen, J. R., Khatri, I. A., Wayne, J. S., and McDowell, C. L. (2016). Bone-Prosthesis Junction for Active Tendon Implants: A Biomechanical Comparison of 2 Fixation Techniques. *The Journal of Hand Surgery*, *41*(4), 526–531.

Tong, H. W., Wang, M., Li, Z. Y., and Lu, W. W. (2010). Electrospinning, Characterization and *in vitro* Biological Evaluation of Nanocomposite Fibers Containing Carbonated Hydroxyapatite Nanoparticles. *Biomedical Materials*, *5*(5), 054111.

Tong, S. Y., Wang, Z., Lim, P. N., Wang, W., and San Thian, E. (2017). Uniformly-Dispersed Nanohydroxyapatite-Reinforced Poly (ε-Caprolactone) Composite Films for Tendon Tissue Engineering Application. *Materials Science and Engineering: C*, *70*, 1149–1155.

Wang, C., and Wang, M. (2017). Electrospun Multicomponent and Multifunctional Nanofibrous Bone Tissue Engineering Scaffolds. *Journal of Materials Chemistry B*, *5*(7), 1388–1399.

Wang, C., Lu, W. W., and Wang, M. (2020). Multifunctional Fibrous Scaffolds for Bone Regeneration with Enhanced Vascularization. *Journal of Materials Chemistry B*, DOI:10.1039/C9TB01520E.

Wang, C., Zhao, Q., and Wang, M. (2017). Cryogenic 3D Printing for Producing Hierarchical Porous and rhBMP-2-loaded Ca-P/PLLA Nanocomposite Scaffolds for Bone Tissue Engineering. *Biofabrication*, *9*(2), 025031.

Wang, M. (2003). Developing Bioactive Composite Materials for Tissue Replacement. *Biomaterials*, *24*(13), 2133–2151.

Wang, M. (2004). Bioactive materials and processing. In *Biomaterials and Tissue Engineering*, Springer, Berlin, Heidelberg, pp. 1–82.

Wang, M. (2006). Deformation and fracture of bioactive particle reinforced polymer composites developed for hard tissue repair. In *Fracture of Nano and Engineering Materials and Structures - Proceedings of the 16th European Conference of Fracture*. Springer, pp. 1029–1030.

Wang, M., and Wang, C. (2019). Bulk properties of biomaterials and testing techniques. In *Encyclopedia of Biomedical Engineering*, Elsevier, the Netherlands, Vol. 1, pp. 53–64.

Wang, M., and Zhao, Q. (2019). Biomedical composites. In *Encyclopedia of Biomedical Engineering*, Elsevier, the Netherlands, Vol. 1, pp. 34–52.

Wang, M., Joseph, R., and Bonfield, W. (1998). Hydroxyapatite-Polyethylene Composites for Bone Substitution: Effects of Ceramic Particle Size and Morphology. *Biomaterials*, *19*(24), 2357–2366.

Wang, M., Leung, L. Y., Lai, P. K., and Bonfield, W. (2004). Effects of Polymer Molecular Weight and Ceramic Particle Size on Flexural Properties of Hydroxyapatite Reinforced Polyethylene. *Key Engineering Materials*, *254*, 611–614.

Wang, M., Porter, D., and Bonfield, W. (1994). Processing, Characterisation, and Evaluation of Hydroxyapatite Reinforced Polyethylene. *British Ceramics Transactions*, 93, 91–95.

Wang, X., Song, G., and Lou, T. (2010). Fabrication and Characterization of Nano-Composite Scaffold of PLLA/Silane modified Hydroxyapatite. *Medical Engineering & Physics*, *32*(4), 391–397.

Yang, W., Yi, Y., Ma, Y., Zhang, L., Gu, J., and Zhou, D. (2016). Preparation and Characterization of Nano Biphasic Calcium Phosphate/Poly-L-lactide Composite Scaffold. *Science and Engineering of Composite Materials*, *23*(1), 37–44.

Yang, W., Yin, G., Zhou, D., Youyang, L., and Chen, L. (2009). Surface-Modified Biphasic Calcium Phosphate/Poly (L-Lactide) Biocomposite. *Journal of Wuhan University of Technology-Mater. Sci. Ed.*, *24*(1), 81–86.

Yang, W., Zhang, C. Y., Xia, L. L., Zhang, T., and Fang, Q. F. (2016). *In vitro* Investigation of Nanohydroxyapatite/Poly (L-lactic Acid) Spindle Composites Used for Bone Tissue Engineering. *Journal of Materials Science: Materials in Medicine*, *27*(8), 130.

Zimel, M. N., Farfalli, G. L., Zindman, A. M., Riedel, E. R., Morris, C. D., Boland, P. J., and Healey, J. H. (2016). Revision Distal Femoral Arthroplasty with the Compress® Prosthesis has a Low Rate of Mechanical Failure at 10 Years. *Clinical Orthopaedics and Related Research®*, *474*(2), 528–536.

Zou, J., Jiang, X. B., Zhang, J., Shu, Y., Chen, X., and Huang, F. R. (2011). Preparation and Characterization of Surface-Modified β-tricalcium Phosphate/Poly (L-Lactide) Biocomposites. *Advanced Materials Research*, *197*, 120–126. (Trans Tech Publications)

Zou, J., Zhou, Z., Ruan, J., and Zhou, Z. (2010). Fabrication of Degradable Bone-Like Substitutes Based on Poly-L-Lactide and β-Tricalcium Phosphate. *Journal of Macromolecular Science, Part B*, *49*(4), 781–790.

13 Advances in Marine Skeletal Nanocomposites for Bone Repair

I. J. Macha

Department of Mechanical and Industrial
Engineering, University of Dar es Salaam,
Dar es Salaam, Tanzania

CONTENTS

13.1 INTRODUCTION

The human skeleton is quite a complex structure consisting of a whopping 206 bones. These bones are connected by a network of tendon, cartilage, and ligament through a self-assembly model. The main physical functions of the skeletal structure are to provide structural support to a human body and protect the internal organs from physical injuries. Physiologically, bones are capable of self-repair in case of injury, play a role in hematopoiesis, and help form blood cells (Wang, Leng, & Gong, 2018). Structurally, bones are natural nanocomposites with a unique architecture that provides excellent functions required by the human body.

Worldwide, about 20–30 million fractures from road traffic accidents and about 8.9 million fractures from osteoporosis occur annually. More than 33% of these fractures require either joint replacements or application of graft materials such as autografts, allografts, and alloplastic (Diwan, Eberlin, & Smith, 2018; Hagen et al., 2012). Autografts and allografts have disadvantages such as limited availability, risk of disease transmission, and lack of osteoconductivity or osteogenesis for allografts;

in addition, there is a need to improve the existing synthetic graft materials. There are challenges, in spite of the advancement in technology, in developing perfect materials for bone repair and regeneration. Using natural biogenic materials has been suggested as an alternative route to provide solutions in tissue engineering and regenerative medicine.

Biogenic materials possess excellent biological properties with, in many cases, relatively weak mechanical properties. Materials for orthopedic applications should be strong enough to bear biomechanical stress, which limits most biogenic materials to be used in their natural structures without any modifications. The development of products for load-bearing applications from these materials should carefully consider mimicking closely the structure and property performance requirements of natural bones. This can be achieved by enhancing our deep understanding of the structure–properties relationship, especially in their nanoscale. On the other hand, a single material can never offer the complex structure and optimal properties of natural bones. To achieve the optimal biological, physical, and mechanical properties, researchers have suggested that a combinatorial approach, where a variety of materials are combined to form composites, is beneficial (Sadagopan & Pitchumani, 1997; Asyraf et al., 2020; Ilyas et al., 2020, 2018c, 2018a, 2018b; Ilyas et al., 2019; Ilyas et al., 2019; R.A. Ilyas et al., 2019; Ilyas & Sapuan, 2020; Sanyang et al., 2018). Native bone possesses a nanocomposite structure of mainly hydroxyapatite (HAp) as reinforcement, embedded in the collagen matrix that provides an excellent performance. Marine skeletons have a unique structure similar to natural bones, yet they don't provide all the required properties and performance factors. The combination of these materials with others, either with natural or synthetic, is still necessary to provide vital components for bone repair and regeneration.

In this chapter, recent advances in marine skeletal nanocomposites for bone repair and regeneration will be comprehensively covered with a focus on structure and compositions. Different combination options for both natural and synthetic polymer matrix and marine-derived hydroxyapatite will also be presented.

13.2 NATURAL BONE STRUCTURE

The adult skeleton comprises 206 different bones, which are grouped into five different categories based, mainly, on their shapes. Their physical, mechanical, and biological functions are related to their physical shapes. These categories are long bones, short bones, flat bones, irregular bones, and sesamoid bones. Physically, long bones such as the femur have their length longer than their width while short bones have their length equal to their width. Flat bones are relatively thin and in most cases curved, the ribs are a good example. Maxillofacial bones have irregular shapes in which sesamoid bones are small and round located in tendons.

At the nanoscale level, microscopically, mature bones can be differentiated into two distinct types—cortical (compact) and trabecular (spongy) bones (Weatherholt,

Fuchs, & Warden, 2012). They possess similar matrix compositions but substantially differ in their functions and structures. Appearing as solid masses, compact bones make up about 80% of the mass of an adult skeleton with their main function being load-bearing and protection. Unlike trabecular bones, cortical bones have a high ratio of a matrix with low porosity. This architecture gives cortical bones great compressive strength for their mechanical roles. Usually, cortical bones are among long bones that form tubular shaft with diaphysis (dense and hard shell) surrounding epiphysis (hollow region) called medullary cavity, which is filled with bone marrow. Trabecular bones have a high pore-size volume ratio in the range of 50–90% of bone volume with less mass per unit volume (Currey, 2001). The interconnected porous structure is vital for vascularization and mineral homeostasis. Trabecular bones have less compressive strength compared to cortical bones, but they provide internal support and contribute to the ability of bones to evenly distribute the load and absorb energy particularly at the joints. Figure 13.1 shows the bone anatomy.

At the microscopic level, trabecular and cortical bones display structural tissue in either lamellar or woven structures. The structural sizes vary between macro- and nanoscale and facilitate body fluid, oxygen, and nutrient transportation. As a composite material, a bone consists of organic and inorganic components. Fibrous protein type I collagen makes up the organic matrix of about 20% of bone wet weight, which gives bone the required flexibility properties.

The inorganic phase of the bone matrix contributes between 65% and 70% of crystalline bone mineral of calcium hydroxyapatite ($Ca_{10}(PO_4)_6(OH)_2$) embedded in collagen for reinforcement that gives bone most of its stiffness (Landis et al., 1996).

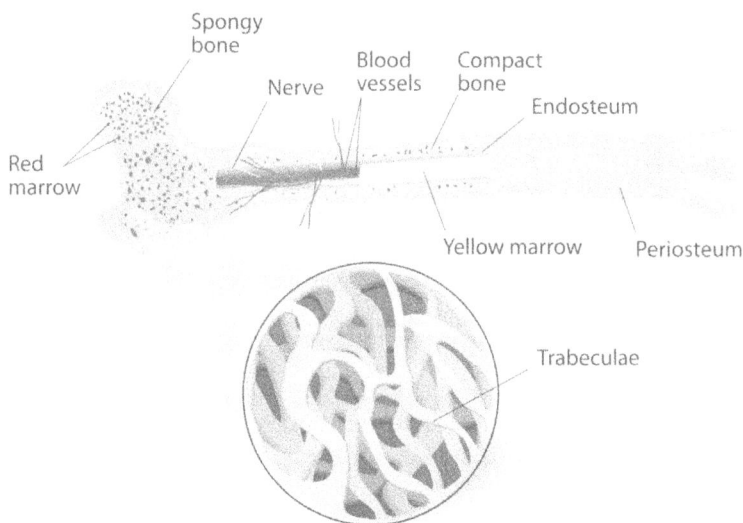

FIGURE 13.1 Bone anatomy (Adapted from Newman 2018).

Bone also contains about 9% water, an essential substance responsible for the transportation of nutrients, waste products, and minerals.

13.3 NANOMATERIALS AND NANOSTRUCTURES

The proper functions of bone are contributed to its hierarchical structure that gives the bone a unique architecture of morphology and sizes ranging from micro to nanostructures. The nanostructural components of bone are designed and coordinated to impart excellent chemical and biological functions of bone. Nanomaterials are materials having size ranges from 1 nm to 100 nm in at least one dimension. These materials have superior properties compared to their respective bulk materials because at the nano level, there is a significant reduction of imperfections within the materials (Khan, Saeed, & Khan, 2019). In terms of physical and chemical interactions, nanomaterials offer a large surface area, which is important for bone cells to grow and function. Many of their properties such as reactivity, toughness, optical properties, and others also depend on their unique small size, shape, and structure. Nanomaterials have gained a wide range of applications in different areas in the development of novel nanocomponents for mechanical, biological, biomedical, and pharmaceutical applications.

The tissue engineering approach focuses on developing 3D structures as substitute bone material with properties equivalent to the native bone for tissue repair and regeneration. Three-dimensional nanostructures with a high level of organization resembling native bone tissue to achieve the desired cell function for the regeneration of damaged tissue are highly needed. From nature, marine skeletons have been researched and used for the development of 3D bone substitutes because their structures and properties are similar to natural bone to a large extent. These 3D nanostructure constructs provide excellent mechanical strength required for the healing of load-bearing bones with a proper mechanotransduction of the mechanical stimuli down to the cell level (Sprio et al., 2011).

13.4 MARINE SKELETAL NANOSTRUCTURES

Marine structures have been researched for several years due to their potential applications, particularly in the biomedical field. Others have expressed environmental concerns regarding the use of marine structures, but the evidence of artificially grown corals, for instance, has overwhelmed (Omori, 2019; Higgins et al., 2019). Coral skeletons are among the marine structures made of calcium carbonate ($CaCO_3$) in forms of calcite or aragonite in platelet structure, with other trace elements such as sodium, magnesium, and strontium. They are considered the hydroxyapatite precursors and bone graft materials (Ben-Nissan, 2005). Coral can be converted to monophasic bioceramic (hydroxyapatite similar to bone minerals) through a two-stage hydrothermal conversion of $CaCO_3$ according to Ben-Nissan (2011) or mechanochemical techniques (Cegla et al., 2014) and used for bone repair and treatment. The calcium phosphate formed depends on the polymorph of the parent $CaCO_3$ used and thus on the species of coral. The coral *Porites* from the aragonite polymorph, under hydrothermal treatment with dibasic ammonium phosphate, converts to HAp, Equation (13.1).

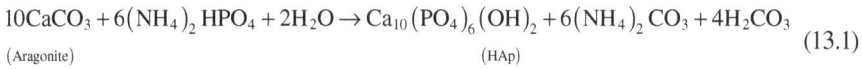

$$10CaCO_3 + 6(NH_4)_2 HPO_4 + 2H_2O \rightarrow Ca_{10}(PO_4)_6(OH)_2 + 6(NH_4)_2 CO_3 + 4H_2CO_3$$

(Aragonite)   (HAp)

$$(13.1)$$

The coral structure is similar to that of cancellous bone, porous structure with interconnected pores (micro, meso, and nano), and has been reported to form chemical bonds with the bone and soft tissue *in vivo* (Ben-Nissan, 2003). Different coral species exist in the marine environment but *Porites* and *Goniopora* possess pore sizes in the range of cancellous bone. Commercially, bioceramic derived coralline named Biocoral (Biocoral Inc., France) in the forms of 3D blocks and granules revealed excellent results for bone augmentation (Titsinides, Agrogiannis, & Karatzas, 2019). It was further reported that the interaction of the primary osteons between the pores via the interconnections allows the propagation of osteoblasts (Heness & Ben-Nissan, 2004). The analysis of osseous reactions of rabbit femoral to coralline derived hydroxyapatite of various pore sizes by radiology and histology showed the substantial production of bone within 500-micron pore size (Kühne et al., 1994). From their studies, Kühne and his colleagues concluded that the pore size of the coralline hydroxyapatite influenced the development of bone in the implants.

13.5 BIOCERAMIC-POLYMER NANOCOMPOSITES

Bone minerals derived from marine skeletons such as hydroxyapatite and tricalcium phosphate are used as bone fillers and implant coatings due to their similarities to natural bone minerals. The scaffolds made of these materials are not desirable for load-bearing applications due to their brittleness. However, their combinations with polymeric biomaterials result in composite materials with improved properties for tissue engineering applications. This paves the way to develop novel therapeutic materials with a similar structure like natural bone but also with the strength, stiffness, and osteoconductivity of hydroxyapatite combined with the flexibility, toughness, and resorbability of the matrix phase (Sahoo et al., 2013). Nie and his co-authors developed chitosan/gelatin calcium phosphate nanocomposite scaffold with the potential to support tissue growth necrotic lesions of rabbit femoral head (Nie et al., 2019). Chitosan is mainly derived from chitin from marine organisms, a biocompatible polymer matrix that offers desirable properties for tissue engineering applications. Nanocomposite scaffolds for bone tissue engineering prepared from collagen as matrix and beta-tricalcium phosphate (β-TCP) displayed promising performance for bone tissue repair and regeneration (Goodarzi et al., 2019). Nanocrystalline hydroxyapatite particles in the needle or spherical shapes can be synthesized and used for reinforcement in polymer nanocomposite development. The challenge pertaining to using nanoparticles is their higher secondary forces such as van der Waals forces. Normally nanoparticles tend to stick together, and may form agglomeration and hinder uniform dispersion within the matrix. Different surface treatments are available and can be used to modify the particles surface chemistry with silane or polyethylene glycol (PEG) to reduce the secondary forces.

Porous nanocomposite scaffolds mimicking the structure of natural bone to provide a natural environment for cell coordination and interaction through extracellular signals, with required mechanical properties for load-bearing applications can be developed through the templating method. To achieve the complex structure of bone, dissolvable compounds such as salt and sugar with different sizes are embedded in the templates and removed later by the leaching process. The properties of the scaffold can be manipulated to suit the type of tissue to be repaired or regenerated. Degradation is one of the major properties in determining the time needed for the tissues to be repaired. The ideal scaffolds degrade slowly while being replaced with natural tissues almost at the same rate to avoid the stress shielding effect for the tissue.

13.5.1 ADVANCED SCAFFOLD FABRICATION TECHNIQUES

Scaffold fabrication techniques such as electrospinning (Bhardwaj & Kundu, 2010), solvent casting (Tjong, 2006), and freeze-drying (Levengood & Zhang, 2014) have less control over pore sizes and distribution, reinforcement arrangement within the matrix, and pore connectivity essential properties for the natural bone structure. Developing scaffolds using these traditional methods may lead to a product with poor nutrients, water, and mineral transportation as well as low cell coordination and migration. Technology advances have offered additive manufacturing processes that create 3D profiles on a layer-by-layer process without using molds as used in traditional manufacturing techniques (Yamamoto et al., 2019). This process provides high precision and controllable designing of the internal and surface structure of scaffold mimicking the unique interconnected pore structure of biological bone tissue. These 3D printing technologies allow the use of printing materials in liquid or solid form. Polymer solutions, for instance containing active biological agents such as cell or cell aggregates, could be used to print 3D construct on a bio substrate (Zhang & Webster, 2009). In 3D printers that use solid feed materials, actually a spool of filament is loaded to the heated nozzle at a temperature about the melting point of the polymer and then extruded through a nozzle. In principle, these machines could have more than one nozzle with different materials for composite manufacturing. On the other hand, filament materials could be developed as composite materials whereby nanoparticles would be embedded in the polymer matrix and used for 3D printing.

13.6   MARINE NANOCOMPOSITES

Marine-derived nanocomposites are a promising alternative route to current practices that use mammalian components such as collagen protein for the development of nanocomposite scaffolds. Marine-derived collagen and chitosan possess favorable properties as the matrix for nanocomposite development. Scaffolds for bone tissue repair and regeneration can be developed from marine-derived gelatin (denatured and insoluble form of collagen), which has superior properties compared to collagen. The additional advantages of gelatin include cell adhesive structure and low immunogenicity among others. Gelatins can be processed as nanoparticles and used as reinforcement with biodegradable polymer for scaffold development (Su & Wang, 2015). Several nanocomposites based on collagen and gelatin as organic phase and a bioactive mineral phase of hydroxyapatite are commercially available and clinically used for bone tissue substitute and regeneration. They also have wider applications as repair materials for cartilage, ocular, periodontal, and skin tissue treatments (Kuttappan, Mathew, & Nair, 2016). Nanocomposites can also be used as drug delivery devices without compromising the ability to repair and regenerate the diseased tissue. Surgical site bacterial infections are one of the challenges facing internal medical implants, and therefore incorporation of antibiotics within these products is an added advantage. The drug-release kinetics could still be controlled by controlling the dissolution or degradation rate of the scaffold. In a recent *in vivo* study of polymer nanohydroxyapatite particles, nanocomposites loaded with alendronate for osteoporosis treatment revealed a complete repair of bone defects and displayed a controlled release of the loaded drug (Wu, Lei, & Liu, 2017). Incorporation of growth factors within the fabricated scaffold to enhance cell interaction for cell proliferation and tissue regeneration has been practiced. It has been reported that composites loaded with BMP-2 displayed improved cell adhesion and proliferation (Sharma et al., 2012).

In vitro and *in vivo* testing of materials before clinical trials are very important. Scaffolding materials should posses several properties, and all these are subjected to vigorous examination for safety reasons. Regulatory bodies such as the US Food and Drug Administration (FDA) are very strict on licensing the manufacturing of medical devices. Bioactivity of the developed scaffold is evaluated using *in vitro* mineralization assays proposed by Kokubo (Kokubo & Takadama, 2006) whereby the materials are immersed in simulated body fluid (SBF) and the bone bioactivity is confirmed by the deposition of HAp on the surfaces of the materials. Cell culture has also been used to test cell interaction with the scaffold. For tissue engineering applications, bone tissue cells like osteoblast or stem cells are used to see how they proliferate and differentiate on the surface of the scaffold. *In vivo* performance is the second stage of material evaluation whereby the performance in terms of cell interaction and bonding with hard and soft tissue can be evaluated.

13.7 CONCLUSION

Marine-derived nanocomposite scaffolds consisting of either marine-derived or synthetic biopolymers and coralline-derived nano-calcium phosphate particles have great potential in bone tissue repair and regeneration applications due to their ability to display performance, structure, and properties similar to natural tissue. Biomimetically, the desirable nanocomposites for tissue repair and regeneration with unique bone architecture can be achieved through 3D printing techniques. This technique has shown several advantages over the traditional methods by producing scaffolds with the interconnected porous structure of nano to microarchitecture. Nanocomposite matrices and reinforcements can easily be derived from marine organisms and skeletons, which are abundantly available with excellent biological properties. Besides, some of these marine organisms such as corals could be grown artificially in the laboratory under specific conditions of temperature and pressure. Research efforts are still needed for the development of marine-based nanocomposite scaffold mimicking the unique structure and morphology of natural bone for bone tissue repair and regeneration. Currently, there are no commercially available marine-derived nanocomposite scaffolds, though they have displayed excellent performance in preclinical studies.

Having a deep understanding of nanocomposite–cell tissue interaction and performances and properties–structures relations, the future for clinical use of these materials will be achieved. A combination of clinically active agents with these scaffolds for the treatment and prevention of different tissue diseases should be considered in the research activities.

REFERENCES

Asyraf, M. R. M., Ishak, M. R., Sapuan, S. M., Yidris, N., and Ilyas, R. A. 2020. Woods and composites cantilever beam: A comprehensive review of experimental and numerical creep methodologies. Journal of Materials Research and Technology.

Ben-Nissan, B. 2003. Natural bioceramics: From coral to bone and beyond. Current Opinion in Solid State & Materials Science, 7(4-5):283–288.

Ben-Nissan, B. 2005. Biomimetics and Bioceramics. Paper read at Learning from Nature How to Design New Implantable Biomaterials: From Biomineralization Fundamentals to Biomimetic Materials and Processing Routes, 2005//, at Dordrecht.

Ben-Nissan, B. 2011. Nanoceramics in biomedical applications. MRS Bulletin, 29(1):28–32.

Bhardwaj, Nandana, and Kundu, Subhas C. 2010. Electrospinning: A fascinating fiber fabrication technique. Biotechnology Advances, 28(3):325–347.

Cegla, Rabea-Naemi Rosa, Macha, Innocent J., Ben-Nissan, Besim, Grossin, David, Heness, Greg, and Chung, Ren-Jei. 2014. Comparative study of conversion of coral with ammonium dihydrogen phosphate and orthophosphoric acid to produce calcium phosphates. Journal of the Australian Ceramic Society, 50(2):154–161.

Currey, J. D. 2001. Bone and Natural Composites: Properties. In Encyclopedia of Materials: Science and Technology, edited by K. H. J. Buschow, R. W. Cahn, M. C. Flemings, et al. Oxford: Elsevier.

Diwan, Amna, Eberlin, Kyle R., and Smith, Raymond Malcolm. 2018. The principles and practice of open fracture care, 2018. Chinese Journal of Traumatology, 21(4):187–192.

Goodarzi, Hamid, Hashemi-Najafabadi, Sameereh, Baheiraei, Nafiseh, and Bagheri, Fatemeh. 2019. Preparation and characterization of nanocomposite scaffolds (collagen/β-tcp/sro) for bone tissue engineering. Tissue Engineering and Regenerative Medicine, 16(3):237–251.

Hagen, Anja, Gorenoi, Vitali, Matthias, P., and Schönermark. 2012. Bone graft substitutes for the treatment of traumatic fractures of the extremities. GMS Health Technology Assessment, 8:13.

Heness, G., and Ben-Nissan, B. 2004. Innovative bioceramics. Materials Forum, 27:104–114.

Higgins, Emily, Scheibling, Robert E., Desilets, Kelsey M., and Metaxas, Anna. 2019. Benthic community succession on artificial and natural coral reefs in the northern Gulf of Aqaba, Red Sea. PLOS ONE, 14(2):e0212842.

Ilyas, R. A., and Sapuan, S. M. 2020. The preparation methods and processing of natural fibre bio-polymer composites. Current Organic Synthesis, 16(8):1068–1070.

Ilyas, R. A., Sapuan, S. M., Atiqah, A., Ibrahim, R., Abral, H., Ishak, M. R., Zainudin, E. S., Nurazzi, N. M., Atikah, M. S. N., Ansari, M. N. M., Asyraf, M. R. M., Supian, A. B. M., and Ya, H. 2019. Sugar palm (Arenga pinnata (Wurmb.) Merr) starch films containing sugar palm nanofibrillated cellulose as reinforcement: Water barrier properties. Polymer Composites, 1–9.

Ilyas, R.A., Sapuan, S. M., Ibrahim, R., Abral, H., Ishak, M. R., Zainudin, E. S., Atikah, M. S. N., Mohd Nurazzi, N., Atiqah, A., Ansari, M. N. M., Syafri, E., Asrofi, M., Sari, N. H., and Jumaidin, R. 2019. Effect of sugar palm nanofibrillated cellulose concentrations on morphological, mechanical and physical properties of bio-degradable films based on agro-waste sugar palm (Arenga pinnata (Wurmb.) Merr) starch. Journal of Materials Research and Technology, 8(5):4819–4830.

Ilyas, R. A., Sapuan, S. M., and Ishak, M. R. 2018a. Isolation and characterization of nano-crystalline cellulose from sugar palm fibres (Arenga Pinnata). Carbohydrate Polymers, 181:1038–1051.

Ilyas, R. A., Sapuan, S. M., Ishak, M. R., and Zainudin, E. S. 2018b. Development and characterization of sugar palm nanocrystalline cellulose reinforced sugar palm starch bionanocomposites. Carbohydrate Polymers, 202:186–202.

Ilyas, R. A., Sapuan, S. M., Sanyang, M. L., Ishak, M. R., and Zainudin, E. S. 2018c. Nanocrystalline cellulose as reinforcement for polymeric matrix nanocomposites and its potential applications: A review. Current Analytical Chemistry, 14(3):203–225.

Ilyas, R A, Sapuan, S. M., Ibrahim, R., Abral, H., Ishak, M. R., Zainudin, E. S., Atiqah, A., Atikah, M. S. N., Syafri, E., Asrofi, M., and Jumaidin, R. 2020. Thermal, biodegradability and water barrier properties of bio-nanocomposites based on plasticised sugar palm starch and nanofibrillated celluloses from sugar palm fibres. Journal of Biobased Materials and Bioenergy, 14(2):234–248.

Ilyas, Rushdan Ahmad, Sapuan, S. M., Ibrahim, R., Abral, H., Ishak, M. R., Zainudin, E. S., Asrofi, M., Atikah, M. S. N., Huzaifah, M. R. M., Radzi, A. M., Azammi, A. M. N., Shaharuzaman, M. A., Nurazzi, N. M., Syafri, E., Sari, N. H., Norrrahim, M. N. F., and Jumaidin, R. 2019. Sugar palm (Arenga pinnata (Wurmb.) Merr) cellulosic fibre hierarchy: A comprehensive approach from macro to nano scale. Journal of Materials Research and Technology, 8(3):2753–2766.

Khan, Ibrahim, Saeed, Khalid, and Khan, Idrees. 2019. Nanoparticles: Properties, applications and toxicities. Arabian Journal of Chemistry, 12(7):908–931.

Kokubo, Tadashi, and Takadama, Hiroaki. 2006. How useful is SBF in predicting in vivo bone bioactivity? Biomaterials, 27(15):2907–2915.

Kühne, J. H., Bartl, R., Frisch, B., Hammer, C., Jansson, V., and Zimmer, M. 1994. Bone formation in coralline hydroxyapatite. Effects of pore size studied in rabbits. Acta Orthop Scand, 65(3):246–52.

Kuttappan, Shruthy, Mathew, Dennis, and Nair, Manitha B. 2016. Biomimetic composite scaffolds containing bioceramics and collagen/gelatin for bone tissue engineering – A mini review. International Journal of Biological Macromolecules, 93:1390–1401.

Landis, William J., Hodgens, Karen J., Arena, James, Song, Min Ja, and McEwen, Bruce F. 1996. Structural relations between collagen and mineral in bone as determined by high voltage electron microscopic tomography. Microscopy Research and Technique, 33(2):192–202.

Levengood, Sheeny K. Lan, and Zhang, Miqin. 2014. Chitosan-based scaffolds for bone tissue engineering. Journal of Materials Chemistry B, 2(21):3161–3184.

Newman, Tim. 2018. Bones: All you need to know. Retrieved from https://www.medicalnewstoday.com/articles/320444#Structure

Nie, Lei, Wu, Qiaoyun, Long, Haiyue, et al. 2019. Development of chitosan/gelatin hydrogels incorporation of biphasic calcium phosphate nanoparticles for bone tissue engineering. Journal of Biomaterials Science, Polymer Edition, 30(17):1636–1657.

Omori, Makoto. 2019. Coral restoration research and technical developments: What we have learned so far. Marine Biology Research, 15(7):377–409.

Sadagopan, D., and Pitchumani, R. 1997. A combinatorial optimization approach to composite materials tailoring. Journal of Mechanical Design, 119(4).

Sahoo, Nanda Gopal, Pan, Yong Zheng, Li, Lin, and He, Chao Bin. 2013. Nanocomposites for bone tissue regeneration. Nanomedicine 8(4):639–653.

Sanyang, M. L., Ilyas, R. A., Sapuan, S. M., and Jumaidin, R. 2018. Sugar Palm Starch-Based Composites for Packaging Applications. In Bionanocomposites for Packaging Applications (pp. 125–147). Springer International Publishing.

Sharma, A., Meyer, F., Hyvonen, M., Best, S. M., Cameron, R. E., and Rushton, N. 2012. Osteoinduction by combining bone morphogenetic protein (BMP)-2 with a bioactive novel nanocomposite. Bone & Joint Research, 1(7):145–151.

Sprio, Simone, Ruffini, Andrea, Valentini, Federica, et al. 2011. Biomimesis and biomorphic transformations: New concepts applied to bone regeneration. Journal of Biotechnology, 156(4):347–355.

Su, Kai, and Wang, Chunming. 2015. Recent advances in the use of gelatin in biomedical research. Biotechnology Letters 37(11):2139–2145.

Titsinides, S., Agrogiannis, G., and Karatzas, T. 2019. Bone grafting materials in dentoalveolar reconstruction: A comprehensive review. Japanese Dental Science Review, 55(1):26–32.

Tjong, S. C. 2006. Structural and mechanical properties of polymer nanocomposites. Materials Science and Engineering: R Reports, 53(3):73–197.

Wang, Huifang, Leng, Yamei, and Gong, Yuping. 2018. Bone marrow fat and hematopoiesis. Frontiers in Endocrinology, 9(694).

Weatherholt, Alyssa M., Fuchs, Robyn K., and Warden, Stuart J. 2012. Specialized connective tissue: Bone, the structural framework of the upper extremity. Journal of Hand Therapy, 25(2):123–132.

Wu, Hongwei, Lei, Pengfei, Liu, Gengyan, et al. 2017. Reconstruction of Large-scale Defects with a Novel Hybrid Scaffold Made from Poly(L-lactic acid)/Nanohydroxyapatite/Alendronate-loaded Chitosan Microsphere: in vitro and in vivo Studies. Scientific Reports, 7(1):359.

Yamamoto, Brennan E, Trimble, A Zachary, Minei, Brenden, and Ghasemi Nejhad, Mehrdad N. 2019. Development of multifunctional nanocomposites with 3-D printing additive manufacturing and low graphene loading. Journal of Thermoplastic Composite Materials, 32(3):383–408.

Zhang, Lijie, and Webster, Thomas J. 2009. Nanotechnology and nanomaterials: Promises for improved tissue regeneration. Nano Today, 4(1):66–80.

14 Magnesium Metal Matrix Composites for Biomedical Applications

V.K. Bommala and M. Gopi Krishna
Department of Mechanical Engineering, Acharya
Nagarjuna University, Guntur, Andhra Pradesh, India

CONTENTS

14.1 INTRODUCTION

Material science is a branch of science that investigates different materials for various applications like automobile, aerospace, and especially in medical implantations. Materials that are used for introduced into the human body environment are called biomaterials. Biomaterials have excellent mechanical properties and must be biocompatible, bio-adhesive, biofunctional, corrosion resistant, and osteoconductive (Gohil et al., 2012). The biomaterial must have tensile strength, ductility, and the ability to absorb strain energy. These properties have been satisfying biomedical applications such as joint replacements, bone plates, wires, screws, rods, dental implants, and cardiovascular stents (Katz et al., 1980). The properties of biomaterials are briefly given in Table 14.1. Biomaterials are further classified into different categories such as metals, ceramics, polymers, and composites in ceramic-based biomaterials; the most commonly used materials are calcium phosphate materials because of their excellent non-toxicity, biocompatibility, and osteoconductive nature in the human body environment (Kalyan et al., 2016).

Ceramic materials are bioactive, bioinert, biodegradable, and exhibit poor mechanical properties due to lack of hardness. These were limitedly developed as compared to metallic and polymeric materials. Polymeric-based biomaterials are most commonly used in bone tissue fixation applications as they can easily develop any type of complex shape and size. The mechanical and chemical properties may change to certain degrees during the sterilization process; surface properties may also be able to be easily modified. These are limited in applications because of their unsatisfactory mechanical properties, and toxic additives

TABLE 14.1
Biomaterial Properties

Mechanical properties	Satisfactory hardness and Young's modulus comparable elastic modulus to bone to overcome stress shielding effect
Biocompatibility	Compatibility in human or animal body environments (Kundu et al., 2013)
Corrosion resistance	High corrosion resistivity to avoid toxicity effects of biocompatibility
Cell viability and cytotoxicity	Nontoxic, not elergetic, maximum cell viability, and does not harm the metabolism

such as plasticizers and stabilizers that are used in the synthesis of polymers. Polymeric biomaterials can be harmful in healing tissue and also damage the human body's metabolism (Ratner et al., 2005).

Metallic implants are generally used to repair fractured tissues and give support to heal the tissue. The commercially available metallic implants like cobalt chromium, stainless steel, and titanium-based alloys and composites are used as metallic implant material to heal tissues (Wu et al., 2013). These materials have outstanding mechanical properties such as hardness, load-bearing capacity, ductility, and high strength; additionally, biometalic implants can be produced in any type of complex shape and size. Mechanical properties and biocompatibility are very imperative factors for the implant materials. However, most metallic implants are not biodegradable, after tissue healing the implants may begin to corrode and damage the human body's metabolism. This is due to chemical reactions, which release metal ions causing a toxic environment in the human body.

Magnesium composites are different when compared to other biomaterials because of their biocompatibility, biodegradability, and mechanical properties. Magnesium composites are used as cardiovascular stents and orthopedic implants because magnesium is one of the essential nutrients for human metabolism (Charyeva et al., 2016). The biodegradable material completely dissolves *In vivo* and *In vitro* conditions after the tissue healing, so there is no need to remove the implant material from the human body. By implanting biodegradable material into the physiological environment, a nontoxic oxide layer may develop without causing any harm to the metabolism. Excess oxides may generate, which will be exerted in the urine system. The major components presented in the biodegradable implant material will not be metabolized in the human body and also show a suitable degradation rate to heal the tissue. The primary drawback of magnesium implants is their low corrosive stability in the physiological environment. Much research has undertaken the challenge to increase its corrosive resistance to heal the tissue.

Magnesium biodegradable biomaterial is classified as pure metal, alloys and metal matrix composites. Pure magnesium and its alloys were already introduced as implant materials and faced a problem in that those corrode quickly before healing the tissue. Developing of metal matrix composites was meant to reduce the degradation rate and increase the mechanical properties to withstand stress shielding effect before heal the tissue. Metal matrix composites used as implant material exhibit the properties of tensile strength, compressive strength, elastic modulus, and corrosive resistance by selecting appropriate reinforcement material. The composite may contain metal matrix and reinforcement to increase its mechanical properties as well as weight to strength ratio. The selection of metal matrix and reinforcement plays a very important role because it should be biodegradable, nontoxic, and biocompatible in the physiological environment.

Magnesium is very light material and has a density of 1.74–2.0 g/cc. Mostly magnesium implants were used as orthopedic implants due to their similarity in properties with human bones. Table 14.2 shows the mechanical properties of magnesium material to the human bone (Xu et al., 2009).Mostly Mg-Ca, Mg-Sn, Mg-Zn,

TABLE 14.2
Mechanical Properties of Composites and Human Bone (Bommala et al., 2019)

Material/Tissue	Density (g/cc)	UTS (MPa)	Yield Stress (MPa)	Elastic Modulus (Gpa)
Cancellous bone	1.0–1.4	1.5–38	—	0.01–1.57
Cortical bone	1.8–2.0	35–283	104.9–114.3	5–23
316L Stainless steel	8.0	450–650	200–300	190
Ti-6Al-4V	4.43	830–1,025	760–880	114
Pure Mg	1.74–1.9	160	90	45
Mg alloy	1.81–2.0	240–250	160	45

Mg-Sr, Mg-Si, and Mg-Zr magnesium alloys are used in biomedical applications. Magnesium-based composites mostly using reinforcement materials of calcium-based ceramics, calcium phosphate particles (CPP), hydroxyapatite (HAP), and tricalcium phosphate (ß-TCP) (Asgar et al., 2009).

14.2 MAGNESIUM COMPOSITE MATERIALS AND APPLICATIONS

14.2.1 MAGNESIUM COMPOSITE MATERIALS

Magnesium-based composite material is used in biomedical applications. The chemical composition matrix material and reinforcement being considered are very crucial factors due to the fact that most material for industrial applications is enormously toxic in the human body environment. The material should possess good mechanical properties and should be biocompatible, especially for biomedical applications.

Generally, a biodegradable biomedical implant must be nontoxic or carcinogenic and it should be compatible with the human body environment. In addition, it is ideal if nutritional minerals like magnesium, zinc, and calcium are present in the implants (see Table 14.3). In order to maintain the mechanical integrity and load-bearing capacity of the implant, the material should possess convenient dissolution rate or measured corrosion rate until healing the surrounding tissues. The load-bearing properties may not be necessary once the healing takes place. Along these lines, pure magnesium, magnesium alloys, and composites are utilized as productive biomedical and biodegradable implants, which are important to control the corrosion rate in a physiological environment (Andrej et al., 1999).

Biodegradable composite material consists of a biodegradable metal matrix and biodegradable reinforcement. Magnesium-based micro composites and Nanocomposites have been developed to achieve the required mechanical properties, corrosion resistance, non-toxicity, and biocompatibility. Many researchers have taken abundant opportunities to develop magnesium biodegradable composites for

TABLE 14.3

Compositions of Alloying Elements through Chemical Analysis for Selection of Matrix Materials (Frignani et al., 2006)

Matrix	Nominal Element Component (wt %)								Trace Elements (Max) (wt%)				
	Al	Zn	Mn	Ca	Li	Nd	Zr	Y	Fe	Cu	Si	Ni	Be ppm
AZ31	3.5	1.4	0.3	—	—	—	—	—	0.003	0.008	1.2	0.001	5–15
AZ91	9.5	0.5	0.3	—	—	—	—	—	0.004	0.025	0.05	0.001	5–15
AM60	6.0	0.2	0.2	—	—	—			0.004	0.008	0.05	0.001	5–15
LAE442	4.0			4.0	2.0				Heavy metallic rare elements				
WE43					3.2	0.5	4.0		Heavy metallic rare elements				

biomedical applications. Magnesium composites have excellent biodegradability and biocompatibility when compared to other metals. The selection of reinforcements plays a very important role in the metal matrix composites to improve its properties. The development of biodegradable composites, both matrix and reinforcement, has to be bioinert, bioactive, and biodegradable while healing the tissue without losing its mechanical integrity.

Mostly ceramic-based reinforcements are used in composites because of their bioactive, bioinert, and biocompatibility in the biological environment. If metallic-based reinforcements are used, they can create a toxic environment due to their lack of biocompatibility in the biological environment. Whenever ceramic-based reinforcements are used in composites, those are called bioceramic composites. The main advantage of bioceramic composites is their low chemical reactivity, corrosion resistance, and biocompatibility. The very early first ceramic reinforcements use in biomedical applications was alumina (Al_2O_3) and zirconium (ZrO_2). To be specific, the important feature of these two reinforcements is their enormously low reaction kinetics. Other ceramic reinforcements exhibit faster reaction kinetics and create toxicity. Bioactive composites come in contact with physiological fluids and have chemical reactions toward tissue to repair or reform the tissue. The reinforcement particle size may improve the grain refinement to increase the properties of the composite. In their current generation, calcium phosphate ceramic reinforcements are focused to generate orthopedic implants and dental implants. HAP is used to improve the bonding between the implant and the orthopedic bone due its outstanding bioactive and biocompatible property. The properties of bioceramic reinforcements are given in Table 14.4.

14.2.2 Applications

Biodegradable magnesium implants have been available in the market since the year 2010with the trade name Magnezix. A powder metallurgy method was used for the first CE-certified biodegradable screw (Seitz et al., 2016). This screw is approved for fixation of bone and fragments, and its mechanical properties were

TABLE 14.4

Properties of Bioceramic Reinforcements

Reinforcement	Characteristics	Applications
Alumina(Al_2O_3)	Bioinert, biocompatible, high hardness, high strength, corrosion resistant, and stress shielding	Femoral head, porous coatings for femoral stems, bone screws, and plates
Zirconia(ZrO_2)	High fracture toughness, high flexural strength, low Young's modulus, biocompatible, bioinert, nontoxic	Artificial knee, bone screws, and plates
Bioglass	Biocompatible, bioactive, brittle, and nontoxic	Artificial bone and dental implants
Hydroxyapatite(Hap)	Bioabsorbable, bioactive, biocompatible, similar composition to bone, good osteoconductive properties	Femoral knee, femoral hip, tibial components, acetubular cup

more appropriate when compared with other implant materials. The ultimate tensile strength of Magnezix is greater than 290 MPa, the elastic modulus is about 45 GPA, and its yield strength is greater than 260MPa. The elongation is up to 8% and pre-clinical studies were conducted from year 2010–2012.

Magnesium-based biomaterials were used to develop cardiovascular stents and achieve necessary angiographic results. The practical time period for coronary stents to full remodeling process of major vessels and degrading with optimal mechanical integrity is from 6 to 12 months (El-Omar et al., 2001).

Magnesium-based biodegradable stents are used in biomedical applications to recover the function of injured vascular arteries as shown in Figure 14.1. Additionally,

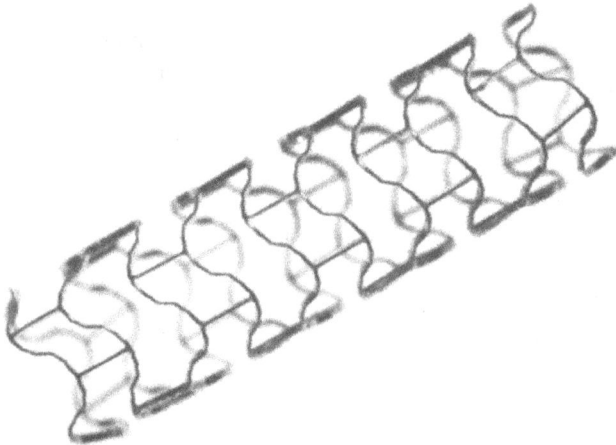

FIGURE 14.1 Cardiovascular stent expanded view (Peuster et al., 2006).

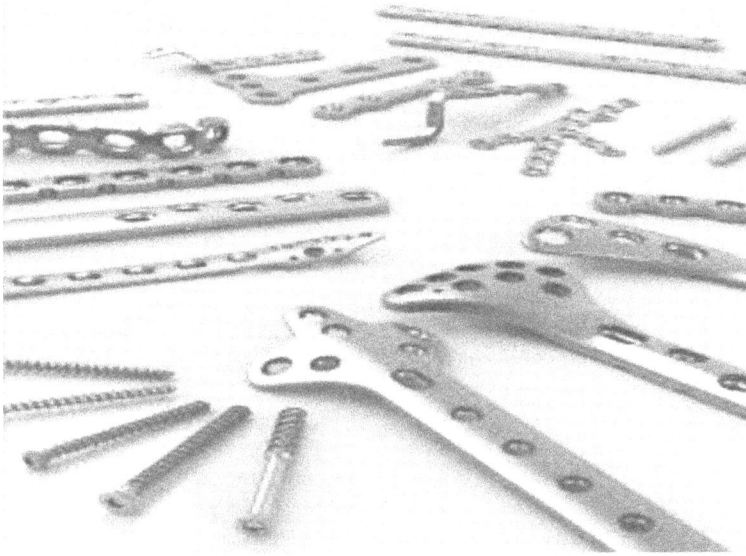

FIGURE 14.2   Magnesium-based biodegradable orthopedic implants (Peuster et al., 2006).

various degradable bone implants for orthopedic applications such as screws (Mg-Ca 0.8 alloy), plates (ZEK 100), nails (LAE442 alloy), laryngeal surgery clips (Mg), wound closing instruments, and dental implants are shown in Figure 14.2. These applications demonstrate magnesium as one of the main important elements in the medical area. In Table 14.5, the applications, advantages, and disadvantages of implant materials are detailed.

14.2.3 TYPES OF REINFORCEMENTS

14.2.3.1 Alumina (Al_2O_3)

Alumina powder has specific characteristics of high hardness and high abrasion resistance. This reinforcement has excellent wear and friction behavior along with surface energy and surface smoothness. Al_2O_3 has a hexagonal structure; the aluminum ions have octahedral interstitial sites, which give the thermal stability of the composite. The abrasion resistance, mechanical strength, and chemical interactions of alumina make it recommended as a ceramic reinforcement for manufacturing of dental and bone implant material composites.

14.2.3.2 Zirconia (ZrO_2)

Zirconia is a biomaterial that has a high mechanical strength, fracture toughness, and excellent biocompatibility. This type of ceramic-based reinforcement gives several advantages of biomedical-based implant materials. Addition of this reinforcement improves the composite strength and thermal stability, as well as reduces the toxicology.

TABLE 14.5

Biomedical Implant Materia Applications, Advantages, and Disadvantages (Wong et al., 2007)

Material	Advantages	Disadvantages	Applications
316L stainless steel	Excellent fabrication property, toughness, easily available and low cost, acceptable biocompatibility	High modulus, poor wear and corrosion resistance	Bone plates, screws, pins, and wires, etc.
Co-Cr alloys	Corrosion resistant, high fatigue strength and wear	Expensive and difficult to manufacture Stress shielding effect, high modulus, toxic due to Cr, Co, and Ni ion release	Bone plates, wires, and hip replacements
Ti alloys	Lower modulus, corrosion resistant, lightweight, and biocompatible	Poor wear resistance, poor ductility	Fixation plates, screws, nails, rods, and wires, total joint replacement
Mg composites	Biocompatibility, biodegradable, light in weight, low stress shielding effect, and low density	Hydrogen evolution during degradation	Bone screws, pins, plates, and stents

14.2.3.3 Carbon

Carbon is an element that exists in a variety of forms such as metals, polymers, and composites that may have carbon aqueous materials, which increase the fatigue strength. However, its intrinsic brittleness and low tensile strength limits its usage in load-bearing applications.

14.2.3.4 Calcium Phosphate Ceramics (CPC)

For 20 years, calcium phosphate salts have often been used as reinforcements in bio-composites for successfully replacing and augmenting bone tissue. Widely used calcium phosphate bioceramics are ß-TCP and HAP. Calcium phosphates may stabilize the phase and improve the grain refinement to increase the corrosion resistance. This type of ceramic-based reinforcement has excellent bioactive, bioinert, and good biocompatibility properties.

14.2.3.5 Other Reinforcements

Si_3N_4 was used as a reinforcement for orthopedic implants to support bone fusion in a spiral surgery and to improve wear and longevity of prosthetic hip and knee joints. It is a hydrophilic negative charged ceramic material, which contains full of nutrients and proteins in the material. It also facilitates bone-cell adherence and integration of the material in the nearest bone. Carbon nanotubes have vast potential in manufacturing and hard tissue implants, scaffolds, micro catheters, and as substrates for neuronal enlargement disorder. Titanium-based reinforcements are used, especially for load-bearing applications due to their biocompatibility, tribological properties, mechanical properties, and corrosion resistance.

14.3 MANUFACTURING METHODS

Generally, the metal matrix composites can develop in three categories. They are solid state, liquid state, and vapor state.

14.3.1 SOLID STATE

In solid state manufacturing methods, the composites are manufactured through solid to solid form. The matrix and reinforcements both are taken in solid form without changing any physical shape and size; these two are bonded together and developed into a new composite material. That is, the two metals of matrix and reinforcement are taken in the powder form and made a composite. The entire solid-state process is called the powder metallurgy method. Powder metallurgy is a term covering a wide range of applications, in which materials or components are made of metallic powders. Using this technique can reduce the usage of metal removal process. That means the direct shape and size of the component or material may directly develop. Therefore, it is drastically reduces the yield losses in the manufacturing process and automatically reduces the manufacturing cost. Powder metallurgy may also be used to make unique materials, which are impossible in other melting or metal forming methods.

Powder compaction is a method of compacting metal powder in a die in the course of the application of high pressure. The tools are seized in the vertical orientation with a punch tool forming the bottom of the cavity. The powder is compacted into a specific shape then removed from the cavity. After making a compaction, it is placed into the furnace to melt and form into a solid object. The parts require minor additional work prior to their final usage, so this process is cost effective.

14.3.2 LIQUID STATE

In liquid state methods, the composites are developed in a liquid state, that is, the solid objects are loaded into furnaces and increased in temperature to the melting point. This method is convenient and low cost compared with other methods. This method is further classified into three categories, which are electroplating, stir casting, and squeeze casting.

14.3.2.1 Electroplating

Electroplating is a procedure that uses an electric current to decrease the dissolved metal cations. This method is primarily used to change the surface properties of an object such as corrosion protection, abrasion lubricity, wear resistance, and aesthetic qualities. The procedure used in electroplating is also called electro deposition. The working method is that one plate is placed as a cathode and whatever metal is to be coated is placed as anode. Both plates are immersed in a solution that is called electrolyte. The electrolyte contains dissolved salts as well as metallic ions that authorize to flow electricity. The power is supplied directly to the anode, and an oxidized metal ion comprises to allow them to dissolve in the solution. The rate of deposition of the anode is equal to the rate of addition of the cathode plate.

14.3.2.2 Stir Casting

Stir casting is a procedure to develop alloys and composites. This is an effective method to develop metal matrix composites, as the reinforcement is uniformly distributed to increase the properties of the base alloy metals. In this manufacturing method, the solid metal matrix is loaded into the furnace and raises its temperature up to its melting point. A vortex is to be created using graphite stirrers in the molten liquid. This uniform distribution of reinforcement may increase the grain refinement to increase the properties of the metal matrix composite. By using this method for development of magnesium matrix composites, an inert environment is required to protect the material from oxidation. The magnesium metal was easily chemical reactive and flammable in nature. To suppress this while making composites, argon gas is used to create an inert environment and to protect from the atmospheric gases the flux can prevent oxidation.

14.3.2.3 Squeeze Casting

Squeeze casting is a procedure by which molten metal solidifies under pressure within blocked dies located between the plates of a hydraulic press. The applied pressure, contact of the molten metal with die surface, produces the rapid heat transfer that yields a pore-free fine grain casting, which increases the mechanical properties of the composites. This process was introduced in the USA in 1960, and it has gained widespread acceptance within the nonferrous casting industries. Magnesium, aluminum, and copper alloys and composites are randomly manufactured with this procedure. This has been widely used to produce high-quality casting because the cooling rate of the casting was increased by applying high pressure during solidification.

14.3.3 Vapor State

In this type of process, vapor deposition is the main consideration of the metal matrix composites. An evaporation method is used for composite fabrication; the reinforcement powders are taken in the vapor form and deposited in the base metal matrix to develop the metal matrix composite. This method is mostly used for nanocomposites to deposit in the small size of components. This method is not suitable for bulk production, nor is it suitable for all types of composites.

14.4 CHARACTERIZATION OF MAGNESIUM COMPOSITES

14.4.1 MICROSTRUCTURE EVOLUTION

Notably affecting corrosion resistance, the microstructure evolution of magnesium composites is the grain size, grain boundary, and phase distribution. Grain refinement leads to transform the density of grain boundaries and increases the mechanical properties as well as the corrosion resistance. Studies revealed the fraction of primary and secondary phases, and grain sizes are the key factors that control corrosion resistance. In metal matrix composites, the base matrix metal has a grain structure that can be refined and reformed to increase the properties of the composite material. Reinforcement is introduced in the metal matrix, which effects grain boundaries based on their shape and size. The size of the reinforcement is decreased, which will increase the grain refinement; whenever grain refinement is done the properties of the composite material increase. A sample with the smallest grain size and largest fraction of secondary phase improves the corrosion resistance because of the secondary phase, which affects galvanic corrosion and suppresses influence of grain size.

14.4.2 MECHANICAL PROPERTIES

Pure magnesium has very similar mechanical properties to bone properties. Later on, similar magnesium alloys were developed for the orthopedic applications. Magnesium alloys have three principal groups, the main group contains pure magnesium; the second group contains aluminum composites such as AZ91, AZ31, and rare earth elements like AE21;andthe third group contains WZ, WE, Mg-Ca, and MZ. The pure elements like Li, Ca, Al, Mn, Zn, Zr, and RE in magnesium combinations may extensively build the mechanical and physical properties of the composite by refining the grain structures, corrosion resistivity, increased strength, machinability, and formability because of the development of intermetallic phases.

Different contaminations usually found in the magnesium combinations are Be, Cu, Fe, and Ni, and a degree of impurities is allowed inside explicit points of confinement during the production of alloy. The plan of adequate levels for Be ranges from 2–4 ppm by weight, Cu is 100–300 ppm, Ni is 20–50 ppm, and Fe is 30–50 ppm. Both Ni and Be refrain from alloying components in biomedical applications considering their carcinogenic nature. The segments Mn, Ca, and Zn are major follow parts for the human body and RE segments showing antagonistic to hostile to cancer-causing properties should be the principal decision for assimilation into a combination. Melody et al. prescribed those especially little measures of uncommon earth components and other alloying metals; for instance, manganese and zinc present in human body condition may increase corrosion resistance. Mn is added to numerous composites to create corrosion resistance and decrease the destructive impacts of contaminations (Polmear et al., 1994).

New innovation has been happening in magnesium metallic materials. Magnesium composite materials have been developed and used as biocompatible, biodegradable implant material because of their nontoxic nature in the human body environment. Magnesium metal matrix material is reinforced with nutritional elements that may improve the mechanical properties as well as corrosion resistance. During the degradation process, the reinforcement acts as a filler material to the implanted tissue and heals the tissue without harm. If apatite-based reinforcements are used in the magnesium composite materials, the HAP or TCP could be increasing the mechanical properties and corrosion resistance, and helping to bond the fractured bone tissues into their original shapes.

It is very important to have enough and suitable mechanical properties for biomedical implants during their lives. Corrosive stability in the implanted structure is a high priority for patient safety because the role of an implant has to support physically damaged tissue throughout the healing process. Biodegradable implants have high mechanical properties and they have to represent best performance. Huge differences between the elastic moduli of implant material to damaged tissue may lead to elastic mismatches and cause stress shielding, especially in metallic biomaterials. Magnesium-based composites face two challenges: stress shielding effects in orthopedic implants and ductility limitation for cardiovascular stent applications. Many researchers have attempted to increase the mechanical properties of magnesium composites, which contain matrix alloys such as Al, Zn, Sr, Zr, Ca, and Mn. The alloy elements also offer the best mechanical properties as biomedical implant materials. The reinforcements are used to incorporate in the matrix, which increases the mechanical properties and corrosion resistance of the biomaterials. Calcium-based reinforcements mostly are used as reinforcements due to their excellent biocompatibility and degradation behavior in the physiological environment.

14.4.3 Biological Properties

Magnesium is one of the most important elements present in the human body where it occupies many number of enzymatic reactions. Magnesium as an element acts as protein, nucleic acid, and stabilizes plasma membranes of cellular activities. The average amount of magnesium present in the adult human body is around 21–28g, and more than 50% of magnesium is presented in the bone tissue. Other soft tissues contain 35–40%, less than 1% is sequestered in the serum. Mg^{2+} ions play a vital role in formative bone frailty. Undeveloped bone tissues contain high concentrations of Mg^{2+} ions, but this concentration changes depending on age. The presence of magnesium in tissue increases elasticity of the bone. Magnesium-based composites are mostly used in orthopedic implants because they have significant effects on osteoblastic cell differentiation. Bone formation, and degradation of magnesium implants show its effect on accelerating on tissue healing. Magnesium presented in the biological environment begins to release hydrogen gas during the degradation process. The high amount of hydrogen gas release can cause tissue healing. The corrosion rate of magnesium should be restricted to decrease risk of gas obstacles during the healing process. By suppression of developed hydrogen gas in the physiological environment, Zn has a nature to control the gas evolution in the human body environment. If the Zn percentage is increased, it creates a toxic environment in the body. This has a wide range of mechanical properties, corrosive resistance, and also increases the osteoblastic immune system and enzymatic reactions. Magnesium composites may contain corrosive resistant, sufficient mechanical properties, and biocompatible and biodegradable matrix and reinforcement materials.

14.5 BIOLOGICAL CORROSION

14.5.1 CORROSION MECHANISMS

Pure magnesium introduced in the human body environment the develops a substantial amorphous layer on its metallic surface which reacted with body fluid and form $MgOH_2$. The rate of oxidation for this layer is 0.01mm/year, whereas the rate of oxidation in saline water is around 0.30mm/year (Ghali et al., 2010).

In magnesium composites, the corrosion rate is controlled by using alloy matrix and reinforcement. Especially in orthopedic applications, pure magnesium and magnesium alloys quickly corrode in the human body environment. To overcome this problem, magnesium composite was developed. Table 14.6 gives the corrosion rate of pure magnesium and magnesium alloy in various physiological solutions.

Body liquids are a mixture of water, dissolved oxygen, proteins, and electrolytic particles, for example, chloride and hydroxide. Magnesium having a negative potential of −2.37V is prone to corrosion and results in free ions relocating from the metal surface to the fringe liquid condition. These particles are synthetic species, for example, metallic oxides, chlorides, hydroxides, and different mixes. In other words, with the assumption that there was no obstruction to oxidation of metal, the reduction was extremely quick to develop hydrogen gas in the human body. As a general rule, the electrochemical response brings about the movement of particles from the metal surface into arrangement, which frames a type of oxide

TABLE 14.6
Corrosion Rates of Pure Mg and Mg Alloys in Various Solutions (Hafekamp et al., 2005)

Material	*In Vitro* Corrosion Rate (mg/cm².h)		*In Vivo* Corrosion Rate (mg/mm².yr)
	Hank Solution	Stimulated Body Fluid	
Pure Mg	0.011	0.038	—
AZ31	0.0065	—	1.17
AZ91	0.0028	—	1.38
LAE 442	—	—	0.39
WE43	—	0.085	1.56

layer on the metal surface. Created on the metal surface, the magnesium hydroxide layer is marginally dissolvable and responds with the chorine particles to make exceptionally solvent chloride and hydrogen gas (Shaw, 2003). The corrosion that occurs on magnesium in an aqueous solution can be communicated in the following conditions. The anodic response is expressed in Equation (14.1), the cathodic response in equation (14.2).

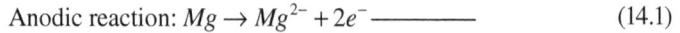

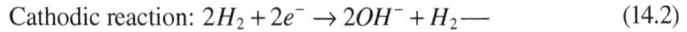

$$\text{Anodic reaction: } Mg \rightarrow Mg^{2-} + 2e^- \text{———————} \tag{14.1}$$

$$\text{Cathodic reaction: } 2H_2 + 2e^- \rightarrow 2OH^- + H_2 \text{——} \tag{14.2}$$

The quick arrangement of hydrogen gas in the physiological condition produces subcutaneous bubbles, which show up in the primary week after presented in the human body and disappear after 2–3 weeks. Song et al. (2007) proposed that the development of hydrogen pace around 0.01 mL/cm²/day can be endured by the human body without a serious threat.

Equation (14.3) gives the general reaction of the process of corrosion:

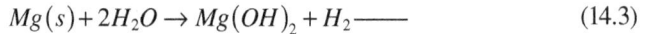

$$Mg(s) + 2H_2O \rightarrow Mg(OH)_2 + H_2 \text{——} \tag{14.3}$$

The rate of corrosion is influenced by the following factors:

- Body fluid pH
- pH variation
- Ion concentration
- Surrounding tissue influences
- Absorption of proteins on the implants.

The factors influence to increase the corrosion resistance. Corrosion resistance of magnesium composites is strongly dependent on various parameters, such as:

- Composition and microstructure
- Mechanical processing and heat treatment
- Kind of surface treatment
- Composition of contacting electrolyte
- Mechanical inputs such as compression or bending process
- Biological environment conditions

14.5.2 TYPES OF BIOLOGICAL CORROSION

When implants are introduced in the human body environment, a chemical reaction takes place and forms oxide layer. In nonferrous metals, the chemical reaction is called oxidation, and in mechanical point, it is called corrosion. When a new material is invented by researchers, its oxidation takes a great deal of time. To reduce this, various techniques are used to determine the oxidation rate of the different media. Characteristic forms of magnesium corrosion in the human body environment are discussed below:

1. Galvanic corrosion
2. Granular corrosion
3. Pitting corrosion
4. Crevice corrosion
5. Fretting corrosion
6. Erosion corrosion
7. Stress corrosion
8. Corrosion fatigue

14.5.2.1 Galvanic Corrosion

Galvanic corrosion is an electrochemical potential of the metal that occurs between two dissimilar metals in the presence of electrolyte, which provides a pathway for electron transfer from one to another metal. Less in size metal becomes anodic, which corrodes and produces corrosion by products in cathodic metal. Considering an example, if screws made of gold are used for attachment of magnesium plate inside the bone during the reconstructive procedure, the magnesium plate will be attacked by the body fluid as shown in Figure 14.3.

Designing a new metallic implant should satisfy electrochemical properties to innovate good implant materials like magnesium-based alloys and composites (Gilbert et al., 1998).

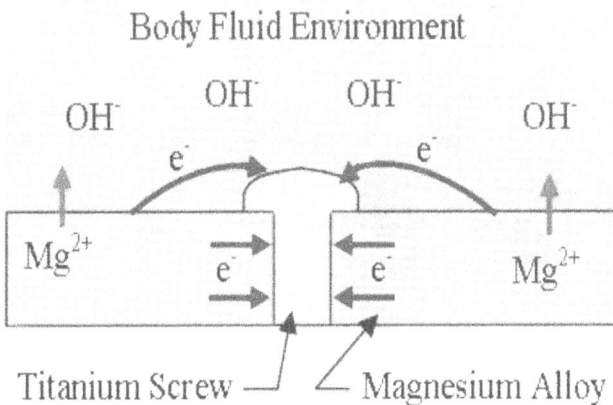

FIGURE 14.3 Galvanic corrosion between two dissimilar metallic materials (Diamant, 1972).

14.5.2.2 Granular Corrosion

In most metals and metallic alloy composites, granular corrosion may happen from the presence of contaminants, which are exhibited in the limits during the process of solidification. Rich galvanic reactions occur in the middle of the metal grid and various polluting influences and incorporations. The subsequent corrosion rate at diverse grain boundary regions exceeds that of the grains and results in an accelerated corrosion rate of the metal matrix. In any case, intergranular corrosion does not happen in magnesium alloys because grains are anodic and their limits are cathodic in nature contrasted with the inside of the grains. The resulting grain boundary corrosion forms nearby grains, which later fall out of the matrix (Han et al., 2007).The reinforcements added into the magnesium alloy matrix materials should be anodic in nature. If they are cathodic, abundant galvanic reactions occur in the boundary grains, which may increase the corrosion rate of the composite.

14.5.2.3 Pitting Corrosion

Pitting corrosion of magnesium material results from the speedy corrosion of little restricted areas that harm the defensive surface oxide layer as shown in Figure 14.4. This sort of corrosion is more insidious than different kinds of corrosion because surface pits are hard to see because of the presence of corroded items. The pits are minute in size; however, they are exceptionally destructive and keep on developing downward and perforating the metal matrix (Zhou et al., 2000). With the underlying nucleation at the surface, the participation of impurities influences the magnesium combination microstructures often assisting in auxiliary corrosion because of the galvanic responses in the materials (Liu et al., 2009). The air inside the pits is extremely damaging with chloride species from the body liquids and Mg^+ particles from anodic suspension incredibly disturbing the circumstance. The pit mouth is exceptionally little and it quickens the corrosion rate in the outside of the grid material because of loss of the particles introduced superficially. During this procedure, the progression of electrons from

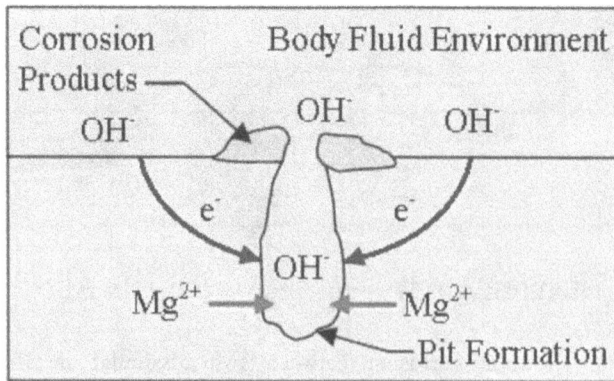

FIGURE 14.4 Pit formation at the surface of a magnesium composite (Greene et al., 1987).

the pit causes the surface to debilitate and the pit entrance becomes cathodic nature. The defensive oxide layer is additionally debilitated by pit development. When pitting begins, magnesium metallic particles are completely infiltrated inside a brief timeframe on account of biomedical inserts and furthermore face an extraordinary decrease in load-bearing resistivity. Another issue related to pit-ting emerges from a confined increment in stress delivered by the pit, which can possibly form cracks (Zeng et al., 2008). Magnesium composite materials are created to conquer the development of the pit on the surfaces in a shorter period. The reinforcements are presented in the network, which lies on the grain limits to ensure the oxide layer just as pits develop.

14.5.2.4   Crevice Corrosion

Crevice corrosion is a local corrosion that happens between the metal and nonmetal segments. For instance, a magnesium composite plate is to be fixed in an area by many screws with a little gap between the screw head and plate. The gap must be adequate to allow a stream of body liquids and forestall any dormant stream as shown in Figure 14.5. The subsequent corrosion cell at that point begins to assault the metal parts of the implant material.

14.5.2.5   Fretting Corrosion

Fretting corrosion occurs in metal parts by direct physical contact with one another within the site of little vibratory surface movements. The small-scale movements are made by daily activities experienced by the human body, which are comparative with the mechanical wear and metallic debris between the surfaces of metallic parts making up the biomedical inserts (Olszewski et al.,1993; Steffan et al., 2003). During the process of corrosion, the metallic particles are delivered, which can frame a

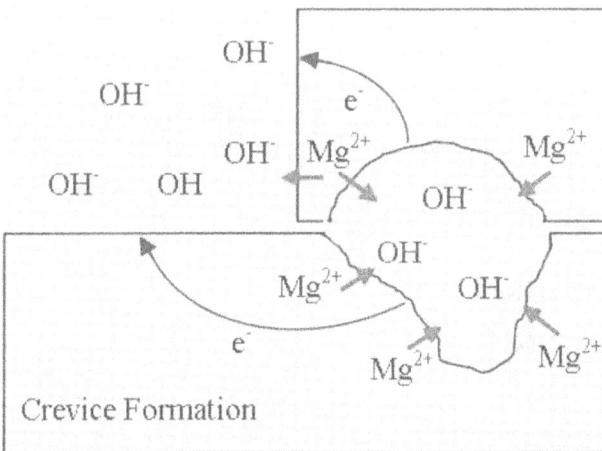

FIGURE 14.5   Crevice corrosion occurring in magnesium material in body fluid conditions (Greene et al., 1987).

wide scope of natural metallic buildings. On account of magnesium composites, the metallic particles discharged during fusing ought to be considered as physiological advantages in light of the fact that these particles can be devoured or consumed by the surrounding tissues, or be broken up and promptly discharged through the kidneys (Venon-Robert et al., 1998).

14.5.2.6   Erosion Corrosion

Erosion corrosion occurs from the exhausting manner of the metal surface or detached layer by the effect of wear debris in the physiological condition encompassed in the implant. The metallic debris impacts the outside of the implant, moving vitality into the area of the contention and plastically twisting the surface (Miller et al., 1993). During the process of deformation, the surface becomes work-hardened to the peak where the following effect surpasses the strain required for surface pitting, cracking, or chip development. All biometals utilized as inserts typically consume at some limited rate when drenched in the unpredictable electrolyte condition of the body; even Ti combinations with the least erosion rate produce corrosion debris (Atrens et al., 2005). This debris may impact the wear behavior and erosion resistance properties of the implant.

14.5.2.7   Stress Corrosion

Electrochemical potential happens among unstressed and focuses on areas of metallic implants under the load conditions; there is an increment in the chemical reactions of the metal. This type of stress initiated corrosion mechanism efficiently increases the corrosion rate (Olson et al., 1987).

14.5.2.8   Corrosion Fatigue

Corrosion fatigue is the product of material being exposed to the combined effects of a cyclic load and corrosive atmosphere (Kainer et al., 2004).Generally, metallic fatigue occurs from cyclic stacking and emptying of a metal part, which originates the formation of cracks on the metal surface and also affects the passive protect layer. Any agglomerations on the surface, such as pores and oxidation pits, can significantly fasten the crack growth rate. Magnesium composite is vulnerable to fatigue corrosion due to the occurrence of ions (chloride) in body fluids. The corrosion promotes the propagation of cracks with cyclic loading and the crack growth considerably increases. The physiological environment body fluid can reduce fatigue life of the magnesium alloys. In magnesium-based composite materials the reinforcements overcome this problem.

14.6   CONCLUSION

Magnesium metal matrix composites are enormously biocompatible and have properties very comparable to natural bone. The development of new bioactive, biocompatible metal matrix composites, with greater physiological as well as mechanical properties, has the potential to enhance the performance of magnesium composite implants to improve their corrosion resistance and wear resistance. Moreover, biomedical studies are required to examine the interface between the material surface and the bounded tissue environment. To examine the long-term effects of the reinforcements being released during the biological corrosion of magnesium metal matrix composites, *In vivo* experimental studies are required. This chapter outlines the potential of magnesium metal matrix composites to fabricate orthopedic implants, the challenges to be overcome, and also some clinical trials that are required to establish the enduring biocompatibility of magnesium-based composites within physiological environments.

REFERENCES

A.W. Asgar and R. Bonan (2009). Worked on biodegradable stents. US Cardiology, vol.6(1), 81–84.

A. Atrens, G. Song, N. Winzer, E. Ghali, W. Deitzel, K.U. Kainer, N. Hort and C. Blawert (2005). Presented a critical review of the stress corrosion cracking of magnesium alloys. Advance Engineering Materials, vol.7, 659–693.

V. K. Bommala, M. Gopi Krishna and T. Rao (2019). Magnesium matrix composites for biomedical applications: A review. Journal of Magnesium Alloys, 7(1), 72–79.

O. Charyeva, O. Dakischew, U. Sommer, C. Heiss, R. Schnettler and K.S. Lips (2016). Studied on biocompatibility of magnesium implants in primary human reaming debris derived cell stem cell in vitro. Journal of Orthopedics, vol.17(1), 63–73.

R.M.E. Diamant (1972). Stated that prevention of corrosion. Applied Chemistry for Engineers. The Pitman Press, Great Britain, Chapter 5, 86–105.

M.M. El-Omar, G. Dangas, I. Iakovou and R. Mehran (2001). Update on in-stent restinosis. Current Interventional Cardiology Report, vol.3(4), 296–305.

A. Frignani, V. Grassi, F. Zucchi, G. Trabanelli and C. Monticelli (2006). Reported on electrochemical behavior of magnesium alloy containing rare earth elements. Journal of Applied Electrochemistry, vol.36, 195–204.

E. Ghali (2010). Stated the corrosion resistance of aluminum and magnesium alloys, understanding, and performance and testing. Chapter 10, John Wiley, 350.

P. Gohil and N.R. Patel (2012). Studied on scope, application & human anatomy significance: A review on biomaterials. International Journal of Emerge, 91–101.

N.D. Greene et al. (1987). Worked on Corrosion Engineering. McGraw-Hill, New York.

H. Hafekamp et al. (2005). Reported on in vivo corrosion of magnesium alloys and associated bone response. Biomaterials, vol.26, 3557–3563.

E. Han, W. Ke and R. Zeng (2007). Reported on corrosion of artificial aged magnesium alloy AZ80 in 3.5 wt pct NaCl solutions. Journal of Materials Science and Technology, vol.23, 353–358.

J.L. Gilbert et al. (1998). Studied on urban corrosion of metal orthopedic implants. Journal of Bone and Joint Surgery, American volume 80, 268–282.

K.U. Kainer et al. (2004). Study on automotive applications of magnesium and its alloys. Journal of Materials Engineering and Performance, vol.13, 7–25.

S.M. Kalyan, A. Tahmasebifar and Z. Evis (2016). Studied on mechanical, electrical and biocompatibility evolution on AZ91D magnesium alloys as a biomaterial. Journal of Alloy Compounds, vol.687, 906–919.

J.L. Katz (1980). Anisotropy of Young's modulus of bone. Nature, vol.283(5742), 106–107.

J. Kundu, F. Pati, Y. Hun Jeong, and D.W. Chow (2013). Studies on Biofabrication, chapter 2, Elsevier, 23–46.

K.W. Miller et al. (1993). Study on material performance and evaluation. Ed RH Jones, ASM International Handbook, Ohio, 251.

D.L. Olson, L.J. Korband and S.C. Dexter (1987). Presented in Corrosion Metals Handbook, ASM International, USA, vol.13, 123–135.

O. Olszewski et al. (1993). Made Research on Fretting wear dependence of hardness ratio and friction coefficient of fretted couple. Wear materials, International conference 9, San Francisco vol.162, 939–943.

M. Peuster, P. Beerbaum, F.W. Bach and H. Hauser (2006). Are resorbable implants about to become a reality? Cardiology in the Young, 16, 107–116.

I.J. Polmear (1994). Studied on magnesium alloys and applications. Material Science and Technology, vol.10, 1–16.

B.D. Ratner (2005). Reported on an introduction to materials in medicine: Biomaterial science, vol.26, 5093.

M. Schlesinger and L.J. Liu (2009). Worked on corrosion of magnesium and its alloys. Corrosion Science, vol.51, 1733–1737.

J.M. Seitz, A. Lucas and M. Kirchner (2016). Reported on magnesium based compression screws. JOM, vol.68, 1177–1182.

B.A. Shaw (2003). Worked on Corrosion Resistance of Magnesium Alloys. ASM Handbook, vol.13a.

G. Song et al. (2007). A critical review of the stress corrosion cracking of magnesium alloys. Advanced Engineering Materials, vol.7, 659–693.

G.L. Song and A. Andrej (1999). Corrosion mechanisms of magnesium alloys. Advanced Engineering Materials, vol.1, 11–33.

I. Steffan, T. Szekeres, C. Lhotka, K. Zweymuller and K. Zhuber (2003). Four year study of cobalt and chromium blood levels in patient managed with two different metal hip replacements. Journal of Orthopedic Research, vol.21, 189–195.

B. Venon-Robert, D.W. Howie, S.D. Roger, D.R. Haynes and S.J. Boyle (1998). Stated variation in cytokines induced by patients from different prosthetic materials. Clinical Orthopedics and Related Research, vol.352, 323–330.

M.H. Wong, K.Y. Chiu, H.C. Man and F.T. Cheng (2007). Characterization and corrosion studies of fluoride conversion coating on degradable Mg implants. Surface Coating Technology, 3, 590–598.

S. Wu, X. Liu, K.W.K. Yeung, H. Guo, P. Li, T. Hu, C.Y. Chung and P.K. Chu (2013). Surface nano-architectures and their effects on mechanical properties and corrosion behavior of Ti-based orthopedic implants. *Surface & Coatings Technology*, 233, 13–26.

L. Xu, G. Yu, F. Pan and E. Zhang (2009). Studied on in vitro and in vivo evaluation of the surface bioactivity of a calcium phosphate coated magnesium alloys. Biomaterials, 2554–2557.

R. Zeng et al. (2008). Reported that progress and challenge for magnesium alloys as biomaterials. Advance Biomaterials, vol.35, B3–B14.

W. Zhou, R. Ambat and N.N. Aung (2000). Studied on evaluation of micro-structural effects on corrosion behavior of AZ91D Mg alloy. Corrosion Science, vol.42, 1433–1455.

Index

293

For Product Safety Concerns and Information please contact our EU
representative GPSR@taylorandfrancis.com
Taylor & Francis Verlag GmbH, Kaufingerstraße 24, 80331 München, Germany